REF
TA
168
2000

HWLCTC

D0217751

BUSINESS/SCIENCE/TECHNOLOGY DIVISION
CHICAGO PUBLIC LIBRARY
400 SOUTH STATE STREET
CHICAGO, IL 60605

DISCARD

Shape and Structure, from Engineering to Nature

Similarities abound in the geometry of flow systems in engineering and in nature. For example, tree-shaped flows are everywhere, in computers, lungs, dendritic crystals, urban street patterns, and communication links.

In this groundbreaking book, Adrian Bejan starts from the design and optimization of engineered systems and discovers a deterministic principle for the generation of geometric form in natural systems. Shape and structure spring from the struggle for better performance in both engineering and nature. This observation leads to *constructal theory*, that is, the thought that the objective and constraints principle used in engineering is also the mechanism from which the geometry in natural flow systems erges. The principle accounts not only for tree-shaped flows but also other geometric forms encountered in engineering and nature – round y spaced internal channels, the proportionality between n in rivers.

 w systems with geometric structure exhibit at least two one with high resistivity (slow, diffusion, walking) that fills f smallest finite scale, and one or more with low resistivity eams, channels, streets). The optimal balancing of the regions nt flow regimes means that the material and channels must be in certain ways. Better global performance is achieved when ution is relatively uniform, this, in spite of the gaping differences he high- and the low-resistivity domains.

 al distribution of imperfection is the principle that generates he system is destined to remain imperfect. The system works best its imperfection (its internal flow resistances) is spread around, so more and more of the internal points are stressed as much as the rdest working points. One good form leads to the next, as the con- ructal principle demands: objective served better, while under the grip of global and local constraints. There is a time arrow to all these forms, and it points toward the better.

From heat exchangers to river channels, the book unites the engineered and the natural worlds. Among the topics covered are mechanical structure, thermal structure, heat trees, ducts and rivers, turbulent structure, and structure in transportation and economics. The numerous illustrations, examples, and homework problems in every chapter make this an ideal text for engineering design courses. Its provocative ideas will also appeal to a broad range of readers in engineering, the natural sciences, economics, and business.

DISCARD

OTHER BOOKS BY ADRIAN BEJAN

Entropy Generation through Heat and Fluid Flow, Wiley, 1982.

Convection Heat Transfer, Wiley, 1984.

Advanced Engineering Thermodynamics, Wiley, 1988.

Convection in Porous Media, with D. A. Nield, Springer-Verlag, 1992.

Heat Transfer, Wiley, 1993.

Convection Heat Transfer, Second Edition, Wiley, 1995.

Thermal Design and Optimization, with G. Tsatsaronis and M. Moran, Wiley, 1996.

Entropy Generation Minimization, CRC Press, 1996.

Advanced Engineering Thermodynamics, Second Edition, Wiley, 1997.

Convection in Porous Media, with D. A. Nield, Second Edition, Springer-Verlag, 1999.

Energy and the Environment, with P. Vadász and D. G. Kröger, eds., Kluwer Academic, 1999.

Thermodynamic Optimization of Complex Energy Systems, with E. Mamut, eds., Kluwer Academic, 1999.

Shape and Structure, from Engineering to Nature

ADRIAN BEJAN

Duke University

CAMBRIDGE
UNIVERSITY PRESS

PUBLISHED BY THE PRESS SYNDICATE OF THE UNIVERSITY OF CAMBRIDGE
The Pitt Building, Trumpington Street, Cambridge, United Kingdom

CAMBRIDGE UNIVERSITY PRESS
The Edinburgh Building, Cambridge CB2 2RU, UK   http://www.cup.cam.ac.uk
40 West 20th Street, New York, NY 10011-4211, USA   http://www.cup.org
10 Stamford Road, Oakleigh, Melbourne 3166, Australia
Ruiz de Alarcón 13, 28014 Madrid, Spain

© Cambridge University Press 2000

This book is in copyright. Subject to statutory exception
and to the provisions of relevant collective licensing agreements,
no reproduction of any part may take place without
the written permission of Cambridge University Press.

First published 2000

Printed in the United States of America

Typeface Sabon 10.25/13 pt.   *System* LaTeX 2_ε [TB]

A catalog record for this book is available from the British Library.

Library of Congress Cataloging in Publication Data
Bejan, Adrian, 1948–
Shape and structure, from engineering to nature / Adrian Bejan.
p.   cm.
Includes index.
ISBN 0-521-79049-2 (hb)
1. Natural organization.   2. Engineering design.   3. Topology Optimization.
4. Constructal Theory.   I. Title.
TA168 .B37   2000
620′.0042 – dc21
00-027314

ISBN 0 521 79049 2 hardback
ISBN 0 521 79388 2 paperback

RO174384524

BUSINESS/SCIENCE/TECHNOLOGY DIVISION
CHICAGO PUBLIC LIBRARY
400 SOUTH STATE STREET
CHICAGO, IL 60605

CONTENTS

LIST OF SYMBOLS

a square root of Biot number, Eq. (8.2)

a_i dimensionless internal contact area of tree of tubes, Eq. (5.56)

a, b dimensionless groups, Eqs. (4.35) and (4.36)

a, b functions, Eqs. (4.20)

A area, m^2

A_i internal contact area of all the tubes of construct of level i, Eq. (5.56), m^2

b square root of Biot number, Eq. (8.4)

b constant, Eq. (10.1)

b, B linear dimensions, m

B base length, m

B volume, Section 11.1, m^3

Be   pressure drop number, Eq. (3.23)

Bi   Biot number, Eq. (10.10)

c specific heat, J kg^{-1} K^{-1}

c_i flow rate of goods, (units) s^{-1}

c_p specific heat at constant pressure, J kg^{-1} K^{-1}

C concentration, kg m^{-3}

C factor depending on the shape of the duct cross section, Eq. (6.7)

C thermal conductance, W K^{-1}

C_f skin friction coefficient, $2\tau/(\rho V^2)$

C_i fluid flow conductance, Table 5.1

C_i total cost rate per unit length, Eq. (11.14), \$ s^{-1} m^{-1}

C_D drag coefficient, Eq. (7.1)

$\hat{C}$ term in Eq. (5.18)

d river depth, m

D diameter, m

D mass diffusivity, m^2 s^{-1}

D_h hydraulic diameter, m

E Young's modulus of elasticity, Pa

f fraction

f friction factor

f function of angle of confluence, Eq. (11.2)

f dimensionless factor, Eq. (7.27)

f length increase factor, L_{i+1}/L_i

f_i flow rate of goods, (units) s^{-1}

f_v pulsating frequency of fires, Eq. (7.13), s^{-1}

F force, N

F functions, Eqs. (4.47) and (10.19)

g gravitational acceleration, m s^{-2}

g price paid, \$ $(unit)^{-1}$

h specific enthalpy, J kg^{-1}

h_{sf} latent heat of solidification, J kg^{-1}

h heat transfer coefficient, W m^{-2} K^{-1}

H dimensionless heat transfer coefficient, Eq. (10.11)

H height, m

i	construct order
I	area moment of inertia, m^4
j	mass flux, kg s^{-1} m^{-2}
k	thermal conductivity, W m^{-1} K^{-1}
k_p	high conductivity, W m^{-1} K^{-1}
k_i	producer–consumer interaction parameter, \$ s^{-1}m^{-1}
k_0	low conductivity, W m^{-1} K^{-1}
$\tilde{k}$	conductivity ratio, k$_p$/k$_0$
K	constant, Eq. (9.12)
K	constant, s$^{-1/2}$, Eq. (10.30)
K	permeability, m^2
KE	kinetic energy, J
K_i	costs, \$ m^{-1} (unit)$^{-1}$
K_0	low permeability, m^2
$K_{1,2,...}$	higher permeabilities, m^2
$\hat{K}$	dimensionless permeability of interstitial material, Table 5.2
L	length, m
m	fin parameter, Eq. (8.27), L^{-1},
m	stream of goods, (units) s^{-1}
m'	stream of goods per unit length, (units) s^{-1} m^{-1}
m, M	mass, kg
$\dot{m}$	mass flow rate, kg s^{-1}
$\dot{m}'$	mass flow rate per unit length, kg s^{-1} m^{-1}
$\dot{m}''$	mass flow rate per unit area, kg s^{-1} m^{-2}
$\dot{m}'''$	mass flow rate per unit volume, kg s^{-1} m^{-3}
M	dimensionless mass flow rate, Eq. (6.14)
M	moment, Nm
$M, \hat{M}$	dimensionless mass flow rates, Eqs. (8.43) and (8.75)
n	exponent, Eq. (10.21)
n	relative index of refraction
Nu	Nusselt number, Section 7.6, Eq. (7.29)
n, N	number of constituents

p	number of elemental cones arranged radially, Fig. 12.7
p	perimeter of contact, m
p, P	pressure, Pa
P	force, N
Pr	Prandtl number, ν/α
q'	heat transfer rate per unit length, W m^{-1}
q''	heat flux, W m^{-2}
q'''	volumetric heat generation rate, W m^{-3}
Q	heat transfer, J
$q, Q, \dot{Q}$	heat current, W
r	flow resistance, Eq. (10.21)
R	thermal resistance, K W^{-1}
R_i	revenue, \$ s^{-1}
r, R	radius, m
Ra	Rayleigh number, Eq. (3.7)
Ra$_p$	porous-medium Rayleigh number, Eq. (7.37)
Ra$_q$	Rayleigh number based on heat source strength, Problem 7.4
Re	Reynolds number, Eq. (3.26)
Re$_l$	local Reynolds number, Eq. (7.6)
s	direction of resultant of pressure forces, Eq. (6.15)
s	stress, Pa
S	spacing, m
St	Stanton number, Section 7.3
S_i	tube cross-sectional areas, Eq. (5.55), m^2
t	time, s
t	thickness, m
t_B	buckling or roll-up time, Eq. (7.11), s
T	temperature, K
T_c	far-field fluid temperature, Eq. (7.14), K
T_h	surface temperature, Section 7.8, K
u, v	velocity components, volume averaged, Eqs. (5.2) and (5.4), m s^{-1}

u, U, U_∞	velocity, m s^{-1}	μ	viscosity, kg s^{-1} m^{-1}
U	overall heat transfer coefficient based on A, W m^{-2} K^{-1}	ν	kinematic viscosity, m^2 s^{-1}
		Π	pressure drop number, or Be, Eq. (3.23)
v, V	velocity, m s^{-1}	ρ	density, kg m^{-3}
V	volume, m^3	τ	dimensionless time, Eq. (10.2)
W	width, m	τ	shear stress, Pa
W	work, J	ϕ	volume fraction
$W, \dot{W}$	power, W	Φ	aggregate integral, Eq. (9.11)
Wo	Womersley number, Eq. (3.40)	Φ	dimensionless group, Eq. (8.20)
x	fraction	$(\ )_{avg}$	average
x, y	cartesian coordinates, m	$(\ )_b$	base, body, breathing
y	dimensionless group, Eq. (4.45)	$(\ )_B$	buckling
		$(\ )_c$	critical, cross-section
α	angle, rad	$(\ )_C$	Carnot, reversible, circumscribed
α	thermal diffusivity, m^2 s^{-1}		
α_i	average porosity, Table 5.2	$(\ )_{down}$	downward component
β	angle, rad	$(\ )_e$	external, expander
β	coefficient of thermal expansion, K^{-1}	$(\ )_f$	fluid, flow cross section
		$(\ )_{hot}$	hot spot
γ	stream of goods per unit area, (units) m^{-2} s^{-1}	$(\ )_H$	high
		$(\ )_i$	construct of level i
δ	boundary layer thickness, m	$(\ )_L$	left, low
δ	geometric aspect ratio, Eq. (5.35)	$(\ )_m$	maximized, minimized
		$(\ )_m$	melting or freezing
δ	thickness, deflection, m	$(\ )_m$	middle
$\Delta(\ )$	difference	$(\ )_m$	porous medium property
ε	dimensionless group, Eq. (4.46)	$(\ )_{mm}$	maximized twice, minimized twice
ε	small dimensionless number	$(\ )_{ma}$	maximum allowable
η	fin efficiency, Eq. (8.24)	$(\ )_{max}$	maximum
η_{II}	second-law efficiency, Eq. (10.7)	$(\ )_{min}$	minimum
		$(\ )_{opt}$	optimum
θ	angle, rad	$(\ )_{out}$	outlet
θ	dimensionless temperature	$(\ )_p$	path, high conductivity
θ	dimensionless time	$(\ )_p$	pulmonary
θ	temperature difference, Eq. (7.24), K	$(\ )_p$	void space, low-resistance ducts
λ	geometric aspect ratio, Eq. (5.34)	$(\ )_{peak}$	peak, largest
		$(\ )_r$	roll
λ	Lagrange multiplier	$(\ )_R$	right
λ_B	buckling wavelength, Eq. (6.1), m	$(\ )_s$	solid, systemic
		$(\ )_t$	transversal

$(\ )_w$ wall

$(\ )_1$ first construct

$(\ )_2$ second construct

$(\ )_0$ reference, elemental volume, or area

$(\ )_\infty$ free stream

$(^-)$ average

$(\tilde{\ })$ dimensionless

$(\hat{\ })$ dimensionless

PREFACE

In this book I propose a topic that has always been and always will be important. The basic idea is that the constrained and purposeful optimizations that engineers perform routinely in the design of thermofluid flow systems can help all of us make better sense of the natural (animate and inanimate) architectures that surround us.

Better sense means a simpler, easier-to-understand, more compact, and general summary of explanations of what we see in nature. Such a summary is called a principle or law. The thirst for a better sense – for rationalizing – has always been the driving force in the historic development of science.

I show that geometric form (shape and structure) springs out of the struggle for better global performance (objective, purpose) subject to global and local constraints. Geometric form is deducible from principle. Optimized geometries such as the tree-shaped flows, the round tubes, and the river cross sections emerge not only in our minds and on our worktables but also in nature, in all flow systems, animate and inanimate. The thought that the same objective and constraints principle is also the mechanism that constructs geometry in natural flow systems is called constructal theory.

There are three aspects of this idea that I pursue in this book. First, to start from principle and to arrive through a mental viewing in the powerful position of predicting geometric forms that appear in nature is to practice *theory*. The time arrow of theory, from principle to nature, runs against the time arrow of empiricism, which begins with nature – the unexplained observation. Empiricism has been the preferred method in the study of naturally organized systems, from river and lung morphology to turbulent eddies and fractal geometry.

The second aspect is useful to us as engineers. Engineering is the science of systems and processes with purpose. It is the science of the useful. By identifying the principle that accounts for geometric form in natural flow we improve our own vision as designers, as creators. For example, nature impresses us with a multitude of tree-shaped flows: lungs, vascularized tissues, river basins and deltas, lightning, botanical trees (canopies, roots, leaves), dendritic crystals, nervous systems, street patterns and urban growth, bacterial colonies, transportation, communication and economic networks, etc. Each tree flow connects an infinity of points (volume, or area) with a single point (source, or sink). The benefit for us is that we may construct similar architectures in engineered systems that require similar volume–point connections.

This is a beautiful example of how, in the end, the theory *returns the favor* to the field that created it, to engineering. We have seen this many times before, most famously in the story of the heat engine and thermodynamic theory. The present theory also contributes to engineering education. Almost without exception, design courses in mechanical and civil engineering are weighted heavily toward the study and optimization of solid structures (e.g., Chap. 2). This book brings the world of flow systems into the same discipline and classroom and under the same deterministic umbrella where geometric form results from principle. The book is recommended as text or supplement for conceptual design courses in all branches of engineering, including transportation, architecture, operations research, spatial economics, and business. For the problems proposed at the end of chapters I wrote a *Solutions Manual*, which can be obtained by writing to Cambridge University Press or directly to me.

The third aspect has to do with the role of engineering in society. Once a noble and revered science (think of Leonardo da Vinci, Sadi Carnot, and the airplane builders during World War II), engineering is now taken for granted. Everywhere we look, from university campus politics to the Nobel prize, engineering ranks either low or not at all on the ladder of respect. The engineering reality is a lot brighter. Scientists of all ages and types have pondered the origin of shape and structure in nature. They have been wondering about our own origins. Zoologists, botanists, and geophysicists speak freely of design, function, necessity, and optimization in their descriptions of natural patterns. Optimization implies an objective or a purpose. All these are engineering concepts. This is why I believe that engineers are destined to play a role in the quest for a rational basis – a principle – for the generation of geometric form in nature.

Many engineers are already proving this through bioengineering. Their designs blur the supposed demarcations between the natural and the artificial. Engineers, with their language and feel for the concepts of objective, constraints, and optimization, are ideally positioned to define the theoretical agenda for life science in this new century.

The preceding observation deserves emphasis. Biologists and physicists are describing what nature is and how it works. What do engineers bring to this apparently full table? Engineers describe how a system changes its configuration in time so that its global performance improves. Becoming better is "what nature is and how it works."

The reader should preview the table of contents, flip the pages, and look at the drawings. The topics assembled in this book cover an extremely wide territory. This great diversity is united by the same principle. Diversity and complexity come from, and are consistent with, principle.

Compact heat exchangers, cracked solids, and swarms of honeybees owe their regularly spaced internal channels to the same principle of global optimization of performance (Chap. 3). Solid bodies that generate heat volumetrically can be cooled from one external sink by means of tree-shaped inserts of much higher conductivity (Chap. 4). Similar trees optimize the flow of fluid between a volume and a point, and vice versa, as in river, respiratory, and circulatory systems (Chap. 5). Travel, transportation, and business links between areas and points become trees when optimized globally for minimum time, minimum cost, and maximum revenue (Chap. 11).

What flows – heat, fluid, people, goods – is not nearly as important as *how* the flow derives its macroscopically visible structure from global objective. The structure is visible as channels, ducts, streets, and fins (Chap. 8) only if the flow possesses at least two regimes with dissimilar resistivities, high and low. The high-resistivity regime (e.g., viscous diffusion, Darcy flow) covers most of the space, as it fills the interstices formed between the smallest channels. The channels and streams are characterized by much lower resistivity (e.g., laminar and turbulent flow in ducts and streams). The same geometric balance between two flow regimes accounts for the eddy of turbulence and the roll of Bénard convection (Chap. 6). Shapeless flow regions of disorganized molecular motion (viscous diffusion) are assigned optimally to lengths of flow with shape (organized molecular motion, streams, eddies), with the global objective of spreading the motion, mixing the fluid, and bringing the entire system to equilibrium faster. The speeding of the approach to equilibrium also accounts for the occurrence of the tree flow structure during volume–point discharges such as lightning (Section 4.8).

Constructal theory is a hierarchical way of thinking that accounts for organization, complexity and diversity in nature, engineering and management. It was first stated in the context of optimizing the access to flow between a point and an area, with application to traffic[*] and the cooling of electronics[†]: "For a finite-size open system to persist in time (to survive) it must evolve in such a way that it provides easier and easier access to the currents that flow through it." The flow path was constructed in a sequence of steps that starts with the smallest building block (elemental area) and continued in time with larger building blocks (assemblies or constructs). The mode of transport with the highest resistivity (slow flow, diffusions, walking, and high cost) was placed at the smallest scales, filling completely the smallest elements. Modes of transport with successively lower resistivities (fast flow, streams, vehicles, and low cost) were placed in the larger constructs, where they were used to connect the area-point or volume-point flows integrated over the constituents. The geometry of each building block was optimized for area-point access. The constructal architecture that emerged was a tree in which every geometric detail is a result – the tree, as a geometric form *deduced* from a principle.

The tree is only the most recent and most complicated geometric form that is being derived from principle. Simpler shapes and much older applications of the principle are the round cross section of tubes for minimum flow resistance and the proportionality between width and depth in rivers of all sizes (Chap. 6). A principle

---

[*] A. Bejan, *J. Adv. Transportation*, Vol. 30, 1996, pp. 85–107.
[†] A. Bejan, *Int. J. Heat Mass Transfer*, Vol. 40, 1997, pp. 799–816.

that has served our understanding so well and for so long deserves to be recognized as law.

The spatial and temporal structures of much more complicated systems, such as power plants and bodies engaged in powered flight, can be rationalized on the same basis (Chaps. 9 and 10). In brief, the optimal *balancing* of the various resistivity regimes means that the material must be distributed or allocated in certain ways. Better global performance is achieved when the distribution is relatively uniform, this, in spite of the gaping differences between the high- and the low-resistivity flow domains.

Optimal distribution of imperfection – this is the principle that generates form. The system is destined to remain imperfect. The system works best when its imperfection (its internal flow resistances) is spread around, so that more and more of the internal points are stressed as much as the hardest working points.

The more we think of engineered systems in this way, the more these systems look and function like animals. The structure of animals has been a puzzle. From the mouse and the salamander, to the crocodile and the whale, animals are correlated by surprisingly accurate power laws between animal body size and other flow and performance parameters. One way to construct a pure theory of structure in living systems is to treat them as energy systems with flows, constraints and, above all, global objective. This is the theoretical line proposed and explored in this book. In Section 10.6, for example, I derive purely theoretically the famous empirical proportionalities between metabolic rate and body size raised to the power 3/4 and between time period (breathing, heart beating) and body size raised to the power 1/4.

The diversity of the leads pursued in this book reflects my most recent research interests. Their breadth may give the impression that in this book I tried to review the field of natural self-organization. This is most definitely not the case. Early on I decided not to review the work of others. I made this decision because of space limitations, the uniqueness of my engineering view on what generates shape and structure everywhere, and because of a lesson learned from my unforgettable MIT professor of mechanics, J. P. Den Hartog (*Mechanics*, Dover, New York, 1961, p. v):

> "It is recorded that Sancho Panza, when he saw his famous master charge into the windmills, muttered in his beard something about relative motion and Newton's third law. Sancho was right: the windmills hit his master just as hard as he hit them".

It is sufficient to note that some of our contemporaries have difficulties with the concept of objective (or purpose, function, design, optimization), even though they themselves rely on it permanently, in thought and way of life.

I believe, nevertheless, that we ought not to suffer ourselves to be daunted by these difficulties; but that, on the contrary, we must look steadfastly into this theory.* I quoted Rudolf Clausius because today we are facing a situation very similar to the one faced by him. To account for coupled thermomechanical behavior

---

* R. Clausius, *Philos. Mag.*, Ser. 4, Vol. 2, 1851, pp. 1–20 and 102–119.

he had to formulate a second principle, the second law, in addition to the conservation of energy. With this new principle came the concept of entropy, which was completely foreign to science. Today the new principle is the construction of geometric form, and the new concept is objective, or purpose.

I wrote this book during 1998–2000, when my research was supported by a grant from the National Science Foundation. In 1999 I also received a 1/2-month summer grant from the Lord Foundation of North Carolina for the purpose of developing a course based on this material. The manuscript was typed by Linda Hayes, with the support of Professor F. Hadley Cocks, Chairman of our Department of Mechanical Engineering and Materials Science. In my life at Duke, I also received significant and steady encouragement from Professor Charles M. Harman, Dean Earl H. Dowell, and President H. Keith H. Brodie. I am very grateful for all this support.

I am also fortunate that in this work I was surrounded by some of my best doctoral students. Their work is exhibited and acknowledged throughout the text. For the color figures, I give special thanks to Marcelo Errera, Majed Almogbel, Maria Neagu, Nicolae Dan, Gustavo Ledezma, Juan Carlos Ordoñez, Mihaela Morega and Alexandru Morega. Professors A. E. Bergles, M. J. Moran, M. O. Coppens, J. de Swaan Arons, and L. Stoicescu offered many helpful comments on what I presented in the final version.

January 2000   Adrian Bejan
                                                                Durham, North Carolina

CHAPTER ONE

NATURAL FORM, QUESTIONING, AND THEORY

1.1 The Great Puzzle: From What Principle can Geometric Form be Deduced?

This book is an invitation to think about a phenomenon that is so prevalent that it is being taken for granted: the macroscopic shapes and structures that generate themselves everywhere in nature. It is about the great puzzle that has been with us from the beginnings of science: From what principle can geometric form be *deduced*? Democritus (approximately 460–370 B.C.) attributed natural geometrical form to "chance and necessity." The doctrine of chance (nondeterminism) has stayed with us ever since, not as an explanation of natural form generation but as an admission of our own failure to rationalize the occurrence of natural self-organization.

Let us start with four empirical observations. First, geometric form is generated in natural systems that are internally "alive" with flows and driving gradients (e.g., temperature, pressure). Such systems are not in equilibrium internally. They are not dead. Second, the geometric shapes that our minds recognize and sort out are not many. Only three shapes cover the live world that is around us and inside ourselves: the tree network of river basins and lungs, the round cross section of blood vessels and bronchial passages, and the "watermelon slice" cross section of rivers. The spherical shape (e.g., a gas bubble in a liquid) is not one of the natural shapes addressed in this book, because it is the shape of systems without internal flows – systems at equilibrium.

The third observation is that natural systems may have the same shape (tree, round, or slice) but are not identical. For example, two bronchial trees are never identical. Similarly, a cut made across a blood vessel never reveals the perfect, mathematical circle. The point is that when presented with one image from the endless diversity of natural flow shapes, the mind knows this image and categorizes it immediately as a tree, round, or slice shape. The tree may be the hardest to describe, but when we see it we know it, and we call it tree.

The fourth observation is the most important. These few natural shapes are everywhere, in both animate and inanimate flow systems. If a single principle – a simple statement – is responsible for the generation of all these shapes, then that

FIGURE 1.1. Tree and roots on the lower Danube (photograph courtesy of Teresa M. Bejan) (see Plate I).

principle manifests itself everywhere. It becomes a law that bridges the gap between physics and biology, or, better said, between two ways of thinking, two fields of vision in the eyes of the same individual.

Let us take a closer look at the shapes with which nature impresses us. Tree-shaped networks are indeed everywhere, in botanical trees, leaves, roots (Fig. 1.1), lungs, vascularized tissues, neural dendrites, river drainage basins, river deltas, urban growth, bacterial colonies, lightning, and dendritic crystals. *Dendron* means tree in Greek. Tree-shaped networks are displayed by flow systems that connect one "root" point (sink, source) to a finite-size area or volume (an infinity of points).

The tree network is not a net. The tree structure is without loops, such that a unique path links the root point to any of the points of the given area of volume.

Natural macroscopic structure is not only spatial but also temporal. In physiology it is known empirically that heart beating and breathing frequencies are sharply defined characteristics of each animal and that they decrease as the body size increases. Many empirical relations have been discovered in physiology, in which they are known as alometric laws. The turbulent wake behind a cylinder in cross flow also has a characteristic, dominant frequency. Again, what principle generates structure over such a wide and diverse spectrum?

The fact that the most common phenomena tend to be taken for granted does not mean that their importance is not being recognized. For example, in the field of river morphology, the recurring geometric features of river drainage basins have been measured extensively and correlated very successfully. We know what these features are, but we do not know where they come from. Bloom [1] wrote that "the techniques of quantitative fluvial geomorphology give an excellent description but no explanation." The same can be said about other successful correlations, e.g., breathing frequencies, the growth of turbulent mixing regions, and the proportionality between width and depth in rivers of all sizes.

More recently, several cases of macroscopic organization in nature have been subjected to the techniques of fractal geometry. The image generated by repeating a number of times an assumed sequence of operations can be designed to resemble a natural tree network. Fractal geometry, like the correlations of river morphology, is descriptive, not predictive. The physics principle from which the fractal algorithm may be deduced is the puzzle. Kadanoff [2] noted the absence of physics in fractal geometry and wrote that "further progress depends upon establishing a more substantial theoretical base in which geometrical form is *deduced* from the mechanisms that produce it."

This book is about the physics principle from which geometric form in natural flow systems can be deduced. It is about the solution to the puzzle stated in the preceding paragraphs. The fully deterministic principle that allows us to anticipate, with eyes closed, the shape and the structure that occurs naturally is theory. If successful, this theory represents the law that extends physics over naturally organized flow systems and into biology.

Proceeding with the eyes closed, without first looking at natural forms, is absolutely essential in the theoretical realm. I became interested in this topic purely by accident, not because I was trying to solve the puzzle. The accident is that a few years ago I reviewed, in book form, two optimization methods that put me in a constructor's frame of mind. One is the geometric optimization of cooling techniques for compact electronic packages, which leads to optimal spacings and shapes in the internal architecture of fixed volumes with fixed flows (heat generation rates) [3]. The other is the method of entropy generation minimization (EGM), which I had recognized as a method first in 1982 [4]. The method has since been expanded in physics under newer names such as 'finite-time' and 'endoreversible' thermodynamics. In the EGM method the thermodynamic performance of the system is improved through physical changes in the design, subject to overall constraints.

I had just proposed to myself the problem of minimizing the thermal resistance between an entire heat generating volume and one point. When I found the geometric-optimization solution, I was astonished to see in front of my eyes, on my piece of paper and under my own pencil, a tree network in which every single feature is a *result*, not an assumption. Only then I thought of the many trees of nature. I drew immediately the conclusion that the natural tree structure is the reflection of the invoked *principle* of global maximization of performance in volume-to-point (or point-to-volume) flow. Because the tree structure is everywhere, the principle that produces it must be acting everywhere.

1.2 The Hardest Questions

Theory grows out of timely and well-placed questions. There is great diversity not only in subjects but also in the way we *formulate* questions. I can imagine two extremes in which formulating questions is easy. In one extreme, we may ask questions about areas of which we are completely ignorant. These are the questions for which even we, the speakers, have no hope. If we all proceed in this direction, our destination can only be the Tower of Babel, because the very act of speaking (asking questions) requires the use of words to which the speaker attaches at least some understanding.

In the other extreme, we may ask questions for which the answers are well known. We do this routinely when we formulate problems for students in introductory courses. In this extreme we teach to the new generations the disciplines that are required by the profession. The beginner is well advised not to linger too much along this path, because too much discipline is poison to the individual's innate creativity.

The hardest questions are in between. They are the most likely to move the boulders that mark the frontiers of knowledge. When we speak we are ignorant, but we also know "just enough" to be able to open our mouths intelligently. These valuable questions are triggered by hunches, or feelings of "aha!" Hunches occur when we meditate in isolation or when we catch a glimpse of a development in a field in which its importance is not recognized – a gemstone among the pebbles of the river bottom, seen by fish. The sample questions formulated in this section are driven by such a hunch – the accident mentioned at the end of Section 1.1.

Why are tree networks everywhere in nature? It is an undisputed fact that natural volume-to-point (or point-to-volume) flows have a common internal tree structure. Channels, or streams, fill the volume partially, and the channels that are closer to the root point are thicker. The interstitial spaces are bathed by a flow with much higher resistance (diffusion), which touches every point of the infinity of points that make up the volume. The tree structure unites the animate with the inanimate, billions and billions of times. What law summarizes this permanent occurrence?

Why cannot channels be smaller than a characteristic size? A common feature of natural tree images is that the channels cannot be smaller than a certain length scale,

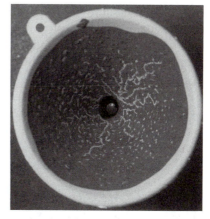

FIGURE 1.2. The formation of the smallest rivers in the drainage basin of a funnel coated with unfiltered coffee sediment. The funnel was held vertically upward, and the photograph was taken at an angle and from above. Note the marriage of shapeless flow (disorganization, diffusion) and flow with tree shape and structure (organization, streams) at the smallest, finite scale. Trees form all around the funnel and are visible from above.

which is why the channels fill the volume only partially. The smallest length scale is a characteristic of the flow medium, e.g., the alveolus in the lung, the smallest capillary in the tissue, the smallest needle of the snowflake, and the smallest rivulet: You may watch the development of the smallest rivers in your own coffee grounds, e.g., Fig. 1.2.

This question is extremely important today as the fractal geometry movement is promoting the claim that natural tree networks are fractal objects. They are not, because according to Mandelbrot's [5] definition of fractal dimension, a fractal is an object generated by repeating ad infinitum an algorithm based on postulated similarity rules. The infinite sequence of stages of branching or coalescence is notably missing from natural tree systems. The fractal algorithm user is forced to interrupt the sequence after only a few steps (the inner cutoff), so that this user may be able to *see* the drawing. If natural structures were truly fractal, "we would be seeing nothing but blurred images and shades of gray" [6]. Similar words of caution have appeared subsequently in *Science* [7]. Why is the inner cutoff a necessary move in the descriptive (not predictive) graphic technique called fractal geometry?

Why bifurcation (pairing), and why does this rule break down at small length scales? Another striking feature of natural tree flows is the bifurcation, dichotomy, or pairing exhibited by the thicker channels. This rule – the integer 2 – breaks down in the direction of smaller scales, such that the number of branches per branching stage increases (e.g., alveoli, the thinnest capillaries, pine needles on the smallest branch). Is the number of branches or tributaries the result of an optimization process?

Is the geometric form of natural flows the result of global minimization of flow resistance, or global maximization of spreading, or access? The tree network is one of the three geometric forms of natural flow systems. The other two – the round cross section of bronchial tubes and the cross section of rivers – can be predicted successfully by minimization of the flow resistance subject to fixed volume (total cross-sectional area). Is the same principle responsible for the natural tree structures and all the features questioned in the preceding paragraphs?

Is the same optimization principle responsible for turbulence? The common features of natural tree networks may hold additional meaning for fluid mechanicists. The coexistence of organized stream flow (channels) with disorganized molecular motion (diffusion) at the smallest macroscopic length scale reminds us of the elementary building block of turbulent flow fields: the smallest eddy. That in natural tree structures streams cannot be smaller than a characteristic length scale reminds us again of the smallest eddy. The finite sequence of stages (branching, or coalescence) that define the tree and fill the space reminds us of the transitions, cascades, and ladders of turbulence.

To start questioning the origin of turbulence along these lines is extremely timely. Classical fluid mechanics asks why a shear flow does not remain laminar forever. The equivalent question for natural trees is why the volume-to-point flow is not always effected by a single flow mechanism – diffusion – with an invisible, quasi-radial flow pattern. Dendritic patterns occur in nature only when flow systems are large and fast enough. The corresponding turbulence question is why diffusion ceases to be the only mechanism when the shear flow becomes sufficiently thick and fast (Chap. 7).

1.3 The Objective and Constraints Principle

The idea pursued in this book is that the spatial and temporal structure exhibited by nature is the result of a global process of optimization subject to global and local constraints. This principle dictates that a finite-size heterogeneous system will undergo changes in shape and structure in order to provide easier access (paths of reduced resistance) to its internal currents. We focus more closely on this statement as the book progresses and as the appropriate terminology is introduced. The principle was named constructal law for the reasons given in Sections 4.1 and 4.5.

The existence of a law of natural pattern formation should not surprise anyone, both in science and on the street. Principles of optimization such as minimum time, minimum flow resistance, and minimum power expenditure have been invoked throughout the history of science and common speech, and continue to be invoked routinely. Leading examples are presented in this book. The deterministic success of these principles has become so routine that success is being taken for granted. Not questioned is the fact that these principles are self-standing: They do not follow from other known laws. Do these principles apply in ad hoc manner – minimum time here, minimum power there – or are they manifestations of a single principle?

It is for this reason that the constructal law deserves scrutiny. News is no longer the deterministic power of the principle but the principle itself.

If this is a book about patterns in nature, then why mix nature with engineering? The reason is that we are part of nature. The engineered, or the artificial, is the realm of the objects that exist only because we exist. They do so as extensions of our own bodies and actions. They expand our individual spheres of activity and influence. We are much bigger than we look. Each of us is connected all over the globe. The principle that operates in nature is the same – has the same structure – as the optimization principles that work inside ourselves, such as in our engineered constructs.

Engineering design begins with a very clear understanding of the *objective* (mission, purpose, function, performance) of the system. This concept also implies *optimization*, because the designed system is expected to function – to perform – in the best manner possible. The word possible means that the designer must recognize the *constraints* that the system must satisfy. The constraints are of two types, global and local.

Well-known examples of global constraints are the mass and the volume of the entire system. The heat current generated inside the volume occupied by an electronic package is another globally constrained quantity, if the electronic components that must be installed in the volume are specified. The end deflection of a cantilever beam is also a global quantity, because to its magnitude contribute the deformations of all the infinitesimal elements of the beam. The values of the global quantities that serve as constraints are obtained when appropriate integrals are performed over all the points of the system volume.

The local constraints are more subtle, but equally important. In a structural member such as a cantilever beam, the stresses felt at every point must not exceed a certain value: the maximum allowable stress s_{ma}. The exact location of the point (or number of points) where s_{ma} occurs is not important. What is important is that the stress at an arbitrary internal point is comparable with, but not greater than, s_{ma}. In this way every point of the infinity of points of the system "works" toward the successful performance of the design, even though some points do not work as hard as others.

An analogous example is the local temperature in a system that generates heat volumetrically (e.g., electric winding, electronics package). The maximum temperature must not exceed a specified level. The number and precise locations of the hot spots are not important. In the end, the local constraints that are satisfied at every point translate into a function (objective) that is met at the system level. The geometry is the unknown. External shape and internal structure are born out of the interplay between satisfying global constraints while meeting the local requirements at every point inside the system, in the pursuit of the system objective.

The starting position occupied by this engineering view is essential to understanding in a deterministic (predictive) sense the occurrence of shape and structure in natural systems. In constructal theory deterministic progress is made *in time*, by proceeding in this direction: from principle (engineering) to nature. The history of science and technology is rich with examples of the opposite flow of ideas, from nature to engineering. Today this classical method is known as biomimetics and continues to be a valuable tool in engineering design. In Fig. 1.3, the hand is a detail from Leonardo da Vinci's *Study of Arms and Hands* (approximately 1474).

Constructal theory proceeds against the method of biomimetics (Fig. 1.3). Through the constructal law, engineering returns biomimetics the favor and lends

Biomimetics

Nature   Engineering

Time

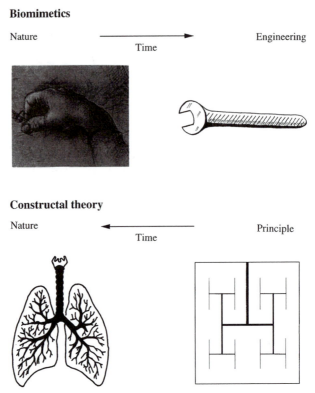

Constructal theory

Nature   Principle

Time

FIGURE 1.3. Proceeding against method, in time: the constructal theory of shape and structure in nature against the empirical method of biomimetics.

the natural scientist a purely theoretical point of view from which nature can be understood and predicted better, i.e., *more simply*. Simplicity is the most desirable feature in science because our physical means for storing, reasoning (connecting), and retrieving information are size constrained. It is not surprising, then, that our physical systems (brain, computer) evolve in the same way as other constrained systems with internal currents. They develop tree networks.

In Fig. 1.4 we take an early look at the constructal principle and why it deserves to be recognized as the law of natural form in flow systems. We do this in juxtaposition to the Fermat principle, which is recognized as law in physics even though its domain of application is only optics [8]. The Fermat law is the statement that when light travels between two given points (A, B) situated in two different media it must choose the path that minimizes the time of travel between the two points (Problem 1.2). From this statement it follows that the angle of refraction is an optimization result, $\sin \beta_i / \sin \beta_r = n_{ir}$, where n_{ir} is the relative index of refraction of the two media (see Fig. P1.2). The same optimal path is deduced if, instead of indexes of diffraction, we use the respective speeds of light propagation through the two media, V_0 and V_1.

In addition to diffraction, the Fermat law covers the much older shortest-path principle of Heron of Alexandria (Problem 1.1). Accordingly, the light ray must

Fermat law

point-point flow

Constructal law

volume-point flow

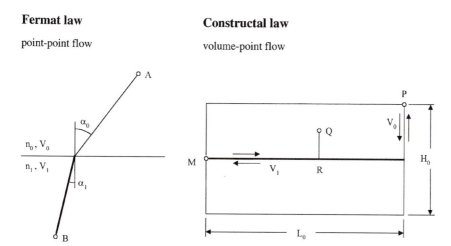

FIGURE 1.4. The generation of geometric form through the minimization of travel time: travel between two points (Fermat law) and between a volume and one point (constructal law).

travel in a straight line through the same medium, and the angle of incidence against a mirror must be equal to the angle of reflection.

In contrast to the point-to-point flow of Fermat and Heron of Alexandria, the constructal law refers to the finite-size system in which the stream flows between one point (M) and each of the points (an infinite number) contained in a volume [6]. In two dimensions the volume may be represented by the rectangular area $A_0 = H_0 L_0$, shown on the right-hand side of Fig. 1.4. The size A_0 is fixed but the shape H_0/L_0 is not. The system is a composite of two media through which movement proceeds at different speeds, $V_0 \ll V_1$. The amount of high-speed material (V_1) is small and fixed – it is much smaller than the rest of the volume, which is occupied by low-speed material (V_0). For brevity, we assume that A_0 is so small that it houses only one channel (strip) of fast material: A_0 is called an elemental system in Chap. 4, in which the problem of volume-to-point flow is considered in greater detail.

The constructal law is the requirement that the access between all the points of A_0 and point M must be maximized. From this requirement follows complete geometric form: the external shape H_0/L_0 and the internal structure of A_0, i.e., the distribution of fast travel through the much larger space covered by slow travel. The optimal internal structure is already shown in Fig. 1.4: The best way to distribute the V_1 material is to shape it as a straight strip and to place it on the longer of the two axes of symmetry of the A_0 rectangle. The maximization of volume-to-point access is accomplished by minimization of the time of travel [9, 10]. This approach and its flow-resistance minimization analog are used throughout this book.

In the constructal law we recognize all the elements of the objective and constraints structure sketched in Section 1.2. The objective of the finite-size system shown on the right-hand side of Fig. 1.4 is to maximize the access between the volume and the point. The global constraints are the fixed system size and the fixed amount of highly permeable material, which must be distributed through the system. The local constraint arises from the fact that some of the points of the given volume will always have easier access to M than other points. To the most distant point in A_0 (point P in Fig. 1.4) belongs the longest travel time to M. This point works hardest: It is the analog of the point of maximum stress in a mechanical structure (Chap. 2) and the hot spot of a thermal structure (Chaps. 3 and 4). By serving the needs of point P in the best way possible, the design also serves the needs of the remaining infinity of points. The design improves by changing its shape and structure so that there is greater uniformity in the way in which all the points work, i.e., in the way that all the points of A_0 can be reached from M.

The optimization of the A_0 construct may be pursued in two ways, with surprisingly identical results [6, 9]. One is the "altruistic" mode in which we perform the optimization to benefit the most disadvantaged traveler – the one who must travel between P and M. By doing this, we can rest assured that all the other travel times (say, between M and the internal point Q) are shorter. The travel time between P and M is $t_0 = H_0/(2V_0) + L_0/V_1$. This time can be minimized with respect to the shape parameter H_0/L_0 subject to the size constraint $A_0 = H_0 L_0$. The result, $(H_0/L_0)_{\text{opt}} = 2V_0/V_1 \ll 1$, is the geometric form of the system.

The alternative approach is to adopt the "egotistical, but socially minded" point of view: Calculate the travel time between an individual point Q and M, average this time over all the points that populate A_0, and then minimize this time with respect to the external shape H_0/L_0. In Chap. 11 we will see that the average time is $\bar{t}_0 = H_0/(4V_0) + L_0/(2V_1)$, which is why the optimal shape is the same as that in the preceding paragraph, $(H_0/L_0)_{\text{opt}} = 2V_0/V_1$.

The agreement between these two approaches suggests that what is good for the most disadvantaged individual – the geometric form – is good for the group as a whole. The urge to organize is an expression of individual need, and the macroscopic form that results is a reflection of that need. The tendency of matter to coalesce, and to form shapes with flow, is so well documented that it often is described as a "property" of matter. In this book we rationalize this behavior through a process of optimization in this same way that the Fermat law allows us to predict the geometric form of the ray of light. The constructal principle asks us to focus our attention on the whole.

Unlike in Fermat's case, in which the break in the trajectory was constrained to slide along a boundary between two semi-infinite media, in volume–point travel the break point (e.g., point R) falls on the centerline of the A_0 rectangle. The centerline is one of the results of the optimization principle, not a postulated constraint. The geometric features that do not move are indicated by the thicker lines in Figs. P1.1–P1.3.

The volume-to-point construct is a bundle containing an infinite number of point-to-point flow paths that obey the Fermat law. The $90°$ bend in the path from Q to M is optimal in the assumed limit where $V_1 \gg V_0$. When this inequality is not strong, the optimal angle between the V_0 and the V_1 portions of each path is obtuse, so that each V_1-long segment (e.g., segment RM) is shorter than in the limit pictured on the right-hand side of Fig. 1.4. This more general situation is discussed in Chap. 11.

PROBLEMS

1.1 Points A and B are connected by the broken line ARB, such that point R is free to move along the base of the drawing. In the principle of Heron of Alexandria, AR is the incident ray of light, RB is the reflected ray, and the base is the mirror. Show that the path ARB is the shortest when $x = 0$, i.e., when the angle of incidence (α_i) is equal to the angle of reflection (α_r).

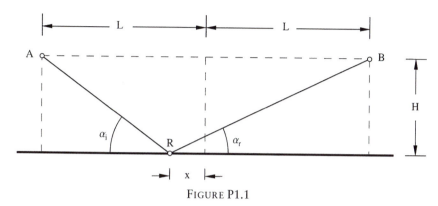

FIGURE P1.1

1.2 Consider the travel along the route ARB shown in Fig. P1.2. The speeds are V_0 along the AR segment and V_1 along the RB segment. Break point R is free to slide along the line that separates the V_0 and the V_1 domains. In the law of Fermat, the two domains are two different media through which a ray of

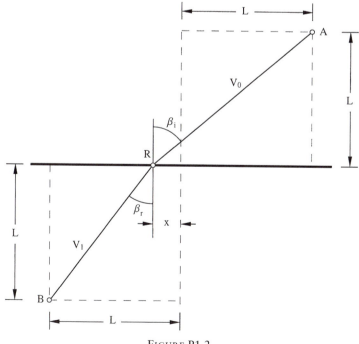

FIGURE P1.2

light connects A to B; the point of refraction (R) is situated on the interface that separates the two media. Show that the time of travel between A and B is minimum when the break in the path is such that

$$\frac{\sin \beta_i}{\sin \beta_r} = \frac{V_0}{V_1}.$$

The angles β_i and β_r are measured relative to the normal to the interface. In optics, the above relation is known as Snell's law.

1.3 Consider the travel between an arbitrary point (Q) in a square territory of side L and a point M situated in the middle of one of the sides. Travel proceeds at two speeds: fast travel (V_1) along one of the axes of symmetry and slow travel (V_0) through the remaining portions of the square. The point of type Q that is situated the farthest from M is point P. Minimize the travel time along the route PRM, where point R is free to slide along the V_1 axis. Show that the optimal location of the point R where the change in speeds occurs is

$$x = \frac{L}{2[(V_1/V_0)^2 - 1]^{1/2}}.$$

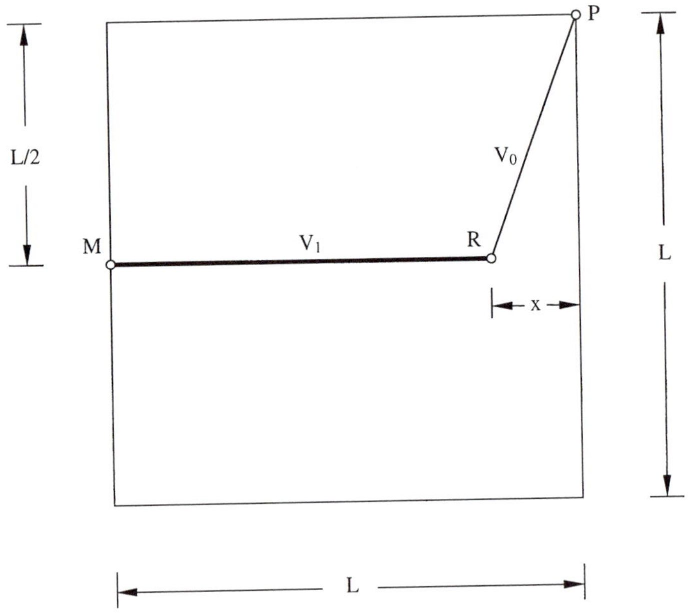

FIGURE P1.3

REFERENCES

1. A. L. Bloom, *Geomorphology*, Prentice-Hall, Englewood Cliffs, NJ, 1978.
2. L. P. Kadanoff, Fractals: where's the physics?, *Phys. Today*, February 1986, pp. 6–7.
3. A. Bejan, *Convection Heat Transfer*, 2nd ed., Wiley, New York, 1995.
4. A. Bejan, *Entropy Generation Minimization*, CRC, Boca Raton, FL, 1996.
5. B. B. Mandelbrot, *The Fractal Geometry of Nature*, Freeman, New York, 1982.
6. A. Bejan, *Advanced Engineering Thermodynamics*, 2nd ed., Wiley, New York, 1997.

7. D. Avnir, O. Biham, D. Lidar, and O. Malcai, Is the geometry of nature fractal?, *Science*, Vol. 279, 1998, pp. 39–40.

8. D. S. Lemmons, *Perfect Form*, Princeton Univ. Press, Princeton, NJ, 1997.

9. A. Bejan, Street network theory of organization in nature, *J. Adv. Transport.*, Vol. 30, 1996, pp. 85–107.

10. A. Bejan and G. A. Ledezma, Streets, tree networks, and urban growth: optimal geometry for quickest access between a finite-size volume and one point, *Physica A*, Vol. 255, 1998, pp. 211–217.

CHAPTER TWO

MECHANICAL STRUCTURE

2.1 Cantilever Beam: Objective and Constraints

In this chapter we review the engineering design method (objective, constraints) and illustrate how shape and structure result from the interplay between the global and local constraints. This chapter is not about flow systems: Mechanical structures are used solely for the purpose of illustrating the objective and constraints principle and how geometric form springs out of the principle. This choice is made for clarity, because we are all most comfortable with discussing mechanical systems. We start with the cantilever beam because it is the best known: It is one of the oldest and most elementary lessons in mechanical engineering. Let us review the design concepts that were mentioned in Chap. 1.

The *objective* of the beam is to support a load, for example force P in Fig. 2.1(a). It must do so with adequate stiffness, i.e., without excessive bending and with a minimum amount of beam material. Stiff beams that require less material perform better.

The principles that govern the behavior of the cantilever beam are classical [1–3]. We review them here because we will invoke them a few times before the chapter is over. With reference to the arbitrarily shaped beam sketched in Fig. 2.1, we assume that the beam is sufficiently long (e.g., with a length/thickness ratio greater than 5) so that its end deflection is due almost entirely to bending. In other words, we are neglecting the deformation that is due to shear. We also assume that the beam material is an elastic solid with Young's modulus E.

The end deflection (δ) is the vertical displacement of the free end $(x = 0)$ relative to the line drawn tangent to the clamped end $(x = L)$. It is given by a general result that is due to Castigliano:

$$\delta = \int_0^L \frac{M}{EI(x)} \frac{\partial M}{\partial P} \, \mathrm{d}x. \tag{2.1a}$$

For the loading shown in Fig. 2.1, the bending moment acting over the $x = \text{constant}$

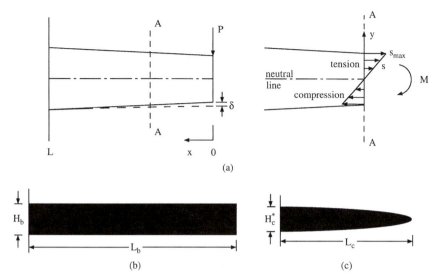

FIGURE 2.1. Two-dimensional cantilever beam: achieving design objective through the shaping of the beam profile.

cross section is $M = Px$, and Eq. (2.1a) becomes

$$\delta = \int_0^L \frac{Px^2}{EI(x)}\, dx. \tag{2.1b}$$

The factor I is the area moment of inertia of the cross section,

$$I = \int_A y^2\, dA, \tag{2.2}$$

where y is the distance measured away from the neutral line (the line of zero bending stress). The units of I are m^4, as shown by the case-specific formulas used in the following examples.

In each cross section the bending moment M is balanced by the collective effect of the bending stresses $s(y)$. Tensile and compressive stresses form above and below, respectively, the neutral line:

$$s = \frac{M}{I} y. \tag{2.3}$$

The maximum stress occurs in the fiber situated farthest from the neutral line:

$$s_{max} = \frac{M}{I} y_{max}. \tag{2.4}$$

The global constraint is the requirement that the designed beam must have a certain stiffness, or spring constant, i.e., a certain deflection δ when P is specified. In the following analysis we will compare several designs that have the same deflection under the same force.

The local constraint is the requirement that all the stresses must not exceed the maximum allowable stress level s_{ma}. We write this requirement as

$$|s(x, y)| \leq |s_{\max}(x)| \leq s_{\mathrm{ma}} \tag{2.5}$$

with the observation that, in general, the maximum stress that occurs in each cross section ($s_{\max}$) depends on longitudinal position, whereas s_{ma} is a constant. The stress $s_{\max}$ depends on x in two ways: through the bending moment ($M = Px$, in our case) and through $I(x)$, which may vary with the cross-sectional shape and dimensions along the beam. Local constraint (2.5) works throughout the beam – in an infinity of beam cross sections – and attains in this way a global character as well.

2.2 External Shape

An infinity of designs may emerge from the purpose and constraints thinking outlined in Section 2.1. The questions before us are the following: Which designs are better and which physical features lead to better designs? In the following analysis we find that one route to better designs is through the selection of the external shape of the system.

One of the simplest designs is the straight beam with the length L_b and constant rectangular cross section $H_b \times B$, where B is the third dimension in the direction perpendicular to the plane of Fig. 2.1(b). The rectangular cross section is characterized by $y_{\max} = H_b/2$ and

$$I_b = \frac{H_b^3 B}{12}. \tag{2.6}$$

By integrating Eq. (2.1) with $M = Px$ we obtain the deflection of the free end, i.e., the global constraint,

$$\delta = \frac{PL_b^3}{3EI_b}. \tag{2.7}$$

Because I_b is constant and $M = Px$, Eq. (2.4) shows that $s_{\max}(x)$ increases linearly from $s_{\max}(0) = 0$ at the free end to $s_{\max}(L) = PL_b y_{\max}/I_b$ at the embedded end. The local constraint amounts to setting $s_{\max}(L) = s_{\mathrm{ma}}$, which is the guarantee that the stress at any other point inside the beam does not exceed the maximum allowable stress,

$$s_{\mathrm{ma}} = \frac{PL_b H_b/2}{I_b}. \tag{2.8}$$

Finally, the goodness of the design of Fig. 2.1(b) is described quantitatively by the amount (e.g., volume) of material that has been used in the beam, for which the necessary dimensions (H_b, L_b) are furnished by Eqs. (2.7) and (2.8):

$$V_b = H_b L_b B = 9 \frac{EP\delta}{s_{\mathrm{ma}}^2}. \tag{2.9}$$

We already know that this design can be improved: We know this from the discussion of how the stresses vary through the beam volume. Most of the beam material does not work nearly as hard as the two lines of length B situated in the plane of the wall $(x = L_b)$ at $y = \pm H_b/2$. One place of improvement is the top and bottom longitudinal fibers of length L_b, along which s_{max} is generally smaller than s_{ma}. The best solution for such places is to vary the beam thickness $H_c(x)$ [Fig. 2.1(c)] in order to force s_{max} to be equal to the constant maximum allowable stress s_{ma} at every x along the beam [see Eq. (2.4)]:

$$s_{ma} = \frac{Px H_c/2}{I_c}, \tag{2.10}$$

where

$$I_c(x) = \frac{H_c^3 B}{12}. \tag{2.11}$$

Equation (2.10) requires that in this new design the group $x\, H_c/I_c$ must be a constant, which means that H_c must vary as $x^{1/2}$:

$$H_c(x) = \left(\frac{6Px}{s_{ma}B} \right)^{1/2}. \tag{2.12}$$

We obtain the deflection of the free end by integrating Eq. (2.1b) in combination with Eqs. (2.11) and (2.12):

$$\delta = \frac{2PL_c^3}{3EI_c^*}, \tag{2.13}$$

where I_c^* is based on the beam thickness at the clamped end:

$$H_c^* = \left(\frac{6PL_c}{s_{ma}B} \right)^{1/2}. \tag{2.14}$$

Because the beam profile is parabolic, the profile area is $(2/3)L_c H_c^*$, for which L_c and H_c^* are furnished by Eqs. (2.13) and (2.14). The required volume is

$$V_c = \frac{2}{3}L_c H_c^* B = 3\frac{EP\delta}{s_{ma}^2}. \tag{2.15}$$

Equations (2.15) and (2.9) tell the whole story: The ratio $V_c/V_b = 1/3$ means that the design of Fig. 2.1(c) requires only one third of the material used in Fig. 2.1(b) while meeting the same global and local constraints. This is astonishing! The savings in beam material are evident in Figs. 2.1(b) and 2.1(c), in which the beam profiles have been drawn to scale. The formulas of this section can be used to show that the relative dimensions of the two designs are $H_c^*/H_b = 2^{-1/3}$ and $L_c/L_b = 2^{-2/3}$.

The design of Fig. 2.1(c) represents an ideal limit, which can be attained only approximately in practice. In this limit the beam cross section is zero at $x = 0$, i.e., right under the end load, where a finite cross-sectional area is demanded by the presence of finite shear stresses over the cross section (neglected in this analysis).

We can pursue the shape optimization by considering alternate global constraints, for example, the beam length L. This is the alternative that appears sometimes in strength-of-materials treatises. The length constraint will be used in Section 2.5. The reductions in solid material when the length is constrained are less dramatic than when the stiffness P/δ is constrained (this section).

The parabolic profile of the cantilever beam of constant strength [Fig. 2.1(c)] is perhaps the oldest result in the theory of slender elastic beams or the field of strength of materials. It was published by Galileo Galilei [4] in 1638, even before Hooke's law was enunciated. This result has a new and important role to play in the present theory, as we will conclude in Chapter 12. The concept of "objective" drove the design in the direction of optimization (less material, improved performance). Out of this evolution sprung the shape of the system – external shape, or *shaping*, as a *mechanism* by which the system achieves its objective.

2.3 Internal Structure

There is another mechanism for improving the system, and we can uncover it by going back to the idea that led us to external shape. When the beam is solid and has a constant cross section [Fig. 2.1(b)], the internal stresses are generally lower than the maximum allowable stress s_{ma}. Notorious for not working hard for the common good is the core of material situated in the vicinity of the neutral line. It makes sense to modify the beam design so that most of this material is eliminated.

In this section we pursue this idea by using a cantilever beam with round cross section. In the simplest design [Fig. 2.2(1)] the beam is solid and its diameter D_1 is constant. This cross section is characterized by $I_1 = \pi D_1^4/64$ and $y_{max} = D_1/2$, so that Eqs. (2.1) and (2.4), in combination with $M = Px$ and $s_{max}(L_1) = s_{ma}$, yield

$$\delta = \frac{PL_1^3}{3EI_1}, \tag{2.16}$$

$$s_{ma} = \frac{32PL_1}{\pi D_1^3}. \tag{2.17}$$

The volume of this cylindrical beam is $V_1 = (\pi/4)D_1^2 L_1$, which, after D_1 and L_1 are eliminated based on Eqs. (2.16) and (2.17), becomes

$$V_1 = 12\frac{EP\delta}{s_{ma}^2}. \tag{2.18}$$

The core material has been eliminated in the competing design shown in Fig. 2.2(2). The beam cross section is constant, and has the shape of an annulus of outer diameter D_2 and thickness t_2, which is considerably smaller than D_2. The relevant bending parameters of this cross section are $I_2 = \pi R_2^3 t_2$ and

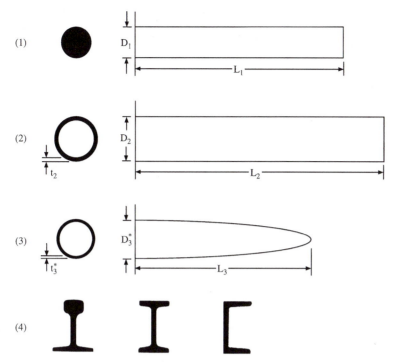

FIGURE 2.2. Solid cantilever beam with round cross-section (1) Achieving design objective through (2) internal structure and (3) combined shape and structure. Practical beam cross-sections that represent improved internal structure (4).

$y_{max} = D_2/2 = R_2$, such that the global and local constraints become

$$\delta = \frac{PL_2^3}{3EI_2},\tag{2.19}$$

$$s_{ma} = \frac{PL_2}{\pi R_2^2 t_2}.\tag{2.20}$$

The volume of the tubular beam is $V_2 = 2\pi R_2 t_2 L_2$, for which L_2 and R_2 are provided by Eqs. (2.19) and (2.20):

$$V_2 = 6\frac{EP\delta}{s_{ma}^2}.\tag{2.21}$$

In conclusion, the tubular design is better than the solid rod design, because $V_2/V_1 = 1/2$: The amount of material has been cut in half! The relative dimensions of the two designs are $D_2/D_1 = (4\varepsilon)^{-2/5}$ and $L_2/L_1 = (4\varepsilon)^{-1/5}$, where $\varepsilon = t_2/R_2$ represents the relative thickness of the annular wall. The cross sections shown in Figs. 2.2(1) and 2.2(2) have been drawn to scale with the assumption that the aspect ratio $\varepsilon = 0.1$. Important to note is that V_2 does not depend on ε, as long as the value of ε is considerably smaller than 1.

As far as structural engineering goes, the tubular beam is a very old result, which was most likely suggested by nature: The stems of reed and bamboo are tubular.

New is the role that this result plays in the present theory. The solid rod and tubular designs satisfy the same global and local constraints. The design evolved from the solid rod and the tubular beam in search of better performance, which is its purpose. During this change the system developed internal structure: an empty core enclosed by a thin cylindrical wall. The new idea is that internal structure, or *structuring*, is a *mechanism* by which the system achieves its objective.

Many other internal structures have evolved in engineering for the same purpose, for example, the I beam and the rail of the railroad track [see Fig. 2.2(4)].

2.4 Shape and Structure, Together

The design achieves its objective even better when it undergoes simultaneously external shaping and internal structuring. The conclusions drawn in Sections 2.2 and 2.3 permit us to shorten the search and focus directly on Fig. 2.2(3). This beam has an annular cross section, and its radius R_3 varies with the longitudinal position so that $s_{max} = s_{ma}$ at every point of the top and bottom fibers of the beam [see Eq. (2.4)]:

$$s_{ma} = \frac{Px}{\pi R_3^2 t_3}. \qquad (2.22)$$

If we assume that the relative thickness of the wall is independent of axial position, $\varepsilon = t_3/R_3 \ll 1$, constant, Eq. (2.22) means that the shape of the beam profile is cubic $(x \sim R_3^3)$, i.e.,

$$R_3 = \left(\frac{Px}{\pi \varepsilon s_{ma}}\right)^{1/3}. \qquad (2.23)$$

The end deflection is derived from Eq. (2.1b), in which $L = L_3$, $M = Px$, and $I = I_3 = \pi R_3^4 \varepsilon$:

$$\delta = \frac{3(\pi \varepsilon)^{1/3} s_{ma}^{4/3} L_3^{5/3}}{5 E P^{1/3}}. \qquad (2.24)$$

The hollow cubic profile of Fig. 2.2(3) was drawn to scale relative to the solid cylinder of Fig. 2.2(1) with the assumption that $\varepsilon = 0.1$. It can be shown that $L_3/L_1 = (5/9)^{3/5}(4\varepsilon)^{-1/5}$ and $D_3^*/D_1 = (5/9)^{1/5}(4\varepsilon)^{-2/5}$. The volume of beam material in Fig. 2.2(3) is obtained by use of the cubic shape of Eq. (2.23):

$$V_3 = \int_0^{L_3} 2\pi R_3 t_3 \, dx = \frac{6(\pi \varepsilon)^{1/3} P^{2/3} L_3^{5/3}}{5 s_{ma}^{2/3}}. \qquad (2.25)$$

After eliminating L_3 between Eqs. (2.24) and (2.25), we find that

$$V_3 = 2\frac{E P \delta}{s_{ma}^2}, \qquad (2.26)$$

which means that the external shaping of the hollow beam reduces its material volume to one third of its original value, $V_3/V_2 = 1/3$. The simultaneous shaping and structuring of the design reduces the required material to one sixth, $V_3/V_1 = 1/6$.

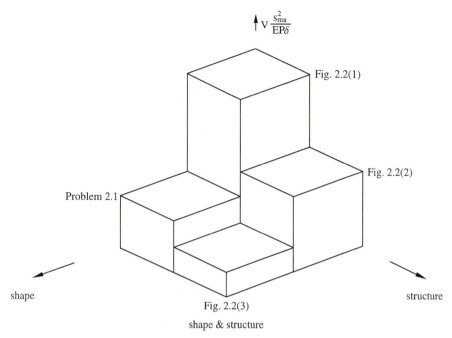

FIGURE 2.3. The interplay between shaping and structuring en route to decreasing the system volume.

These are dramatically large payoffs. They result from relatively simple implementations of the objective and constraints philosophy discussed in Chap. 1. The geometry of the system – its external shape and internal structure – is the result of the principle of optimization subject to global and local constraints. Shape and structure are the mechanisms – the paths, the means – for achieving the design objective.

It is only now, at the very last moment in this presentation, that we look at nature (see the lower part of Fig. 1.3). Tree trunks and branches are always thicker near the embedded end (see Fig. 2.1). The internal structure of long stems (reed, bamboo) and bird bones is similar to that of Fig. 2.2(2). The same principle that gave us the designed system on the engineering side of Fig. 1.3 serves now as *pure theory* on the nature side of the same figure.

The three designs analyzed based on Fig. 2.2 are summarized in Fig. 2.3. Plotted on the vertical is the dimensionless measure of the goodness of the design: the volume of beam material, $Vs_{ma}^2/(EP\delta)$. Problem 2.1 covers the fourth possibility, which is the shaping of the external surface of the solid beam such that the maximum stress is constant along the top and bottom fibers of the beam. Dramatic improvements are registered through shaping, structuring, and shaping and structuring together. The least material is required by the design in which shaping and structuring are applied together.

Although both shaping and structuring lead to improvements in the global performance, the volume change caused by the implementation of one method is greater when the method is implemented alone. In other words, the net impact of one method depends on whether the other method has been applied already. This point is made graphically in Fig. 2.3 and is found in many other instances of system design [5]. Applying more methods at the same time is accompanied by increased costs, not just decreasing improvements. This trend is commonly known as the law of diminishing returns.

It is an amazing coincidence that the round cross section of the thin-walled tube also offers minimal resistance to internal fluid flow (Section 6.3). Large stiffness in bending and low resistance in duct flow call for the same geometry. This coincidence is responsible for the prevalence of the round hollow cylinder as a geometric feature in both nature and engineering.

2.5 Column in End Compression

The cantilever lesson of Figs. 2.1 and 2.2 has been learned many times in engineering. The open endedness of this chapter is illustrated with a second example from mechanical structures with optimal shape: the strut, i.e., the elastic column subjected to end-to-end compression, Fig. 2.4(a). The length of the column L is fixed and serves as the global constraint. We do not have a local constraint in this example. The largest compressive force that the column can support without undergoing any lateral deformation is the first (lowest) critical load P. According to Euler's buckling theory for straight columns with constant cross section [1–3], the first critical load is

$$P = \frac{\pi^2 E I_0}{L^2},\tag{2.27}$$

where E is Young's modulus of elasticity of the column material and I_0 is the area moment of inertia of the cross section. If the constant cross section is round and has

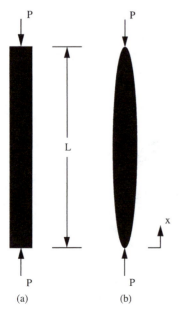

FIGURE 2.4. Slender columns supporting the same critical compressive load: (a) constant cross section, (b) optimally varying cross section.

the diameter D_0, the area moment of inertia is $I_0 = \pi D_0^4/64$ and the total volume of beam material is $V_0 = (\pi/4) D_0^2 L$.

The objective of the beam is to support P across the spacing L. Can this objective be fulfilled by a better design, i.e., by a beam that requires less material? In search of an answer, the designer and the design proceed in the direction traced by Galileo Galilei's constant-strength beam (Fig. 2.1). The better beam will have a variable cross section; its diameter will be larger in the middle section, because there the bending moment is larger than near the two ends. The optimal shape of a solid column with hinged ends has been determined based on variational calculus [6–8] and is given parametrically by

$$A_1(\theta) = \frac{4V_1}{3L} \sin^2 \theta, \qquad (2.28)$$

$$x(\theta) = \frac{1}{\pi}\left(\theta - \frac{1}{2}\sin 2\theta\right), \qquad 0 \leq \theta \leq \pi. \qquad (2.29)$$

In this formulation $A_1[\theta(x)]$ is the variable cross section of the column, where x is the longitudinal position measured from one of the ends. It can be shown that the first critical load for such a column [8] is

$$P = \frac{\pi E V_1^2}{3L^4}, \qquad (2.30)$$

where V_1 is the total volume of column material. This optimal shape has been drawn to scale in Fig. 2.4(b).

FIGURE 2.5. Optimally shaped columns in end compression over the 1992 Genova exhibit celebrating the 500th anniversary of Columbus' discovery of America.

How much material has been saved in going from Fig. 2.4(a) to Fig. 2.4(b)? The answer becomes clear when we put Eq. (2.27) in the same form as Eq. (2.30) by noting that $I_0 = \pi D_0^4/64 = V_0^2/(4\pi L^2)$. We obtain

$$P = \frac{\pi E V_0^2}{4L^4}, \tag{2.31}$$

which, next to Eq. (2.30), shows that $V_0^2/4 = V_1^2/3$ or that $V_1/V_0 = (3/4)^{1/2} = 0.87$. The optimal shaping of the column produces a 13% reduction in material, which is evident when we compare the two black silhouettes shown in Fig. 2.4.

Such reductions are not negligible in applications in which weight is an important issue – a penalty to the greater installation that uses the column in end compression. Think of the cranes mounted on the decks of ships for loading and unloading the cargo stored inside the hull. This image was illustrated with beauty in the harbor of Genova, Italy, in the summer of 1992 (Fig. 2.5): See the struts (hollow, most likely) holding the tent over the exhibit celebrating 500 years since Columbus' discovery of America. Unwittingly, the artist paid homage not only to Columbus but also to another great visionary, Galileo Galilei, who came 100 years later from a city (Pisa) not far from Genova.

2.6 The Concept of "Better"

The objective and constraints principle sculpts the limited amount of material into a form with purpose. The functions of the system are spoken out by its shape. For example, if the mechanical system has functions in addition to the stiffness in bending discussed in this chapter, the system architecture develops additional features that reflect these additional objectives.

The wing feather of the goose (Fig. 2.6) beautifully illustrates this conclusion. The feather must be stiff in bending, and for this reason its stem has the two features highlighted in this chapter: The bonelike material is structured into a thin

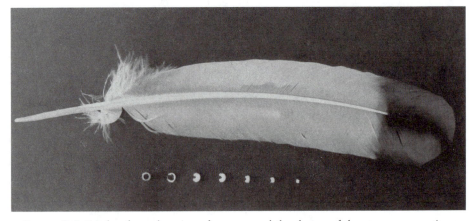

FIGURE 2.6. Feather from the wing of a goose and the shapes of the stem cross section.

wall, and the transversal dimension of the cross section decreases smoothly away from the root. Support for this explanation is offered by the additional observation that the geometry of Fig. 2.2(3) is found at both ends of the stem of the feather. The hollow and tapered tube occurs on both sides of the plane of implantation into the wing. Both parts are subjected to bending that is due to distributed transversal loading. The kidney-shaped cross section is found only in the part that works in air. Perhaps this additional feature (the kidney shape) can be reasoned in the same way as all the other features discussed in this book. Many other natural structures can be reasoned on the basis of mechanical design principles [9–11].

Natural shapes appear to be perfect – perfectly smooth as a Brancusi sculpture – because the less-than-perfect designs were melted one by one into perfection over a very long period of time. To our eyes and minds, seen through our extremely narrow time slit, it seems as if nature has "smoothed over" its previous mistakes. None of the preceding principle-generated forms were mistakes. Similarly, none of the simple structural forms that began this chapter were bad ideas. One good form leads to the next, as the constructal principle demands: objective served better, while under the grip of global and local constraints.

There is a time arrow to all these forms, and it points toward the *better*. Better from what point of view? To meet what objective? The concept of better (or optimal, shortest, fastest, cheapest, etc.) is defined by the constructal law in the same way that the concepts of temperature, energy and entropy are defined, respectively, by the zeroth, first, and second laws of thermodynamics.

PROBLEMS

2.1 Consider the constant-strength design of a *solid* straight beam of length L, round cross section, and variable diameter $D(x)$. As in the three examples of Fig. 2.2, the beam is embedded at one end and loaded with the transversal force P at the other end. The distance x is measured from the loaded end toward the embedded end. The end deflection δ caused by P is specified. The maximum stresses are set equal to the maximum allowable stress s_{ma} (constant).

Show that the beam shape must be such that D is proportional to $x^{1/3}$. Derive expressions for the beam length and the largest diameter [$D^* = D(L)$] as functions of the specified material and constraints (E, s_{ma}, δ, P). Determine the volume of this solid constant-strength beam and show that it is equal to 1/3 of the volume required by the corresponding design with constant diameter, Fig. 2.2(1).

2.2 The horizontal beam with variable thickness $H(x)$ shown in Fig. P2.2 connects a body of weight F and temperature T_0 to a support of temperature T_L. The beam geometry is slender and two dimensional, with the width B measured in the direction perpendicular to the figure. The distance L and the amount of beam material are fixed. The thermal conductivity k and modulus of elasticity E of this material are known constants.

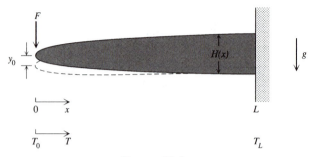

FIGURE P2.2

The function of the beam is to support the weight F as rigidly as possible while impeding the transfer of heat from T_L to T_0. It is a mechanical support and, simultaneously, a thermal insulation. The designer is interested in arranging the beam material [the thickness $H(x)$] in such a way that the tip deflection y_0 and the end-to-end heat transfer rate q are minimized.

One possibility is to shape the beam cross section such that the thickness H increases as x^n as x increases, where the exponent n is a number between 0 and 1. Determine analytically y_0 and q as functions of n and comment on how the shape of the beam cross section (n) affects the success of the design [12, 13].

2.3 A low-temperature apparatus T_0 (Fig. P2.3) is supported from room temperature T_L through a vertical bar in tension. This bar must be strong enough to carry the weight of the apparatus F, slender enough to allow the smallest heat transfer rate q to flow from T_L to T_0, and rigid enough to fix the position of the apparatus on the vertical line. The rigidity requirement means that the total elastic elongation ΔL caused by the longitudinal force F must be as small as possible.

At certain temperatures below room temperature, the thermal conductivity and modulus of elasticity of structural materials vary roughly as

$$k(T) \cong \bar{k}[1 + b(2\theta - 1)],$$
$$E(T) \cong \bar{E}[1 - c(2\theta - 1)],$$

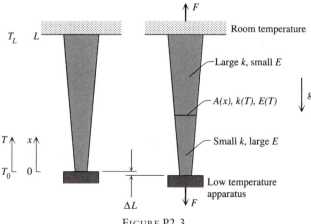

FIGURE P2.3

where $\bar{k}$ and $\bar{E}$ are the temperature-averaged values,

$$\bar{k} = \frac{1}{T_L - T_0} \int_{T_0}^{T_L} k \, dT, \qquad \bar{E} = \frac{1}{T_L - T_0} \int_{T_0}^{T_L} E \, dT,$$

b and c are two positive numbers less than 1, proportional to dk/dT and $-dE/dT$, and

$$\theta = \frac{T - T_0}{T_L - T_0}.$$

Note that k decreases and E increases toward lower temperatures. These characteristics of structural materials suggest [12, 13] that a tapered bar $(0 < a < 1)$,

$$A(x) = \bar{A}\left[1 + a\left(2\frac{x}{L} - 1\right)\right] \quad (0 < a < 1),$$

$$\bar{A} = \frac{1}{L}\int_0^L A(x) \, dx \quad \text{(constant)},$$

might exhibit a smaller elongation in tension than the bar with x-independent geometry $(a = 0)$. It can be argued that the overall elongation is reduced when more material is positioned near the warm end, where E is smaller, than near the cold end. Note further that when the L-averaged cross-sectional area is held as fixed, the total amount (weight) of the structural material used is fixed, regardless of the shape chosen for the bar.

Evaluate the merit of this proposal by determining the effect of taper (a) not only on the total elongation ΔL but also on the heat leak q. It is advisable to conduct the analysis by using the dimensionless heat transfer rate

$$\tilde{q} = \frac{q}{\bar{k}\bar{A}(T_L - T_0)/L}$$

and the dimensionless total elongation

$$\delta = \frac{\Delta L}{L}\frac{\bar{A}\bar{E}}{F}.$$

Determine δ and $\tilde{q}$ as functions of geometry (a) and choice of structural material (b, c), and comment on how you might design a rigid support in tension so that it can serve as thermal insulation at the same time.

1. J. P. Den Hartog, *Strength of Materials*, Dover, New York, 1961.
2. L. Stoicescu, *Strength of Materials*, Galati University Press, Galati, Romania, 1987.
3. J. Case and A. H. Chilver, *Strengths of Materials and Structures*, 2nd ed., Edward Arnold, London, 1971.
4. G. Galilei, *Discorsi e Dimostrazioni Matematiche*, Leiden, The Netherlands, 1638, in I. Todhunter and K. Pearson, *A History of the Theory of Elasticity and of the Strength of Materials*, Dover, New York, 1960, Vol. 1.
5. A. Bejan, *Advanced Engineering Thermodynamics*, 2nd ed., Wiley, New York, 1997, p. 409.

6. T. Clausen, *Bulletin Physico-Mathematiques et Astronomiques*, 1849, pp. 279–294; summarized in I. Todhunter and K. Pearson, *A History of the Theory of Elasticity and of the Strength of Materials*, Dover, New York, 1960, Vol. 2.

7. J. B. Keller, The shape of the strongest column, *Arch. Rational Mech. Anal.*, Vol. 5, 1960, pp. 575–585.

8. J. F. Wilson, D. M. Holloway, and S. B. Biggers, Stability experiments on the strongest columns and circular arches, *Exp. Mech.*, Vol. 11, 1971, pp. 303–308.

9. S. A. Wainwright, W. D. Biggs, J. D. Currey, and J. M. Gosline, *Mechanical Design in Organisms*, Princeton Univ. Press, Princeton, NJ, 1976.

10. S. A. Wainwright, *Axis and Circumference*, Harvard Univ. Press, Cambridge, MA, 1988.

11. M. French, *Invention and Evolution: Design in Nature and Engineering*, 2nd ed., Cambridge Univ. Press, Cambridge, UK, 1994.

12. A. Bejan, Heat transfer as a design-oriented course: mechanical supports as thermal insulators, *Int. J. Mech. Eng. Educ.*, Vol. 22, No. 1, 1994, pp. 29–41.

13. A. Bejan, *Heat Transfer*, Wiley, New York, 1993.

CHAPTER THREE

THERMAL STRUCTURE

3.1 Cooling Electronics: Objective and Constraints

In this chapter we begin to discuss flow systems, which is the main focus of this book. Our objective continues to be the illustration of how objective and constraints thinking takes us, almost without notice, to geometric form in engineering. The flows in this chapter carry energy and fluid: the heat generated in a constrained volume and the coolant used for the purpose of removing the heat current.

Consider designing a package of electronic components, which must fit inside a given volume – the *global* constraint. The *objective* of the design is to install as much circuitry as possible in this volume. In rough terms, this is equivalent to installing as much heat generation rate q as possible, because electrical components generate heat. The highest temperatures in the package (the hot spots) must not exceed a specified value T_{max}: This is the *local* constraint. If the local temperature T rises above this allowable ceiling, the electronic functioning of the local component is threatened. In sum, the thermal design is better when q is larger. In other words, the design is better when the global thermal-conductance ratio $q/(T_{max} - T_0)$ is larger, where T_0 is the initial (reference, sink) temperature of the coolant that absorbs q.

3.2 Volume Cooled by Natural Convection

Internal structure (characteristic graininess, internal spacings) results from the pursuit of the design objective subject to global and local constraints. To illustrate, in this section we assume that the heat generation rate q is spread almost uniformly over the given volume. The electronic circuitry is mounted on equidistant vertical boards of height H, filling a space of height H and horizontal dimensions L and W. For simplicity, we assume that the configuration is two dimensional with respect to the ensuing flow, as shown in Fig. 3.1. Not specified is the board-to-board

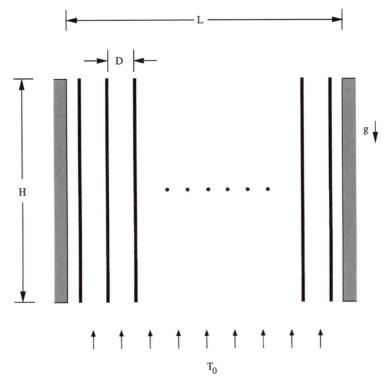

FIGURE 3.1. Two-dimensional volume that generates heat and is cooled by natural convection.

spacing D or the number of vertical parallel-plate channels,

$$n = \frac{L}{D}. \tag{3.1}$$

Determining the spacing D that maximizes q can be a laborious task. We will show, however, that with a little ingenuity the problem can be solved on the back of an envelope. The method, outlined first in 1984 [1] and refined in Refs. 2 and 3, begins with the assumptions that (1) the flow is laminar, (2) the board surfaces are sufficiently smooth to justify the use of heat transfer results for natural convection over vertical smooth walls, and (3) the maximum temperature T_{max} is closely representative of the temperature at every point on the board surface. The method consists of two steps. In the first, we identify the two possible extremes: the small-D limit and the large-D limit. In the second step, the two extremes are intersected to locate the D value that maximizes q.

Small Spacings. When D becomes sufficiently small, the channel formed between two boards becomes narrow enough for the flow and heat transfer to be fully developed. According to the first law we have $q_1 = \dot{m}_1 c_p (T_{max} - T_0)$ for the heat transfer rate extracted by the coolant from one of the channels of spacing D, where T_0 is the inlet temperature and T_{max} is the outlet temperature. The mass flow rate is $\dot{m}_1 = \rho(DW)U$, where the mean velocity U can be estimated by the replacement

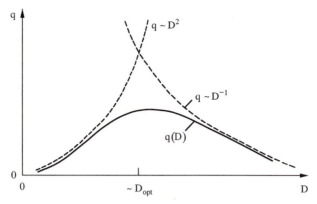

FIGURE 3.2. The intersection of asymptotes method: the maximization of the global thermal conductance of a stack of vertical plates in natural convection.

of $\Delta p/L$ with $\rho g \beta (T_{max} - T_\infty)$ in the Poiseuille flow solution [2]:

$$\rho g \beta (T_{max} - T_0) = f \frac{4}{D_h} \frac{1}{2} \rho U^2. \tag{3.2}$$

For laminar flow between parallel plates we have $f = 24/(U D_h / \nu)$ and $D_h = 2D$; hence

$$U = g\beta (T_{max} - T_0) D^2/(12\nu), \tag{3.3}$$

$$\dot{m}_1 = \rho D W U = \rho W g \beta (T_{max} - T_0) D^3/(12\nu). \tag{3.4}$$

The rate at which heat is removed from the entire package is $q = nq_1$, or

$$q = \rho c_p \, W L g \beta (T_{max} - T_0)^2 D^2/(12\nu). \tag{3.5}$$

In conclusion, in the $D \to 0$ limit the total heat transfer rate varies as D^2. This trend is indicated by the small-D asymptote plotted in Fig. 3.2.

Large Spacings. Consider next the limit in which D is large enough that it exceeds the thickness of the thermal boundary layer that forms on each vertical surface: $\delta_T \sim H \mathrm{Ra}_H^{-1/4}$, where $\mathrm{Ra}_H = g\beta H^3 (T_{max} - T_0)/(\alpha \nu)$. In other words, the spacing is considered large when [1, 2]

$$D > H \left[\frac{g\beta H^3 (T_{max} - T_0)}{\alpha \nu} \right]^{-1/4} \qquad (\mathrm{Pr} \ge 0.5). \tag{3.6}$$

In this limit the boundary layers are distinct (thin compared with D), and the center region of the board-to-board spacing is occupied by fluid of temperature T_0. The number of distinct boundary layers is $2n = 2L/D$ because there are two for each D spacing. The heat transfer rate through one boundary layer is $\bar{h} H W (T_{max} - T_0)$ for which the heat transfer coefficient $\bar{h}$ is furnished by [2]

$$\frac{\bar{h} H}{k} = 0.517 \mathrm{Ra}_H^{1/4}, \tag{3.7}$$

where $Ra_H = g\beta(T_{max} - T_0)H^3/(\alpha\nu)$. The rate of heat transfer extracted from the entire package is $2n$ times larger than $\bar{h}HW(T_{max} - T_0)$:

$$q = 2\frac{L}{D}HW(T_{max} - T_0)\frac{k}{H}0.517\,Ra_H^{1/4}. \tag{3.8}$$

Equation 3.8 shows that in the large-D limit the total heat transfer rate varies as D^{-1} as the board-to-board spacing changes. This second asymptote has been added to Fig. 3.2.

Optimal Spacings. What we have determined so far are the two asymptotes of the actual (unknown) curve of q versus D. Figure 3.2 shows that the asymptotes intersect above what would be the peak of the actual $q(D)$ curve. The trade-off between the two extremes is illustrated in color in Fig. 3.3, and is explained in the caption. It is not necessary to determine the actual $q(D)$ relation: The optimal spacing D_{opt} for maximum q can be estimated as the D value where Eqs. (3.5) and (3.8) intersect [1, 2]:

$$\frac{D_{opt}}{H} \cong 2.3\left[\frac{g\beta(T_{max} - T_0)H^3}{\alpha\nu}\right]^{-1/4}. \tag{3.9}$$

This D_{opt} estimate is within 20% of the optimal spacing deduced based on much lengthier methods, such as the maximization of the $q(D)$ relation [4] and the finite-difference simulations of the complete flow and temperature fields in the package (for a review of this body of work see Ref. 5).

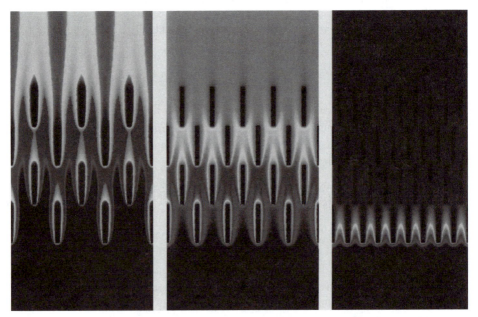

FIGURE 3.3. Three designs for the internal structure of a fixed volume with fixed total heat generating rate and a stream of coolant flowing vertically. The objective is to maximize the global thermal conductance between the volume and the stream, which is equivalent to minimizing the red areas (hot-spot temperatures) that occur inside the volume (see Plate II).

An estimate of the maximum heat transfer rate can be obtained by substitution of D_{opt} into Eq. (3.8) [or Eq. (3.5)]:

$$q_{max} \approx 0.45 k (T_{max} - T_0) \frac{LW}{H} \mathrm{Ra}_H^{1/2}. \tag{3.10}$$

The approximation sign is a reminder that the peak of the actual $q(D)$ curve falls under the intersection of the two asymptotes (Fig. 3.2). This result also can be expressed as the maximum volumetric rate of heat generation in the $H \times L \times W$ volume:

$$\frac{q_{max}}{HLW} \approx 0.45 \frac{k}{H^2} (T_{max} - T_0) \mathrm{Ra}_H^{1/2}. \tag{3.11}$$

In conclusion, if the heat transfer mechanism is laminar natural convection, the maximum volumetric rate of heat generation is proportional to $(T_{max} - T_0)^{3/2}$, $H^{-1/2}$, and the property group $k(g\beta/\alpha\nu)^{1/2}$.

The assumption that the heat generation rate q is spread on equidistant vertical plates as tall as the volume was made for the sake of illustrating in the simplest terms how internal structure results from the purpose and constraints principle. Optimal internal structure is a characteristic of all systems that share the same purpose and constraints. The generality of this result is supported by several more recent studies in which the stack of H-tall plates was replaced with arrays of internal heating elements of other shapes.

For example, when the fixed volume is filled with many equidistant parallel staggered plates, which are considerably shorter than the volume height [$b \ll H$ in Fig. 3.4(a)], there exists an optimal horizontal spacing (D) between neighboring plates [6]. In laminar natural convection, the optimal spacing scales with the height of the fixed volume, not with the height of the individual plate, e.g.,

$$\frac{D_{opt}}{H} \cong 0.63 \left(\frac{Nb}{H}\right)^{1.48} \mathrm{Ra}_H^{-0.19}, \tag{3.12}$$

where N is the number of plate surfaces that face one elemental channel [e.g., $N=4$ in Fig. 3.4(a)]. The dimensionless group (Nb/H) is of the order of 1 and represents the relative contact area present along the boundaries of the elemental channel. The similarities between relations (3.12) and (3.9) are important, because they point to the robustness of the design principle that generates internal spacings. The corresponding correlation for the maximum thermal conductance is [6]

$$\frac{q_{max}}{k(T_{max}-T_0)LW/H} \cong 1.92 \exp\left(-0.7\frac{Nb}{H}\right) \mathrm{Ra}_H^{0.43}, \tag{3.13}$$

where W is the dimension perpendicular to the plane of Fig. 3.4(a). The form and numerical estimates based on relation (3.13) are similar to those of relation (3.11). Experiments and numerical simulations show that correlations (3.12) and (3.13) for vertical staggered plates in natural convection are accurate to within 6% in the range $10^3 \leq \mathrm{Ra}_H \leq 5 \times 10^5$ and $0.4 \leq (Nb/H) \leq 1.2$.

When the volume is heated by an array of horizontal cylinders of diameter D [Fig. 3.4(b)], the optimal cylinder-to-cylinder spacing S scales once again with the overall height of the volume [7]. In laminar natural convection the optimal spacing

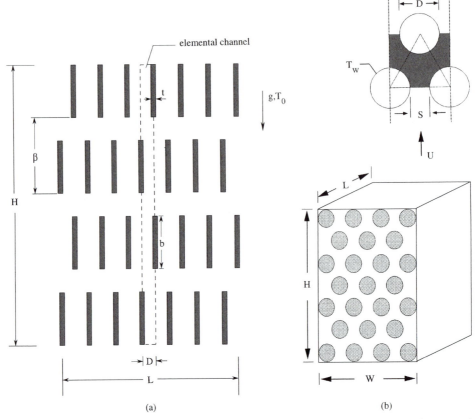

FIGURE 3.4. Natural convection cooling of a volume heated uniformly: (a) array of staggered plates [6], and (b) array of horizontal cylinders [7].

is correlated closely by a formula that resembles relation (3.9):

$$\frac{S_{opt}}{H} \cong 2.72 \left(\frac{H}{D}\right)^{1/12} Ra_H^{-1/4} + 0.263 \frac{D}{H}, \tag{3.14}$$

where the second term on the right-hand side is a small correction factor. The correlation for the corresponding maximum thermal conductance is

$$\frac{q_{max} D^2}{k(T_{max} - T_0)HLW} \cong 0.448 \left[\left(\frac{H}{D}\right)^{1/3} Ra_D^{-1/4}\right]^{-1.6}, \tag{3.15}$$

where $Ra_D = g\beta(T_{max} - T_0)D^3/(\alpha\nu)$. Relations (3.14) and (3.15) agree to within 1.7% with numerical simulations and experiments performed in the range $Pr = 0.72$, $350 \leq Ra_D \leq 10^4$, and $6 \leq H/D \leq 20$.

In conclusion, the maximization of the overall thermal conductance subject to the global and local constraints generates a characteristic internal structure – an optimal relative positioning of the solid components that fill the given volume. This is the principle that generates internal geometric form.

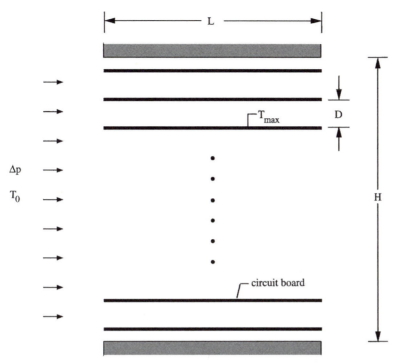

FIGURE 3.5. Two-dimensional volume that generates heat and is cooled by forced convection.

3.3 Volume Cooled by Forced Convection

Consider now the analogous problem of installing the optimal number of heat generating boards in a space cooled by forced convection [8, 9]. As shown in Fig. 3.5, the swept length of each board is L and the transverse dimension of the entire package is H. The width of the stack W is perpendicular to the plane of the figure. We retain the assumptions (1)–(3) listed under Eq. (3.1). The thickness of the individual board is again negligible relative to the board-to-board spacing D, so that the number of channels is

$$n = \frac{H}{D}. \tag{3.16}$$

The pressure difference across the package, Δp, is assumed constant and known. This is a good model for electronic systems in which several packages and other features (e.g., channels) receive their coolant *in parallel* from the same plenum. The plenum pressure is maintained by a fan, which may be located upstream or downstream of the package.

Small Spacings. When D becomes sufficiently small, the channel formed between two boards becomes narrow enough for the flow and the heat transfer to be fully developed. In this limit, the mean outlet temperature of the fluid approaches the board temperature T_{max}. The total rate of heat transfer from the $H \times L \times W$ volume

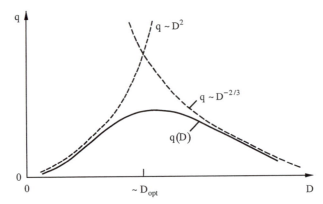

FIGURE 3.6. The maximization of the global thermal conductance of a stack of parallel plates in forced convection.

is $q = \dot{m}c_p(T_{max} - T_0)$, where $\dot{m} = \rho HWU$. The mean velocity through the channel, U, is known from the Hagen–Poiseuille flow solution [2]:

$$U = \frac{D^2}{12\mu} \frac{\Delta p}{L}. \tag{3.17}$$

The corresponding expression for the total heat transfer rate is

$$q = \rho HW \frac{D^2}{12\mu} \frac{\Delta p}{L} c_p(T_{max} - T_0). \tag{3.18}$$

In this way we conclude that the total heat transfer rate varies as D^2. This trend is illustrated by the small-D asymptote in Fig. 3.6.

Large Spacings. Consider next the limit in which D is large enough that it exceeds the thickness of the thermal boundary layer that forms on each horizontal surface. In this case it is necessary to determine the free-stream velocity U_0 that sweeps these boundary layers. Because the pressure drop Δp is fixed, a force balance on the $H \times L \times W$ control volume reads

$$\Delta p HW = \bar{\tau}(2n)LW, \tag{3.19}$$

where $\bar{\tau}$ is the wall shear stress averaged over L, namely, $\bar{\tau} = 0.664\,\rho U_0^2\,Re_L^{-1/2}$ for $Re_L \leq 5 \times 10^5$ [2]. Equation (3.19) yields

$$U_0 = \left(\frac{\Delta p H}{1.328 n L^{1/2}\rho\nu^{1/2}}\right)^{2/3}. \tag{3.20}$$

The heat transfer rate through a single board surface is $q_1 = \bar{h}LW(T_{max} - T_0)$, where the heat transfer coefficient averaged over L is $\bar{h} = 0.664\,(k/L)\,Pr^{1/3}\,Re_L^{1/2}$ when $Pr \geq 0.5$ [2]. The total heat transfer rate from the entire package is $q = 2nq_1$ or, after the $\bar{h}$ and U_0 expressions given above are used,

$$q = 1.21 kHW(T_{max} - T_0)\left(\frac{PrL\Delta p}{\rho\nu^2 D^2}\right)^{1/3}. \tag{3.21}$$

In conclusion, in the large-D limit the total heat transfer rate varies as $D^{-2/3}$ as the board-to-board spacing changes. This trend is shown in Fig. 3.6.

Optimal Spacings. The intersection of the two $q(D)$ asymptotes, Eqs. (3.18) and (3.21), yields an estimate for the board-to-board spacing for maximum global thermal conductance:

$$\frac{D_{\text{opt}}}{L} \cong 2.7 \left(\frac{\Delta p L^2}{\mu \alpha} \right)^{-1/4}. \tag{3.22}$$

This spacing increases as $L^{1/2}$ and decreases as $\Delta p^{-1/4}$ with increasing L and Δp, respectively. Relation (3.22) underestimates by 12% the more exact value obtained by locating the maximum of the actual $q(D)$ curve [8] and is adequate when the board surface is modeled either as uniform flux or isothermal. It has been shown [10] that relation (3.22) holds even when the board thickness is not negligible relative to the board-to-board spacing.

At this stage it is useful to introduce the dimensionless group called Be in Refs. 11 and 12 (or Π_L in Ref. 9):

$$\text{Be} = \frac{\Delta p L^2}{\mu \alpha}, \tag{3.23}$$

so that Eq. (3.22) becomes

$$\frac{D_{\text{opt}}}{L} \cong 2.7 \, \text{Be}^{-1/4}.$$

The manner in which the design parameters influence the maximum rate of heat removal from the package can be expressed as

$$q_{\max} \approx 0.6k(T_{\max} - T_0)\frac{HW}{L}\text{Be}^{1/2}, \tag{3.24}$$

which is obtained by setting $D = D_{\text{opt}}$ in Eq. (3.18) or Eq. (3.21). Once again, the approximation sign is a reminder that the actual $q_{\max}$ is as much as 20% smaller because the peak of the $q(D)$ curve is situated under the point where the two asymptotes cross in Fig. 3.6. The maximum volumetric rate of heat generation in the $H \times L \times W$ volume is

$$\frac{q_{\max}}{HLW} \approx 0.6\frac{k}{L^2}(T_{\max} - T_0)\text{Be}^{1/2}. \tag{3.25}$$

The similarity between the forced convection results [relation (3.22) and approximation (3.25)] and the corresponding results for natural convection cooling [relation (3.9) and approximation (3.11)] is worth noting. The role played by the Rayleigh number Ra_H in the free convection case is played in forced convection by the dimensionless group Be (or Π_L).

The optimal internal spacings belong to the volume as a whole, with its purpose and constraints, not to the individual element (L-long plate in Fig. 3.5) on which heat is being generated. The robustness of this conclusion becomes clear when we

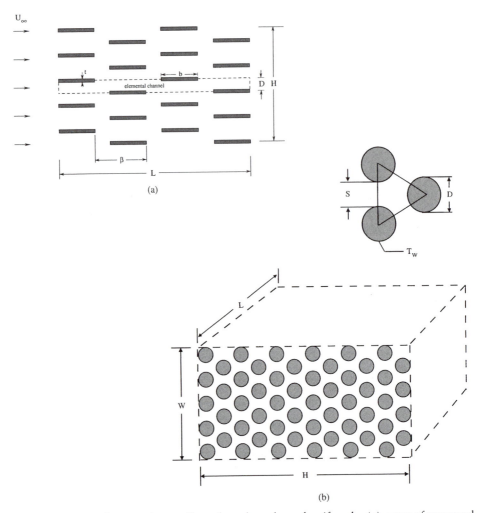

FIGURE 3.7. Forced convection cooling of a volume heated uniformly: (a) array of staggered plates [13], (b) array of horizontal cylinders [14, 15], and (c) square pins with impinging flow [16].

look at other elemental shapes for which optimal spacings have been determined. A volume heated by an array of staggered plates in forced convection [Fig. 3.7(a)] is characterized by an internal spacing D that scales with the swept length of the volume L [13]:

$$\frac{D_{opt}}{L} \cong 5.4 \, Pr^{-1/4} \left(Re_L \frac{L}{b} \right)^{-1/2}. \tag{3.26}$$

In this relation the Reynolds number is $Re_L = U_\infty L / \nu$. The range in which this correlation was developed based on numerical simulations and laboratory experiments is $Pr = 0.72$, $10^2 \leq Re_L \leq 10^4$, and $0.5 \leq (Nb/L) \leq 1.3$.

Similarly, when the elements are cylinders in cross flow [Fig. 3.7(b)], the optimal spacing S is influenced the most by the longitudinal dimension of the volume. The

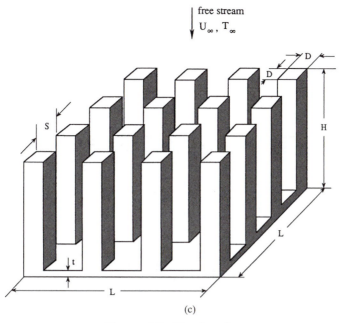

(c)

FIGURE 3.7. (Continued)

optimal spacing was determined based on the method of intersecting the asymptotes [14, 15]. The asymptotes were derived from the large volume of empirical data accumulated in the literature for single cylinders in cross flow (the large-S limit) and for arrays with many rows of cylinders (the small-S limit). In the range $10^4 \leq \tilde{P} \leq 10^8$, $25 \leq H/D \leq 200$, and $0.72 \leq \mathrm{Pr} \leq 50$, the optimal spacing is correlated to within 5.6% by

$$\frac{S_{\mathrm{opt}}}{D} \cong 1.59 \frac{(H/D)^{0.52}}{\tilde{P}^{0.13}\,\mathrm{Pr}^{0.24}}, \qquad (3.27)$$

where $\tilde{P}$ is a dimensionless pressure drop number based on D [review Eq. (3.23)]:

$$\tilde{P} = \frac{\Delta p D^2}{\mu \nu}. \qquad (3.28)$$

When the free-stream velocity U_∞ is specified (instead of Δp), we may transform relation (3.27) by noting that, approximately, $\Delta p \sim (1/2)\rho U_\infty^2$:

$$\frac{S_{\mathrm{opt}}}{D} \cong 1.7 \frac{(H/D)^{0.52}}{\mathrm{Re}_D^{0.26}\,\mathrm{Pr}^{0.24}}. \qquad (3.29)$$

This correlation is valid in the range $140 < \mathrm{Re}_D < 14{,}000$, where $\mathrm{Re}_D = U_\infty D/\nu$. The minimized global thermal resistance that corresponds to this optimal spacing is

$$\frac{T_D - T_\infty}{\dot{Q}D/(kLW)} \cong \frac{4.5}{\mathrm{Re}_D^{0.9}\,\mathrm{Pr}^{0.64}}, \qquad (3.30)$$

where T_D is the cylinder temperature and $\dot{Q}$ is the total rate of heat transfer from the HLW volume to the coolant (T_∞). If the cylinders are arranged such that their centers form equilateral triangles [see Fig. 3.7(b)], the total number of cylinders present in the bundle is $HW/[(S + D)^2 \cos 30°]$. This number and the contact area based on it may be used to deduce from relation (3.30) the volume-averaged heat transfer coefficient between the array and the stream.

Optimal spacings emerge also when the flow is three dimensional, as in an array of pin fins with impinging flow [Fig. 3.7(c)]. The flow is initially aligned with the fins, and later makes a 90° turn to sweep along the baseplate and across the fins. The optimal spacings are correlated to within 16% by [16]

$$\frac{S_{\text{opt}}}{L} \cong 0.81 \, \text{Pr}^{-0.25} \, \text{Re}_L^{-0.32},\tag{3.31}$$

which is valid in the range $0.06 < D/L < 0.14$, $0.28 < H/L < 0.56$, $0.72 < \text{Pr} < 7$, $10 < \text{Re}_D < 700$, and $90 < \text{Re}_L < 6000$. Note that the spacing S_{opt} is controlled by the linear dimension of the volume L. The corresponding minimum global thermal resistance between the array and the coolant is given within 20% by

$$\frac{T_D - T_\infty}{\dot{Q}/(kH)} \cong \frac{(D/L)^{0.31}}{1.57 \, \text{Re}_L^{0.69} \, \text{Pr}^{0.45}}.\tag{3.32}$$

The global resistance refers to the entire volume occupied by the array (HL^2) and is the ratio between the fin–coolant temperature difference ($T_D - T_\infty$) and the total heat current $\dot{Q}$ generated by the HL^2 volume. In the square arrangement of Fig. 3.7(c) the total number of fins is $L^2/(S + D)^2$.

All the correlations that have been developed for the more complicated arrays shown in Figs. 3.4 and 3.7 have been tested extensively, numerically and experimentally. Along with the simple analyses that led to Figs. 3.2 and 3.6, they provide a very clear image of the origin of internal structure in optimized systems subject to constraints.

3.4 The Method of Intersecting the Asymptotes

The intersection of asymptotes method proposed in Refs. 1 and 2 and illustrated in the Sections 3.2 and 3.3 should not be confused with the method of *matched* asymptotic expansions. Matched asymptotes is an old technique of applied mathematics, in which the asymptotes are indeed fused (aligned, spliced) together through the appropriate choice of free parameters. Classical examples from fluid mechanics and heat transfer are the boundary-layer solutions of Blasius [17] and Pohlhausen [18].

The intersection of asymptotes is quite different: It is the competition, clash, or collision of the asymptotes, not the match. It is the sharp intersection of two vastly dissimilar trends (e.g., Figs. 3.2 and 3.6) that determines the spacings for optimal internal structure subject to global constraints. This method provides a much more direct route to structure optimization than the tedious and questionable (model-dependent) development of the complete behavior of the system versus its varying internal structure. Not every design is important; not every point of the rounded solid-line curves sketched in Figs. 3.2 and 3.6 deserves to be determined.

The important designs are in the close vicinity of the intersection of the two competing trends. Optimal architecture is the visible (physical) statement made by the *balance*, equipartition, or equilibrium between competing geometries. We will reach the same conclusion many times as we examine the other flow shapes and structures exhibited in this book.

3.5 The Balance between Stream-Travel Time and Diffusion Time

There is yet another way to see balance, or equipartition, as the mechanism through which the constrained system maximizes its performance. Think of the fluid that enters and bathes volumes, such as those of Figs. 3.1 and 3.5, and ask how the *entire* flow region "works the hardest" toward meeting the global objective of the system. As in the examples of Chap. 2, the solution is to eliminate the lazy flow regions, namely those regions in which the fluid does not cool the walls. In Figs. 3.1 and 3.5 this means that we must make sure that none of the fluid leaves the system as cold as it entered. We must also eliminate the flow regions in which the entire channel cross section is choked with fluid that is practically as hot as the walls. Each ministream works the hardest when it is allowed to leave the system as soon as its fluid packets – all of them – have interacted thermally with the wall. This physical configuration is achieved when the time of longitudinal travel from inlet to exit is equal to the time of transversal diffusion across each channel.

Let us see how well this time balance works in predicting the optimal spacing between vertical plates with natural convection (Fig. 3.1). The longitudinal fluid travel time along each channel is

$$t_1 \sim \frac{H}{v}, \tag{3.33}$$

where v is the vertical velocity scale for laminar boundary-layer natural convection (Ref. 2, pp. 163 and 174):

$$v \sim \frac{\alpha}{H} \mathrm{Ra}_H^{1/2}. \tag{3.34}$$

Combined, approximations (3.33) and (3.34) yield the stream-flow (convection) time scale

$$t_1 \sim \frac{H^2}{\alpha} \mathrm{Ra}_H^{-1/2}. \tag{3.35}$$

The time of transversal thermal diffusion (t_2) can be estimated based on the viscous diffusion solution discussed in more detail above Eq. (7.9). Replacing v with α and y with $D/2$, we write

$$\frac{D/2}{2(\alpha t_2)^{1/2}} \sim 1 \tag{3.36}$$

and note that the factor of 2 in the denominator is a peculiarity of the position (shape) of the knee in the error-function profile to which Eq. (7.9) refers. Setting the t_2 scale of approximation (3.36) equal to the t_1 scale of approximation (3.35), we obtain

$$\frac{D}{H} \sim 4\,\text{Ra}_H^{-1/4}. \tag{3.37}$$

This simple result is remarkable, because it reproduces the form and order of magnitude of the D_{opt} estimate obtained by means of the intersection of asymptotes method, relation (3.9).

The same time-matching argument can be used to anticipate the optimal spacing between parallel plates cooled by laminar forced convection (Fig. 3.5). The transversal diffusion time scale is the same as in approximation (3.36). The longitudinal travel time is

$$t_1 \sim \frac{L}{U}, \tag{3.38}$$

for which U is furnished by Eq. (3.17). Setting $t_1 \sim t_2$, we arrive at

$$\frac{D}{L} \sim 3.7\,\text{Be}^{-1/4}. \tag{3.39}$$

This estimate matches the intersection of asymptotes result of relation (3.22) in every essential respect.

In this section we invoked the equality between longitudinal (convection) time and transversal (diffusion) time and rederived even faster the optimal internal spacings that formed the subject of this chapter. This equality of communication times joined together two vastly different forms of transport, organized motion (streams, convection) with disorganized motion (molecular agitation, diffusion). The reader will encounter this time-balance principle several times in this book, particularly in tree networks for minimum-time traffic (Chap. 11) and the transition to turbulence in all shear flows and Bénard convection (Chap. 7).

3.6 Optimal Longitudinal Flow Pulsations

More evidence that the equality of the convection and diffusion times generates optimal physical structure is provided by the discovery of optimal pulsating flow, as a means of maximizing the mass and heat transfer between two fluid reservoirs at different temperatures [19–22]. In this section we discuss only heat transfer (Fig. 3.8). The fluid sloshes back and forth between the T_1 and the T_2 reservoirs, with the time period t_p. The lateral heat transfer between the bundle of channels and the surroundings is negligible.

During the first half of the cycle, when the flow is from T_1 to T_2, the fluid coats the walls with boundary layers of temperatures comparable with T_1. During the return stroke, the T_2 fluid sweeps the channel, and its job is to make the closest thermal contact possible with the T_1 boundary layers created during the first stroke.

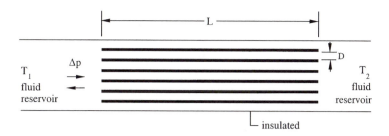

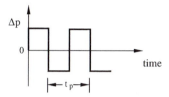

FIGURE 3.8. Insulated stack of parallel two-dimensional channels with periodic fluid flow between two end reservoirs at different temperatures.

The energy picked up by the T_2 fluid is then transported to the T_2 reservoir during the third stroke.

Once again, we face the best use of a constrained space problem that formed the object of this entire chapter. From the argument that started Section 3.5, we know that each fluid column that fills a channel does its best when the time of residence in the channel matches the time of transversal thermal diffusion across the channel. This balance yields the optimal channel spacing [approximation (3.39)]. This result is new. The existence of this optimal geometry has not been recognized yet in the field of Refs. 19–22, in which the channel geometry is assumed given and the narrow-channel limit with Hagen–Poiseuille flow is always postulated.

Recognized has been the existence of an optimal sloshing frequency ω, which maximizes the time-averaged heat transfer rate from T_1 to T_2. This result has been reported as an optimal Womersley number,

$$\mathrm{Wo} = a\left(\frac{\omega}{\nu}\right)^{1/2},\qquad(3.40)$$

where a is the half spacing of one channel, $a = D/2$. Kurzweg [22] developed optimal Wo results for parallel channels with walls as thick as the fluid channels and wall thermal conductivity equal to the fluid thermal conductivity. He correlated the optimal results found for $\mathrm{Pr} = 0.01$, 10, and 1000 by writing approximately

$$\mathrm{Wo} \cong \pi^{1/2}\,\mathrm{Pr}^{-1/2}.\qquad(3.41)$$

Constructal theory anticipates not only the optimal slenderness of each channel, which is new, but also the optimal frequency found in relation (3.41). The optimal sloshing period t_p is twice the time of viscous diffusion across one channel, t_2, approximation (3.36). Writing $t_p = 2t_2$ and $\omega = 2\pi/t_p$, we find that Womersley number definition (3.40) that corresponds to t_p becomes

$$\text{Wo} \sim 3.5\,\text{Pr}^{-1/2}. \tag{3.42}$$

This result reproduces the essential features of correlation (3.41). It also provides a purely theoretical foundation for the *existence* of optimal frequencies in this class of heat and mass transfer enhancement techniques.

The same principle anticipates many other cases of temporal organization (optimal frequencies) in animate and inanimate flow systems, as we will see in Chap. 10.

Theory also reveals the proper dimensionless groups – the most concise terminology for expressing and correlating empirical data. I presented many examples of this in my book on convective heat transfer [1, 2]. To the pulsating flows of this section, constructal theory contributes the prediction that the proper dimensionless group for frequency is Wo $\text{Pr}^{1/2}$ [see approximation (3.42)]. The definition of this theoretical group is $a(\omega/\alpha)^{1/2}$, which is the same as replacing ν with α in Eq. (3.40).

3.7 From Constructal Principle to Internal Structure

We are entitled to disregard for a moment the reasoning that generated the geometries presented in this chapter. We may act as natural scientists, cut open the "animal" – the electronics package, heat exchanger, or electric winding – and marvel at the spatial organization that we see. We marvel also at how much alike are the internal structures of contemporary systems (e.g., two competing personal computers) in the same way that a zoologist marvels at the similarity in bronchial tube sizes and numbers in two animals that have nearly the same mass. In this way we *observe* structure, and, just like the natural scientist, we may go on and measure, correlate, and catalog (classify) the dimensions and patterns that we see.

In contrast with this empirical course, the reasoning exposed in this chapter had nothing to do with observations or empiricism. We invoked consistently a single principle of purpose (conductance maximization) and constraints, and from this principle we *deduced* the optimal internal structure. We practiced theory. The discovery is that geometric form (internal structuring) emerges as the *mechanism* through which the global system achieves its objective.

We are now at that late moment in time – the last point along the bottom arrow in Fig. 1.3 – when we look at nature. Natural systems also develop internal spacings. Bees control the hot-spot temperatures in their swarm by opening nearly parallel vertical channels through which the ambient air cools the swarm. As shown in Fig. 3.9, the bees construct wider channels when the ambient air is warm and the required air flow rate is larger [23, 24]. In Section 3.8 we see that cracks form at surprisingly regular intervals in volumetrically shrinking solids [25]: The principle outlined in this chapter becomes the backbone of a much awaited theory capable of predicting these natural occurrences.

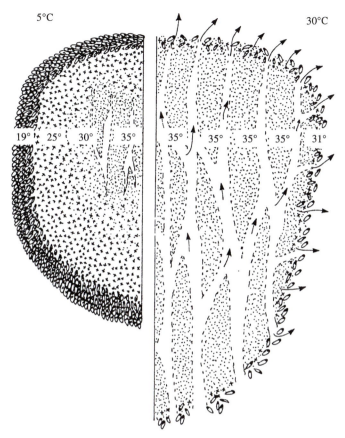

FIGURE 3.9. The regulation of temperature in a swarm of honeybees [23]. The left-hand side shows the structure of the swarm cluster at a low ambient temperature. The right-hand side is for a high ambient temperature and shows the construction of almost equidistant ventilation channels with a characteristic spacing. Also indicated are the heat transfer from the swarm (arrows), areas of active metabolism (crosses), areas of resting metabolism (dots), and local approximate temperatures.

3.8 Cracks in Shrinking Solids

The formation of cracks in solids is an old and busy field that so far has been dug mainly by materials scientists, physicists, and chemists. The challenge that persists is to predict the origin of such patterns, that is, to explain why they are necessary. During the past decade it has become fashionable to describe cracking patterns in terms of fractal images. This tool is pleasing, but not predictive.

Let us think freely about the most common example of patterned cracks. Wet soil exposed to the sun and the wind becomes drier, shrinks superficially, and develops a network of cracks (Fig. 3.10). The loop in the network has a characteristic length scale. The loop is round, more like a hexagon or a square, not slender. The loop is smaller (i.e., cracks are denser) when the wind blows harder, that is, when the drying rate is higher.

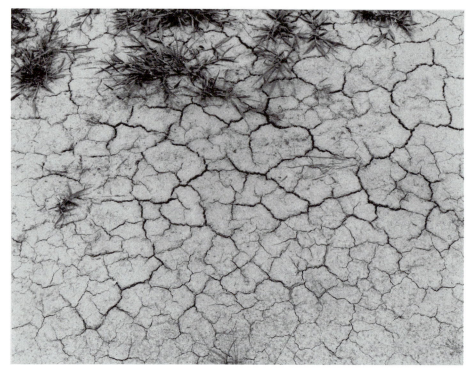

FIGURE 3.10. Pattern of cracks on the ground [25].

These unexplained characteristics of mud cracks are hints that their pattern is another natural occurrence of access optimization: the maximization of the mass transfer rate from the system (wet soil) to the ambient or the minimization of the overall drying time. In view of the analogy between mass transfer and heat transfer, we can explore this theoretical route by considering the thermal analog sketched in Fig. 3.11. A one-dimensional solid slab of thickness L is initially at the high temperature T_H and has the property of shrinking on cooling. The coolant is a single-phase fluid of temperature T_L.

The question is how to maximize the thermal contact between the solid and the fluid or how to minimize the overall cooling time. This objective makes it necessary to allow the fluid to flow through the solid. In Fig. 3.11 the cracks are spaced uniformly, but their spacing R is arbitrary. The channel width D increases in time, as each solid piece R shrinks. The fluid is driven by the pressure difference ΔP, which is maintained across the solid thickness L. The imposed ΔP is an essential aspect of the channel spacing selection mechanism. For example, in the air cooling of a hot solid layer the scale of ΔP is set at $(1/2)\rho_f U_\infty^2$, where ρ_f and U_∞ are the density and the free-stream velocity, respectively, of the external air flow.

To examine the effect of the channel spacing R on the time needed for cooling the solid, we consider the two asymptotes $R \to 0$ and $R \to \infty$. The approach is the same as in the geometric optimization of electronic packages (Section 3.4). In other words, electronic packages emerge as patterns of heat generating blocks

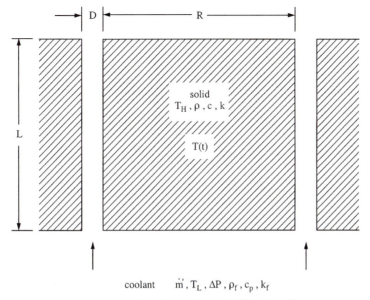

FIGURE 3.11. Channels in a shrinking solid cooled by forced convection [25].

separated by optimal cooling channels for the same reason that optimal patterns of cracks occur in nature. This observation reinforces the commonality of natural and synthetic patterns under constructal theory.

When the number of channels per unit length is large, the spacing R is small and so is the eventual shrinkage that is experienced by each R element. This means that when $R \to 0$ we can expect $D \to 0$ and laminar flow through each D-thin channel, such that the channel mass flow rate is $\dot{m}' = \rho_f D U \sim \rho_f D^3 \Delta P / (\mu L)$. In the same limit, R is small enough so that the solid conduction is described by the lumped thermal capacitance model. The solid piece R is characterized by a single temperature T, which decreases in time from the initial level T_H to the inlet temperature of the fluid T_L. This cooling effect is governed by the energy balance $\rho c R L (dT/dt) = -q'$, where ρ and c are the density and the specific heat, respectively, of the solid. The cooling effect (q') provided by the flow through the channel is represented well by $q' = \dot{m}' c_p (T - T_L)$, where c_p is the specific heat of the coolant. We obtain the order-of-magnitude statement $\rho c R L (\Delta T / t) \sim \dot{m}' c_p \Delta T$, where ΔT is the scale of the instantaneous solid excess temperature $T - T_L$. Finally, by using the scale, we find the cooling time scale:

$$t \sim \frac{\rho c}{\rho_f c_p} \frac{\mu R L^2}{D^3 \Delta P} \qquad (R \to 0) \qquad (3.43)$$

In the opposite limit, R is large and the shrinkage (the channel width D) is potentially very large in proportion to R. The fluid present at one time in the channel is mainly isothermal at the inlet temperature T_L. The cooling of each solid side of the crack is ruled by one-dimensional thermal diffusion into a semi-infinite medium. The cooling time in this regime is the same as the time of thermal diffusion over the distance R,

$$t \sim \frac{R^2}{\alpha} \qquad (R \to \infty), \qquad\qquad (3.44)$$

where $\alpha = k/(\rho c)$ and k is the thermal conductivity of the solid.

To summarize, in the limit $R \to 0$, the cooling time is proportional to R/D^3 or R^{-2} because we expect a proportionality between D and R, namely, $D/R \sim \beta \Delta T \ll 1$, where $\Delta T \sim T_H - T_L$ and β is the coefficient of thermal contraction of the solid. In the opposite limit, $R \to \infty$, the cooling time is proportional to R^2. Put together, these proportionalities suggest that the cooling time possesses a sharp minimum with respect to R or the channel density. Intersecting the two asymptotes, we find that the optimal crack distance R_{opt} for fastest cooling is of the order of

$$R_{\mathrm{opt}} \sim \left[\frac{k}{k_f} \frac{\alpha_f v L^2}{U_\infty^2 (\beta \Delta T)^3} \right]^{1/4}. \qquad\qquad (3.45)$$

This result is promising for two fundamental reasons, in addition to the practical aspect of knowing how to extract heat or mass from the heart of a solid in the fastest way possible. One reason is that the optimal crack distance decreases as the external pressure (or flow) is intensified. This effect is in accord with numerous observations that mud cracks become denser when the wind speed increases. This result, in association with the theoretical view that natural cracks occur such that the cooling speed is maximized, is the first time that the effect of wind speed on crack density is predicted. The second reason is that the R_{opt} result predicts a higher density of cracks (a smaller R_{opt}) as the solid excess temperature ΔT increases. This trend too is in agreement with observations, and it is being predicted for the first time.

An important geometric aspect of the R_{opt} scale is that the optimal distance between consecutive cracks must increase as $L^{1/2}$. This is relevant to predicting the length scale of the lattice of vertical cracks formed in a horizontal two-dimensional surface cooled (or dried) from above, under the influence of external forced convection. As the air flow direction changes locally from time to time and as the material (its graininess) is such that cracks may propagate in more than one direction, we arrive at the problem of cooling a two-dimensional terrain (area A, when seen from above) with cracks of length L and associated area elements of width R_{opt}.

Figure 3.12 shows the two extremes in which L may find itself in relation to R_{opt}. First, when L is considerably shorter than R_{opt} it is impossible to cover the area A exclusively with patches of size $L \times R_{\mathrm{opt}}$. The reason is that when two cracks of length L are joined at an angle, the elemental area $\sim L^2$ trapped between them is too small to accommodate the amount of ideally cooled solid material. When L is considerably longer than R_{opt}, any lattice of cracks will fail to cover the area A completely. Now the trapped elemental area ($\sim L^2$) is considerably larger than the amount of ideally cooled solid ($\sim L R_{\mathrm{opt}}$): Most of the interior of the area element of size L^2 would require a cooling time that is considerably longer than the minimum time determined in the preceding analysis.

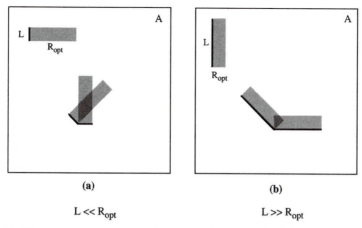

(a)   (b)

$L \ll R_{\mathrm{opt}}$ $L \gg R_{\mathrm{opt}}$

FIGURE 3.12. The two extremes in covering a two-dimensional solid (A) with cracks (L) and optimally cooled volume elements [25].

In conclusion, the fastest way possible to cool the entire solid A is to cover the A cross section with $L \times R_{\mathrm{opt}}$ elements, in which $L \sim R_{\mathrm{opt}}$. The optimal pattern is one with relatively round or square loops, not slender loops. Combining $L \sim R_{\mathrm{opt}}$ with the R_{opt} expression, we find the optimal length scale of the loop in the network of cracks that will minimize the cooldown time: $R_{\mathrm{opt}} \sim (\alpha_f \nu k / k_f)^{1/2} / [U_\infty (\beta \Delta T)^{3/2}]$. Once again, in agreement with observations, we see that the lattice length scale R_{opt} must decrease as the wind speed and the initial excess temperature increase.

PROBLEMS

3.1 The optimal spacing between horizontal tubes in a fixed volume cooled by natural convection can be determined based on the method of Section 3.2. Consider the bundle of horizontal cylinders shown in the middle panel of Fig. 3.4(b). The overall dimensions of the bundle (H, L, W) and the cylinder diameter (D) are fixed. Natural convection heat transfer q occurs between the cylinder surfaces (T_w) and the surrounding fluid reservoir (T_∞). The objective is to select the number of cylinders in the bundle or the cylinder-to-cylinder spacing (S) such that the overall thermal conductance between the bundle and the ambient $q/(T_w - T_\infty)$ is maximized.

For the sake of concreteness, assume that the cylinders are staggered and that their centers form equilateral triangles. Other arrays can be treated similarly. The analysis consists of estimating the overall thermal conductance in the two asymptotic regimes (large S, small S), and intersecting the two asymptotes to locate the optimal spacing for maximum conductance.

The height-averaged thickness (spacing) of the flow channel may be approximated by $\bar{S} = S(S + 2D)/(S + D)$, where S is the smallest spacing (middle panel of Fig. 3.4(b)). The average heat transfer coefficient from a single cylinder is given by $\bar{h}D/k = c \, \mathrm{Ra}_D^{1/4}$, where c is a constant of the order of 0.5 and $\mathrm{Ra}_D = g\beta(T_w - T_0)D^3/(\alpha \nu)$.

3.2 Determine the optimal spacing of cylinders in cross flow. The fixed volume $H \times L \times W$ shown in Figure 3.7(b) contains a bundle of parallel cylinders of diameter D and temperature T_w and is bathed by a cross flow of temperature T_0 and velocity U_∞. Maximize the heat transfer q between the bundle and the surrounding fluid by selecting the cylinder-to-cylinder spacing S or the number of cylinders in the bundle.

That an optimal number of cylinders must exist can be expected based on the following argument. If the volume $H \times L \times W$ contains only one cylinder, then perhaps two or more cylinders will transfer more heat to the surrounding flow. This trend continues until the cylinders become so numerous that they almost touch and the bundle becomes impermeable to the free stream. The optimal spacing $\bar{S}$ emerges as a trade-off between the limit of dense cylinders (not enough coolant) and that of sparse cylinders (not enough heat transfer area).

This argument also suggests the analysis you should construct. Begin with the calculation of the heat transfer rate in the two asymptotic regimes noted above. For the large-S limit, assume that the average heat transfer coefficient $\bar{h}$ is given by $\bar{h}D/k = c(U_\infty D/\nu)^{1/2}$, where c is an empirical constant of the order of 0.5. For the small-S limit, show that the H-averaged transversal dimension of each flow channel, $\bar{S}$, is adequately represented by $\bar{S} = S(S+2D)/(S+D)$, where S is the smallest spacing, Fig. 3.7(b). Finally, determine the optimal cylinder-to-cylinder spacing by intersecting the two asymptotic estimates obtained for the total heat transfer between the bundle and the free stream.

REFERENCES

1. A. Bejan, *Convection Heat Transfer*, Wiley, New York, 1984, p. 157, problem 11.
2. A. Bejan, *Convection Heat Transfer*, 2nd ed., Wiley, New York, 1995.
3. A. Bejan, G. Tsatsaronis, and M. Moran, *Thermal Design and Optimization*, Wiley, New York, 1996.
4. A. Bar Cohen and W. M. Rohsenow, Thermally optimum spacing of vertical, natural convection cooled, parallel plates, *J. Heat Transfer*, Vol. 106, 1984, pp. 116–123.
5. A. Bejan, Geometric optimization of cooling techniques, in S. J. Kim and S. W. Lee, eds., *Air Cooling Technology for Electronic Equipment*, CRC, Boca Raton, FL, 1996, Chap. 1.
6. G. A. Ledezma and A. Bejan, Optimal geometric arrangement of staggered vertical plates in natural convection, *J. Heat Transfer*, Vol. 119, 1997, pp. 700–708.
7. A. Bejan, A. J. Fowler, and G. Stanescu, The optimal spacing between horizontal cylinders in a fixed volume cooled by natural convection, *Int. J. Heat Mass Transfer*, Vol. 38, 1995, pp. 2047–2055.
8. A. Bejan and E. Sciubba, The optimal spacing of parallel plates cooled by forced convection, *Int. J. Heat Mass Transfer*, Vol. 35, 1992, pp. 3259–3264.
9. A. Bejan, *Heat Transfer*, Wiley, New York, 1993.
10. S. Mereu, E. Sciubba, and A. Bejan, The optimal cooling of a stack of heat generating boards with fixed pressure drop, flow rate or pumping power, *Int. J. Heat Mass Transfer*, Vol. 36, 1993, pp. 3677–3686.
11. S. Petrescu, Comments on the optimal spacing of parallel plates cooled by forced convection, *Int. J. Heat Mass Transfer*, Vol. 37, 1994, p. 1283.

12. S. Bhattacharjee and W. L. Grosshandler, The formation of a wall jet near a high temperature wall under microgravity environment, *ASME HTD*, Vol. 96, 1988, pp. 711–716.

13. A. J. Fowler, G. A. Ledezma, and A. Bejan, Optimal geometric arrangement of staggered plates in forced convection, *Int. J. Heat Mass Transfer*, Vol. 40, 1997, pp. 1795–1805.

14. A. Bejan, The optimal spacings for cylinders in crossflow forced convection, *J. Heat Transfer*, Vol. 117, 1995, pp. 767–770.

15. G. Stanescu, A. J. Fowler, and A. Bejan, The optimal spacing of cylinders in free-stream cross-flow forced convection, *Int. J. Heat Mass Transfer*, Vol. 39, 1996, pp. 311–317.

16. G. Ledezma, A. M. Morega, and A. Bejan, Optimal spacing between pin fins with impinging flow, *J. Heat Transfer*, Vol. 118, 1996, pp. 570–577.

17. H. Blasius, Grenzschichten in Flüssigkeiten mit kleiner Reibung, *Z. Math. Phys.*, Vol. 56, 1908, p. 1; also NACA TM 1256.

18. E. Pohlhausen, Der Wärmeaustausch zwischen festen Körpern und Flüssigkeiten mit kleiner Reibung und kleiner Wärmeleitung, *Z. Angew. Math. Mech.*, Vol. 1, 1921, pp. 115–121.

19. P. C. Chatwin, On the longitudinal dispersion of passive contaminant in oscillatory flows in tubes, *J. Fluid Mech.*, Vol. 71, 1975, pp. 513–527.

20. E. J. Watson, Diffusion in oscillatory pipe flow, *J. Fluid Mech.*, Vol. 133, 1983, pp. 233–244.

21. U. H. Kurzweg and L. D. Zhao, Heat transfer by high-frequency oscillations: a new hydrodynamic technique for achieving large effective thermal conductivities, *Phy. Fluids*, Vol. 27, 1984, pp. 2624–2627.

22. U. H. Kurzweg, Enhanced heat conduction in oscillating viscous flows within parallel-plate channels, *J. Fluid Mech.*, Vol. 156, 1985, pp. 291–300.

23. B. Heinrich, The mechanisms and energetics of honeybee swarm temperature regulation, *J. Exp. Biol.*, Vol. 91, 1981, pp. 25–55.

24. T. Basak, K. K. Rao, and A. Bejan, A model for heat transfer in a honey bee swarm, *Chem. Eng. Sci.*, Vol. 51, 1996, pp. 387–400.

25. A. Bejan, Y. Ikegami, and G. A. Ledezma, Constructal theory of natural crack pattern formation for fastest cooling, *Int. J. Heat Mass Transfer*, Vol. 41, 1998, pp. 1945–1954.

CHAPTER FOUR

HEAT TREES

4.1 The Volume-to-Point Flow Problem

In this chapter we review the recent application of the objective and constraints principle in the cooling of electronics and how the naturally shaped tree emerged as its chief and surprising result. The engineering system is the same as that in Chap. 3 (volume filled with heat generating components; the same purpose and constraints), except that this time the generated heat current cannot be washed away by a stream of coolant. Why we must search for another escape route for the heat current is explained next.

Smaller, faster, smarter, and cheaper: The research frontier in the cooling of electronics is being pushed in the direction of smaller and smaller package dimensions. There comes a point at which miniaturization makes convection cooling impractical, because the ducts through which the coolant must flow take up too much space. The only way to channel the generated heat out of the electronic package is by conduction. The conduction path will have to be very effective (of high thermal conductivity, k_p), so that the temperature difference between the hot spot (the heart of the package) and the heat sink (on the side of the package) will not exceed a certain value. Conduction paths also take up space. Designs with fewer and smaller paths are better suited for the miniaturization evolution. The new fundamental problem that I proposed in Ref. 1 is the following:

> "Consider a finite-size volume in which heat is being generated at every point and which is cooled through a small patch (heat sink) located on its boundary. A finite amount of high conductivity (k_p) material is available. Determine the optimal distribution of k_p material through the given volume such that the highest temperature is minimized."

The discovery made in Ref. 1 is purely geometric: Every portion of the given volume can have its shape optimized such that its resistance to heat flow is minimal. This unique principle applies at any volume scale and to other forms of transport (fluid, electric charge, mass species). The volume-to-point path was determined in a sequence of steps consisting of shape optimization and subsequent construction (assembly, grouping). The time arrow of this construction, which points from small to large, is very important. It starts from the smallest building block (elemental system) and proceeds toward larger building blocks (assemblies, constructs). It was shown that determinism vanishes if the direction is reversed, from large to small. To emphasize the link between determinism and the direction from small to large and as a reminder that theory runs against fractal thinking, the geometric optimization principle proposed in Ref. 1 was named constructal.*

4.2 Elemental Volume

Consider a volume V that generates heat at the uniform volumetric rate q''', which is collected over the volume and led to a heat sink on its boundary. For simplicity, assume that the heat-flow geometry is two dimensional over the area A; hence $V = AW$, where W is the third dimension of the volume. Most of A is occupied by the material of low conductivity k_0, which generates q'''. A small fraction of A (namely A_p) is to be covered by high-conductivity material.

We regard A as a collection of subsystems of various sizes $(A_0, A_1, A_2, \ldots,)$ and begin with the smallest such system (A_0), which is shown in the upper part of Fig. 4.1. The defining feature of this elemental system is that its heat current $(q_0 = q''' A_0 W)$ is collected by a single k_p blade. The other important feature is that the size A_0 is known and fixed: For example, in electronics this smallest size is dictated by manufacturing constraints. The size A_0 is fixed, but the shape H_0/L_0 may vary. The objective of the following analysis is to determine the optimal shape such that the overall resistance between A_0 and the heat sink [the D_0 patch located at the origin $(0, 0)$] is minimal.

An analytical solution is possible in the limit $\tilde{k} \gg 1$, where $\tilde{k} = k_p/k_0$ is a known constant of the conducting composite. The requirement to minimize resistance is the reason why the D_0-thin blade is placed on the longer of the two axes of symmetry of A_0 (note that $H_0 > L_0$ in Fig. 4.1). When $\tilde{k} \gg 1$, heat flows vertically through the k_0 material and horizontally through the k_p material. This allows us to develop the expression for the hot-spot temperature $T(L_0, \pm H_0/2)$ as a sum of two terms. First, the temperature drop through k_0 material from the farthest corner $(L_0, H_0/2)$ to the insulated (right) end of the blade $(L_0, 0)$ is of the order of $\Delta T_{k_0} \sim q''' H_0^2/k_0$. This scale comes from writing the Fourier law for the vertical heat current $q''' H_0$ that is generated between the two indicated points, $q''' H_0 \sim k_0 \Delta T_{k_0}/H_0$. The next temperature drop occurs along the k_p blade, from the insulated end $(L_0, 0)$ to the heat-sink end $(0, 0)$: Its order of magnitude is $\Delta T_{k_p} \sim q''' H_0 L_0/k_p$, which comes from the Fourier law written for the same heat current, $q''' H_0 \sim k_p \Delta T_{k_p}/L_0$. The overall temperature drop from the hot spot to the heat sink is $\Delta T_0 = \Delta T_{k_0} + \Delta T_{k_p} \sim q''' H_0(H_0/k_0 + L_0/k_p)$, where H_0 and L_0 are related through the size constraint $A_0 = H_0 L_0$. The expression for ΔT_0 can be derived

---

* From the Latin verb *constrŭĕre* (to build), which survives as *construire* in French, Italian, and Romanian and as *construir* in Spanish and Portuguese.

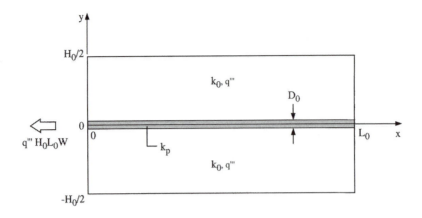

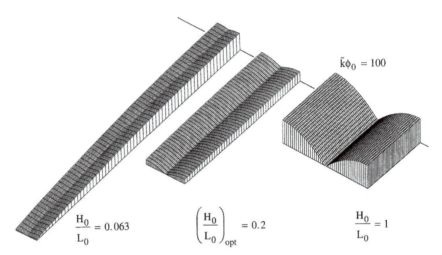

FIGURE 4.1. Elemental volume and the minimization of the volume-to-point resistance by varying the external shape [1]. The lower drawing was made by Dr. G. A. Ledezma.

exactly and arranged in the following dimensionless form [1]:

$$\frac{\Delta T_0 k_0}{q''' A_0} = \frac{1}{8} \times \frac{H_0}{L_0} + \frac{k_0 H_0}{2 k_p D_0} \times \frac{L_0}{H_0}. \tag{4.1}$$

Next we note that the ratio D_0/H_0 is another constant of the conductive composite, because it accounts for the proportion in which k_p material is allocated to A_0, namely, $A_{p0}/A_0 = D_0 L_0/(H_0 L_0)$. We call this constant $\phi_0 = D_0/H_0 \ll 1$. With this, Eq. (4.1) shows clearly that the overall volume-to-point resistance can be minimized with respect to the shape of the elemental system (H_0/L_0):

$$(H_0/L_0)_{opt} = 2(\tilde{k}\phi_0)^{-1/2} < 1, \tag{4.2}$$

$$\left(\frac{\Delta T_0 k_0}{q''' A_0}\right)_{min} = \frac{1}{2(\tilde{k}\phi_0)^{1/2}}. \tag{4.3}$$

Additional results are $H_{0,opt} = (2 A_0)^{1/2}(\tilde{k}\phi_0)^{-1/4}$ and $L_{0,opt} = (A_0/2)^{1/2}(\tilde{k}\phi_0)^{1/4}$. An important feature of Eq. (4.3) is that at the optimum the resistance through the k_0 material matches the resistance through the k_p material. In other words, the two terms have the same value on the right-hand side of Eq. (4.1). This principle of *equipartition* of resistance [2] applies at every volume scale.

The lower part of Fig. 4.1 shows how the peak temperature of the elemental volume responds to changes in the shape ratio H_0/L_0 when the elemental size $H_0 L_0$ and elemental heat current $q''' H_0 L_0 W$ are fixed. The longer dimension of the rectangular base is L_0. The vertical dimension is proportional to the local temperature. The three drawings correspond to the case $\tilde{k}\phi_0 = 100$, for which the optimal shape is $(H_0/L_0)_{\text{opt}} = 0.2$ [see Eq. (4.2)]. This optimal design is the middle drawing: Here the peak temperature is lower than when H_0/L_0 is larger or smaller than the optimal value. The middle drawing also shows that at the optimum the overall temperature difference in the y direction is equal to the overall temperature difference in the longitudinal x direction, in accordance with the resistance equipartition principle.

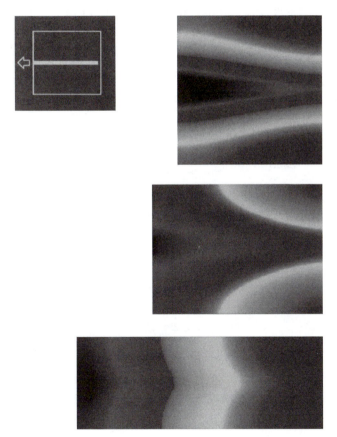

FIGURE 4.2. Deriving optimal shape from the minimization of global resistance between a volume and one point. Three competing designs are shown. The volume and the heat generation rate are fixed. The aspect ratio of the rectangular domain is variable. The resistance is proportional to the peak temperature difference, which is measured between the hot spots (red) and the heat sink (blue). The middle shape minimizes the areas covered by red and has the smallest volume–point resistance (see Plate III).

The same process of deducing shape from principle is illustrated in color in Fig. 4.2.

4.3 First Construct and Growth

As we proceed toward larger scales, we can assemble elemental volumes only of size A_0, which have to be connected so that their heat currents can be collected into a single stream (Fig. 4.3). At our disposal we have a lot more than just two materials (k_0, k_p) with which we started the elemental optimization (Section 4.2). Now we have elemental volumes of fixed size and optimal shape that can be used as building blocks (pieces of a mosaic) to fill the given volume. The optimization product achieved at the lower scale is preserved (memorized) as we proceed toward the next larger scale: Memory is a defining feature of the constructal sequence.

This first construct (A_1) contains n_1 elemental volumes; $A_1 = n_1 A_0$. The number n_1 (or the shape H_1/L_1) is a degree of freedom. The elemental heat currents are collected by a new blade of k_p material and thickness D_1.

When n_1 is large, the first construct (Fig. 4.3) conducts heat in almost the same way as the elemental volume (Fig. 4.1). The only difference is that the k_0 regions of Fig. 4.1 are now replaced with a composite material. The effective thermal conductivity of this composite in the vertical direction is $k_1 = k_p D_0/H_0$. The hot-spot temperature difference ΔT_1 occurs between the far corner $(L_1, H_1/2)$ and the heat sink $(0, 0)$. Nothing has changed except the notation in Eq. (4.1), which now is

$$\frac{\Delta T_1 k_1}{q''' A_1} = \frac{1}{8} \times \frac{H_1}{L_1} + \frac{k_1 H_1}{2 k_p D_1} \times \frac{L_1}{H_1}. \tag{4.4}$$

The geometric optimization too is a repeat of what went on at the elemental level, although for more clarity we use a somewhat different terminology than in the original communication [1]. First, it is more convenient to rewrite Eq. (4.4)

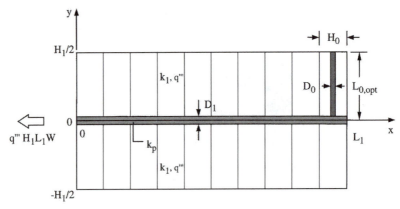

FIGURE 4.3. The first construct: a large number of elemental volumes connected to a central high-conductivity channel [1].

such that it shows explicitly the effect of n_1:

$$\frac{\Delta T_1 k_0}{q''' A_1} = \frac{1}{4 n_1 (\tilde{k} \phi_0)^{1/2}} + \frac{n_1}{2^{3/2} \tilde{k} (\tilde{k} \phi_0)^{1/4} \tilde{D}_1}, \tag{4.5}$$

where $\tilde{D}_1 = D_1 / A_0^{1/2}$. The trade-off role played by n_1 is clear. The total current $q''' A_1$ increases as the size of the construct (A_1, or n_1) increases. Note the behavior of the overall resistance that must be overcome by this current. In the beginning, as the size of the assembly increases from a small n_1 value, the overall resistance decreases: Resistance minimization is achieved through *growth*. This trend ends at a certain (optimal) value of n_1, when the second term of Eq. (4.5) takes over and the resistance starts to increase. The optimal size is represented by

$$n_{1,opt} = \frac{(\tilde{k} \tilde{D}_1)^{1/2}}{2^{1/4} (\tilde{k} \phi_0)^{1/8}}. \tag{4.6}$$

The corresponding minimal value of the first-assembly resistance is

$$\left(\frac{\Delta T_1 k_0}{q''' A_1} \right)_m = \frac{1}{2^{3/4} \tilde{k}^{7/8} \phi_0^{3/8} \tilde{D}_1^{1/2}}. \tag{4.7}$$

A second resistance-minimization route is available when the volume fraction of k_p material allocated to A_1 is fixed, $\phi_1 = A_{p1} / A_1$, where $A_{p1} = L_1 D_1 + n_1 \phi_0 A_1$. This constraint is

$$\phi_1 = \frac{\tilde{D}_1}{2^{1/2} (\tilde{k} \phi_0)^{1/4}} + \phi_0. \tag{4.8}$$

The resistance-minimization equation (4.7) subject to constraint (4.8) yields $\phi_{0,opt} = \phi_1 / 2$ and $\tilde{D}_{1,opt} = 2^{-3/4} \tilde{k}^{1/4} \phi_1^{5/4}$, which can also be written as

$$\left(\frac{D_1}{D_0} \right)_{opt} = \left(\frac{\tilde{k} \phi_1}{2} \right)^{1/2} \gg 1. \tag{4.9}$$

Working these results back into Eqs. (4.6) and (4.7), we obtain

$$n_{1,opt} = \left(\frac{\tilde{k} \phi_1}{2} \right)^{1/2} \gg 1, \tag{4.10}$$

$$\left(\frac{\Delta T_1 k_0}{q''' A_1} \right)_{mm} = \frac{1}{\tilde{k} \phi_1}, \tag{4.11}$$

where the subscript *mm* is a reminder that the resistance has been minimized twice. It can be verified that at this optimum the first-assembly shape and dimensions are exactly the same as those in Ref. 1, namely $(H_1 / L_1)_{opt} = 2$, $H_{1,opt} = (2 A_1)^{1/2}$, and $L_{1,opt} = (A_1 / 2)^{1/2}$. The temperature field in a first construct with only four elemental volumes is illustrated in Fig. 4.4.

Examined from the point of view of the natural sciences, Eq. (4.10) represents an important step: Growth, assembly, or aggregation emerges as a *mechanism* of global resistance minimization. The number optimized in Eq. (4.10) represents *optimal* growth when the optimized geometry has been preserved (memorized) at the elemental level. Growth, assembly, or aggregation is no longer a "natural"

FIGURE 4.4. The temperature field in a first construct containing four elemental volumes. Note the formation of hot spots (red) in the corners situated the farthest from the heat sink (blue) (see Plate IV).

behavior taken for granted. It is not accepted empirically as "self"-organization. It is not haphazard. It is not an assumed "property," as in Holland's theory [3]. In constructal theory, aggregation is a result – a consequence of the deterministic principle called constructal law – the optimization of access to internal currents, subject to global constraints. This law is stated more explicitly in Section 4.5.

4.4 Second and Higher-Order Constructs

To fill an even larger volume, we connect a large number of optimized first constructs and arrive at the second construct shown in Fig. 4.5. The second construct can be optimized similarly, i.e., two times: for overall size ($n_2 = A_2/A_1 \gg 1$) and for internal distribution of high-conductivity material. The second degree of freedom is represented by the thickness of the new central fiber D_2, which collects the heat currents contributed by the n_2 constituents. The volume-to-point thermal resistance of the second assembly is $\Delta T_2/(q''' A_2)$, where ΔT_2 is measured between the upper right-hand corner (the hot spot) and the midpoint of the left-hand side of the A_2 rectangle (the heat sink). The overall resistance is [1]

$$\left(\frac{\Delta T_2 k_0}{q''' A_2} \right) = \frac{H_1}{2\tilde{k}\phi_1 L_2} + \frac{A_2}{2\tilde{k} D_2 H_2}, \qquad (4.12)$$

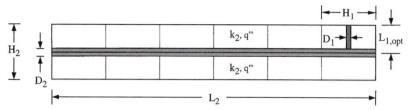

FIGURE 4.5. Second construct containing a large number of first constructs [1].

or, in terms of the number of constituents n_2,

$$\left(\frac{\Delta T_2 k_0}{q''' A_2}\right) = \frac{1}{n_2 \tilde{k} \phi_1} + \frac{n_2 A_1^{1/2}}{2^{3/2} \tilde{k} D_2}. \tag{4.13}$$

The first minimization of this resistance is performed with respect to n_2, and the results are

$$n_{2,\mathrm{opt}} = \frac{2^{3/4} \tilde{D}_2^{1/2}}{\phi_1^{1/2}}, \tag{4.14}$$

$$\left(\frac{\Delta T_2 k_0}{q''' A_2}\right)_m = \frac{2^{1/4}}{\tilde{k} \phi_1^{1/2} \tilde{D}_2^{1/2}}, \tag{4.15}$$

where $\tilde{D}_2 = D_2/A_1^{1/2}$. Once again, we see that optimal growth means resistance minimization. The second minimization is made with respect to $\tilde{D}_2$ when the volume fraction of k_p material is constrained: $\phi_2 = A_{p2}/A_2$, constant. The ϕ_2 constraint can also be written as

$$\phi_2 = 2^{-1/2} \tilde{D}_2 + \phi_1. \tag{4.16}$$

The results of this second optimization step are $\phi_{1,\mathrm{opt}} = \phi_2/2$ and $\tilde{D}_{2,\mathrm{opt}} = 2^{-1/2} \phi_2$. These, in combination with the preceding results, lead to

$$\left(\frac{D_2}{D_1}\right)_{\mathrm{opt}} = 2, \tag{4.17}$$

$$n_{2,\mathrm{opt}} = 2, \tag{4.18}$$

$$\left(\frac{\Delta T_2 k_0}{q''' A_2}\right)_{mm} = \frac{2}{\tilde{k} \phi_2}. \tag{4.19}$$

It can be verified that Eqs. (4.17)–(4.19) match the results reported originally [1], in which the optimization was performed with respect to shape (H_2/L_2) instead of size (n_2). It is in Eq. (4.18) that the integer 2 (dichotomy, bifurcation, pairing) becomes a feature of the optimized architecture as a *result* of the resistance minimization principle, not as an *assumption* as in the tree network analyses of physiology and river morphology (for a review, see Ref. 4). Next, the pairing result $n_{2,\mathrm{opt}} = 2$ recommends a revision – a more exact remake – of the analysis of Eqs. (4.12)–(4.19), because that analysis was based on the assumption that n_2 is

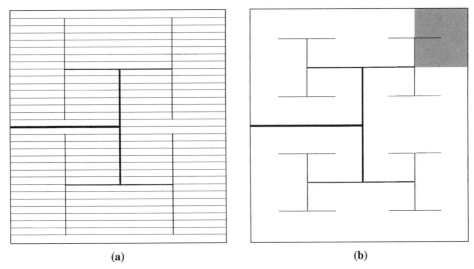

FIGURE 4.6. The fourth and the eighth constructs: note that (a) the fourth construct is the same as the darkened corner of (b) the eighth construct [1].

considerably greater than 1. The revised results for the second assembly with $n_2 = 2$ can be found in Ref. 1.

In summary, the constructal sequence is one in which the volume-to-point flow grows over an increasingly larger space. This growth occurs in steps (assemblies, constructs), as illustrated by the higher-order constructs of Fig. 4.6. Each step contributes external shape and internal structure, which are complete results of the constrained minimization of resistance. Table 4.1 summarizes all the external and internal geometric features of the first six constructs. The architecture of each new assembly is such that it allows the flow to continue its outward expansion with minimal resistance.

4.5 Constructal Law

We reached again that late moment when, with the principle-generated form already in front of us, we open our eyes to nature to admire its forms. The image generated in Fig. 4.6 is much more complex than a simple stick drawing that connects a square area to one point. In this deductive process small came first and large later. This sequence began with diffusion in the interstices (k_0) and continued with channeled flow through the k_p blades. The shape of the area – the tree canopy – is an optimization result, not an assumption. The tree links become thicker toward the root point. Pairing (or bifurcation) is a feature of the tree links at levels higher than the second construct. This rule breaks down at smaller scales, where the tributaries (or branches) are considerably more numerous. The tree links fill the space only partially, as the smallest links cannot be smaller than the interstices covered by diffusion.

TABLE 4.1. The Main Parameters of the First Six Optimized Assemblies [1]

Assembly Level i	Assembly Height Factor H_i/H_{i-1}	Assembly Shape[a]	Path-Width Factor D_i/D_{i-1}	Path-Length Factor L_i/L_{i-1}	$\dfrac{\Delta T_{i,\min}}{q''H_0^2/k_0}$	$\phi_i\dfrac{H_0}{D_0}$
0		r			$\dfrac{1}{4}$	1
1	$\left(\dfrac{D_0 k_p}{H_0 k_0}\right)^{1/2}\gg 1$	R	$\left(\dfrac{D_0 k_p}{H_0 k_0}\right)^{1/2}\gg 1$	1	$\dfrac{3}{8}$	2
2	1	S	$\dfrac{4}{3^{1/2}}=2.31$	1	$\dfrac{3}{8}\left(1+\dfrac{1}{3^{1/2}}\right)$	$2\left(1+\dfrac{1}{3^{1/2}}\right)$
3	2	R	2	1	$\dfrac{3}{8}\left(1+\dfrac{2}{3^{1/2}}\right)$	$2\left(1+\dfrac{2}{3^{1/2}}\right)$
4	1	S	2	2	$\dfrac{3}{8}\left(1+\dfrac{4}{3^{1/2}}\right)$	$2\left(1+\dfrac{4}{3^{1/2}}\right)$
5	2	R	2	1	$\dfrac{3}{8}\left(1+\dfrac{6}{3^{1/2}}\right)$	$2\left(1+\dfrac{6}{3^{1/2}}\right)$
6	1	S	2	2	$\dfrac{3}{8}\left(1+\dfrac{10}{3^{1/2}}\right)$	$2\left(1+\dfrac{10}{3^{1/2}}\right)$

[a] S, square; R, rectangle (aspect ratio = 2); r, rectangle (aspect ratio $\ll$ 1).

This tree drawing matches in every category the description of any natural tree: Review the questions formulated in Section 1.2. It is astonishing not that the high-conductivity channels form a tree, but that every feature of this tree is deterministic, i.e., the result of a single principle. This conclusion runs against the doctrine that natural flow structures are nondeterministic – consequences of chance and necessity. The discovery is not the tree but the principle that generated it.

This principle is raised to the rank of *law* by the structure of all the tree networks anticipated by constructal theory. This law can be stated as follows [1]:

> "For a finite-size system to persist in time (to live), it must evolve in such a way that it provides easier access to the imposed currents that flow through it."

This statement has two parts. First, it recognizes the natural tendency of imposed currents to construct shapes and structures, i.e., paths of optimal access through constrained open systems. The second part accounts for the improvements of these paths, which occur in an identifiable direction that can be aligned with time itself.

This formulation of the constructal law refers to a system with imposed steady flow, as in the heat-flow configuration analyzed until now. If the system discharges itself to one point in unsteady fashion, then the constructal minimization of volume-to-point resistance is equivalent to the minimization of the time of discharge or the maximization of the speed of approach to internal equilibrium (uniformity, zero flow). This configuration is discussed in Section 4.8.

If the volume is unbounded, the constructs compound themselves and continue to spread indefinitely. Complexity continues to increase in time because growth is a path to resistance minimization (Section 4.3). Examples are the jet injected into a fluid reservoir and the dendritic crystal growing into a subcooled liquid (Section 7.10). The reader should flip ahead through the book to see the wide diversity of the structured phenomena that are covered by constructal theory.

4.6 Tapered Channels and Optimal Angles

The heat trees of Fig. 4.6 do not look entirely natural: This is due to the simplifying features that were introduced tacitly at the start. For example, the high-conductivity blades were always drawn with constant thickness and perpendicular to their tributaries. These features served their purpose: They kept the number of geometric degrees of freedom to a minimum, and in this way they made possible the analytical exposition of the geometric-optimization method.

Constructal trees look more and more natural if their freedom to provide easier access to their internal currents is expanded. If the thickness of the elemental channel is allowed to vary arbitrarily with the longitudinal position [$D_0(x)$ in Fig. 4.1], the optimal shape for minimum global volume-to-point resistance at the elemental level corresponds to the power law $D_0 \sim x^{1/2}$, as shown in Fig. 4.7. The optimal profile of the two-dimensional channel looks like the pine needle or the needle of the snowflake: It is rounded at the tip, and its thickness increases at a decreasing rate as the distance from the tip increases. This result was obtained by variational

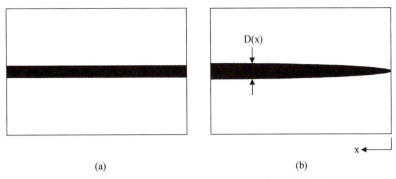

(a)   (b)

FIGURE 4.7. Two-dimensional high-conductivity channel (black) with fixed volume and length: (a) uniform thickness, and (b) optimal thickness function for minimal volume-to-point resistance at the elemental level [5].

calculus subject to the constraint that the volume of the high-conductivity material is fixed [5].

In the elemental system of Fig. 4.1 it was assumed that the high-conductivity channel stretches all the way across the elemental volume. When this simplifying assumption is not made, we find numerically that there is an optimal spacing between the tip of the k_p channel and the adiabatic boundary of the elemental volume [6]. Figure 4.8 shows five cases of optimized elemental volumes with spacings at the tips, i.e., with k_0 material all around the tips of the k_p blades. Incidentally,

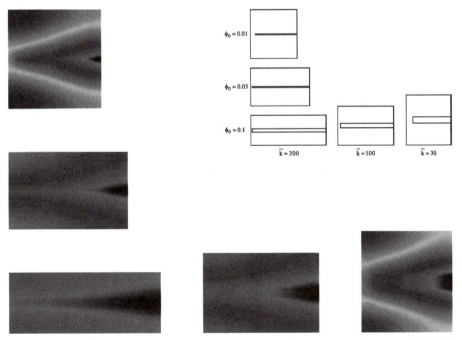

FIGURE 4.8. The optimal shapes of elemental volumes with spacings at the tips of the high-conductivity channels: the effect of varying $\tilde{k} = k_p / k_0$ and $\phi_0 = D_0 / H_0$ [6]. Note that the optimal shape becomes more slender and the global resistance decreases (the red disappears) as the volume (ϕ_0) and conductivity (k_p) of the central channel increase (see Plate V).

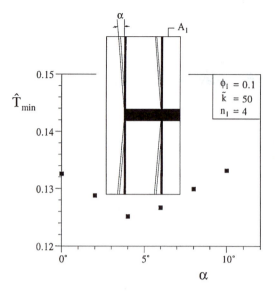

FIGURE 4.9. The optimization of the angle of confluence between tributaries and their central stem in a first construct ($\phi_1 = 0.1$, $\tilde{k} = 50$, $n_1 = 4$) [5].

this figure shows how the optimized elemental volume – the smallest building block of the constructal design – changes as the proportions of the composite material change: The elemental shape becomes more slender as either $\tilde{k}(= k_p / k_0)$ or $\phi_0(= D_0 / H_0)$ increases.

When the angle formed between each tributary channel and its central stem is allowed to vary, numerical calculations of the two-dimensional heterogeneous conduction field show that there exists an optimal angle for minimal volume-to-point resistance at the construct level [5]. This effect is illustrated for a first construct in Figs. 4.9 and 4.10, in which, for simplicity it was assumed that all the tributaries are tilted at the same (variable) angle. The optimal inclination is similar to that of tributaries in nature: Pine needles, rivulets, and bronchial ramifications point away from the thicker support (stem).

Figure 4.9 also shows that the volume-to-point resistance of the construct ($\hat{T}_{min}$) decreases only marginally (by 5.8%) as the angle α changes from the perpendicular position ($\alpha = 0°$) to the optimal position ($\alpha \cong 4°$). Similarly, the improvement derived from the optimal needle shape [Fig. 4.7(b)] relative to the simpler constant-thickness shape [Fig. 4.7(a)] is only a 6% reduction in the elemental resistance. The optimized tip spacings of Fig. 4.8 produce reductions of 20% in the global elemental resistance.

These relatively unimportant improvements tell a very important story: The tree design is *robust* with respect to various modifications in its internal structure. This means that the global performance of the system is relatively insensitive to changes in some of the internal geometric details. Trees that are not identical have nearly identical performance.

The robustness of the tree design sheds light on why natural tree flows are never identical geometrically. They do not have to be, if the maximization of global performance is their guiding principle. The ways in which the details of natural trees may differ from case to case are without number because, unlike in the constructions presented in this chapter, the number of degrees of freedom of the emerging form is not constrained. Local details differ from case to case because of unknown and

FIGURE 4.10. The optimization of the angle between the first-construct stems and the stem of a second construct (see Plate VI). In this example the second construct has four first constructs, and each first construct has eight elemental volumes. When the stems are perpendicular ($\alpha = 0°$, the third frame) the hot spots are concentrated in the farthest corners relative to the heat sink (blue). When the angle is too large (the first frame), the hot spots jump to the corners that are on the same side as the heat sink. In the optimal configuration (the second frame), the hot spots are spread most uniformly around the periphery of the system. More points work the hardest in this configuration. We pursue this idea further in the shapes of constant resistance constructed in Chap. 12.

unpredictable local features such as the heterogeneity of the natural flow medium and the history and lack of uniformity of the volumetric flow rate that is distributed over the system. Marvelous illustrations of this element of chance are provided by the seemingly irregular river drainage basins of Chap. 6. The point is that the global performance is predictable, and the principle that takes the system to this level of performance is deterministic.

4.7 Three-Dimensional Heat Trees

The optimization principle illustrated so far in two dimensions continues to apply in three-dimensional flow between a volume and one point [7]. The actual formulas for the optimized results are different (optimal shapes, numbers of constituents), but the optimization opportunities remain the same. For example, if the elemental volume is a cylinder of diameter H_0, length L_0, and fixed volume (Fig. 4.11), and if the volume fraction occupied by k_p material is fixed, $\phi_0 = (D_0/H_0)^2$, the external shape can be optimized for minimal volume-to-point resistance [7]:

$$\left(\frac{H_0}{L_0}\right)_{opt} = \frac{\pi}{4}\left(\frac{2bk_0}{ak_p}\right)^{1/2},$$

$$a(\phi_0) = \frac{1}{16}(-1 - \ln \phi_0 + \phi_0), \qquad b(\phi_0) = \frac{8}{\pi^2}\left(\frac{1}{\phi_0} - 1\right), \qquad (4.20)$$

$$T_{hot,min}\frac{k_0^{2/3}k_p^{1/3}}{q'''V_0^{2/3}} = \frac{3}{4}2^{-1/3}\pi^{-2/3}(-1 - \ln \phi_0 + \phi_0)^{2/3}\left(\frac{1}{\phi_0} - 1\right)^{1/3}.$$

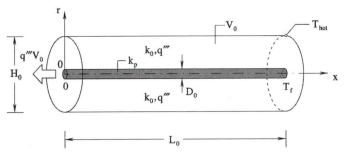

FIGURE 4.11. Cylindrical elemental volume with uniform volumetric heat generation and axial high-conductivity fiber [7].

In the limit $\phi_0 \ll 1$, the ratio $(H_0/L_0)_{\text{opt}}$ varies as $(\phi_0 k_p/k_0)^{-1/2}$, which is the same trend as in Eq. (4.2). In the same limit, the minimized hot-spot temperature increases monotonically as ϕ_0 decreases, which is the expected trend when the amount of k_p material decreases. The analysis leading to Eqs. (4.20) is analogous to that shown in Section 4.2 and is valid when $(k_0/k_p) < \phi_0 < 1$.

The first and higher-order constructs may be visualized as buds or pine cones on a stem, as shown in Fig. 4.12, or as coral branches, as shown in Fig. 4.13. The cylindrical constituents do not fit smoothly in a cylindrical aggregate, which is why the analysis based on Fig. 4.12 can be only approximate. Its purpose is to guide us to the optimization opportunities that exist and to more exact methods of geometric

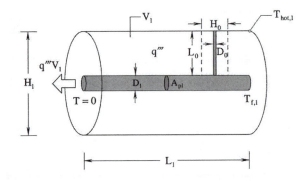

FIGURE 4.12. Cylindrical first construct containing a large number of elemental cylinders [7], and a similar structure encountered in pine cones (Courtesy of Col. Stefan Florin Varga).

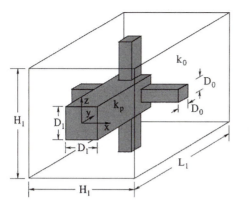 

FIGURE 4.13. Parallelepipedic first construct containing four elemental cylinders [7] and branches of coral (Courtesy of Col. Stefan Florin Varga).

optimization. Beginning with the first construct, the optimization opportunities at every level of assembly are two: the external shape (or number of constituents) and the proportion in which the high-conductivity material is allocated to the central fiber of the assembly and all the fibers of the constituents.

A more exact method of geometric optimization is the full numerical simulation of heat conduction in the three-dimensional heterogeneous body [7]. The selection of parallelepipedic shapes and numerical grids is helpful, so that the constituents fit smoothly in the construct. One drawback is that the number of constituents that can be assembled is limited to two or four, as in the first construct shown in Fig. 4.13. The numerical work developed in Ref. 7 shows that three-dimensional trees exhibit all the features encountered in two-dimensions, including the robustness of the tree design and performance.

4.8 Time-Dependent Discharge from a Volume to One Point

The universality and robustness of the tree form for volume-to-point flow are strengthened by more recent results on time-dependent conduction [8]. The discharge of a volume to one point is the fundamental problem type that covers many natural phenomena, for example, dielectric breakdown (lightning) and the formation of the river basin after a sudden rain. This was studied in a carefully designed time-dependent analog of the steady-state problem of Ref. 1. The same two-dimensional configuration and simplifying assumptions were kept. At time $t = 0$ the volume is isothermal (T_i) and placed in sudden contact with a heat-sink point (T_0) on its boundary.

Instead of minimizing the volume-to-point resistance at every volume scale, we minimized consistently the time of discharge (the cooldown time). The latter was defined as the time when the hot-spot temperature inside the volume $[T_{\max}(t) - T_0]$ drops to a fraction (e.g., 10%) of the original temperature difference $(T_i - T_0)$. The search for optimal heat-flow paths was conducted analytically and based on full numerical simulations of time-dependent conduction.

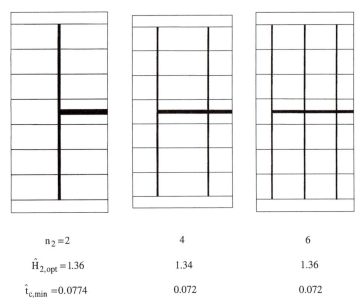

$n_2 = 2$	4	6
$\hat{H}_{2,opt} = 1.36$	1.34	1.36
$\hat{t}_{c,min} = 0.0774$	0.072	0.072

FIGURE 4.14. Second-order constructs for the time-dependent discharge from a two-dimensional volume to one point [8].

The optimized geometric features of the resulting structure are consistently the same as those of the steady-state configuration. For example, at the elemental level we found that the optimal external shape is nearly the same as that in Eq. (4.2), namely,

$$\left(\frac{H_0}{L_0}\right)_{opt} = \frac{2.11}{(\tilde{k}\phi_0)^{1/2}}. \qquad (4.21)$$

Several second-construct designs are illustrated in Fig. 4.14. They show that the overall performance is relatively insensitive to some of the internal details of the channel arrangement, provided that the arrangement is a tree and that it has been optimized for minimum discharge (cooldown) time, $\hat{t} = t\alpha_0 / A_2$. In the same figure, n_2 is the number of first constructs assembled in the second construct and $\hat{H}_2 = H_2 / A_2^{1/2} = (H_2 / L_2)^{1/2}$. Very similar designs have emerged from the numerical optimization of the steady volume-point flow in two dimensions [5].

The fact that the minimization of steady-state resistance (Sections 4.2–4.4) and the minimization of time of approach to equilibrium (this section) lead to the same internal flow structure should not surprise anyone. When the internal resistances of an isolated nonequilibrium system are minimized, the internal currents find it easiest to bring the system to the state of internal uniformity, and equilibrium is attained in minimum time. The existence of the equilibrium end state is proclaimed by the second law of thermodynamics. The geometric minimization of the time of approach to equilibrium – the generation of internal structure to achieve this objective – is an entirely different idea. It is part of the domain of the constructal law, and it covers many natural flow structures, from lightning to river basins (Chap. 6) and turbulence (Chap. 7).

4.9 Constructal Design: Increasing Complexity in a Volume of Fixed Size

In this section an alternative way of looking at the volume-to-point flow is proposed, namely, the reverse of the growth described in Sections 4.1–4.4. This time the size of the volume (A) is fixed, although its external shape may vary in order to accommodate optimally the internal structure that we will determine: *the design*. The volume fraction of high-conductivity material $\phi_0 = A_p/A$ is also fixed. The design objective is to distribute this material over A such that the resistance from A to the heat sink M is minimal. The design approach described in this section was proposed in Ref. 9.

We contemplate using several designs for the internal structure of the same volume, A. In Fig. 4.15 each is indicated by the subscript $i = 0, 1, 2, \ldots$, which represents, in order, the elemental design, the first-construct design, the second-construct design, etc. These designs have the same size ($A = A_0 = A_1 = A_2 = A_3$), and are indicated by the labels put in the lower-left-hand corner of each frame. More specifically, by first-construct design we mean that the A_1 frame of Fig. 4.15 has the same internal structure as that of Fig. 4.3, even though that structure is not shown in detail in Fig. 4.15. Similarly, the A_2 frame of Fig. 4.15 has the same internal tree structure as that of Fig. 4.5.

Next, we optimize the internal dimensions of each design (A_i, ϕ_i) and then see how well they perform against each other. In other words, given the system size (A) and amount of k_p material (ϕ), which design – what degree of internal complexity – is the best for minimizing the volume-to-point flow resistance?

Elemental Design. The elemental design is represented by a single high-conductivity blade of thickness D_0, and by the external shape H_0/L_0. The resistance minimization analysis for this geometry is the same as that in Section 4.2. The optimal external shape is given by Eq. (4.2) with ϕ in place of ϕ_0. This shape is independent of the size A_0. The optimal thickness of the k_p blade is $D_{0,\text{opt}} = \phi_0 H_{0,\text{opt}} = \phi_0(2A)^{1/2}(\tilde{k}\phi_0)^{-1/4}$. The minimized overall thermal resistance is shown in Eq. (4.3), where ΔT_0 is the excess temperature registered in the hot-spot corners (see Fig. 4.15).

First-Construct Design. Consider next the design in which the k_p material is distributed between one central blade (D_1, L_1) and n_1 equidistant side blades (D_0, L_0), as shown in the second frame of Fig. 4.15. When $n_1 \gg 1$, the temperature drop along the D_1 blade is equal to $q''' H_1 L_1^2/(2k_p D_1)$, in accordance with the analysis given in Ref. 1 [see also the discussion above Eq. (4.1)]. The temperature drop from the hot-spot corner to the right end of the D_1 blade is sustained by an elemental design of size $A_0 = A_1/n_1$ and is given by Eq. (4.3). Adding these two temperature differences, we obtain the overall resistance:

$$\frac{\Delta T_1 k_0}{q''' A_1} = \frac{1}{2n_1(\tilde{k}\phi_0)^{1/2}} + \frac{n_1^{1/2}}{2^{3/2}\tilde{k}^{5/4}\phi_0^{1/4}\hat{D}_1}, \qquad (4.22)$$

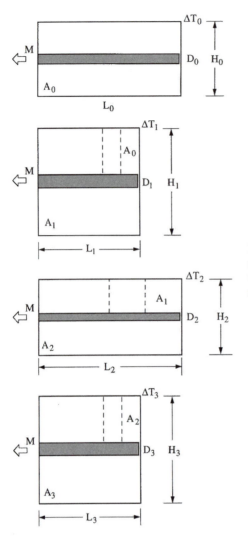

FIGURE 4.15. Constructal design: the minimization of resistance through the increase in the internal complexity of a volume of fixed size [9]. Note that $A_0 = A_1 = A_2 = A_3$.

where $\hat{D}_1 = D_1 / A_1^{1/2}$ and $\phi_0 = D_0 / H_0$. It is worth noting that $\hat{D}_1$ is not the same as $\tilde{D}_1$ of Eq. (4.5). Neither is Eq. (4.22) the same as Eq. (4.5). The main difference this time is that A_1 is fixed and A_0 varies, unlike in Eq. (4.5) in which A_0 was fixed and A_1 (or n_1) varied. Only the shape of A_0 is fixed, in accordance with the elemental design (Section 4.2) or Eq. (4.2).

The first-construct design has 2 degrees of freedom: the number of side blades n_1 and the relative size of the two blade thicknesses. The second degree of freedom is represented by either $\hat{D}_1$ or ϕ_0. By minimizing expression (4.22) with respect to n_1, we find

$$n_{1,\text{opt}} = 2\tilde{k}^{1/2}\phi_0^{-1/6}\hat{D}_1^{2/3} \tag{4.23}$$

and the corresponding resistance minimum,

$$\left(\frac{\Delta T_1 k_0}{q''' A_1}\right)_m = \frac{3}{4\tilde{k}\phi_0^{1/3}\hat{D}_1^{2/3}}. \tag{4.24}$$

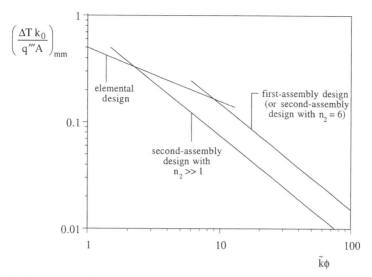

FIGURE 4.16. Constructal design: the minimized thermal resistances of the elemental, first-construct, and second-construct designs [9].

Next, the minimization with respect to $\hat{D}_1$ (or ϕ_0) is subjected to the k_p material constraint $\phi_1 = A_{p1}/A_1 = \text{constant}$, which assumes the form

$$\phi_1 = \hat{D}_1^{4/3}\phi_0^{-1/3} + \phi_0. \tag{4.25}$$

The results of minimizing expression (4.22) with respect to this second degree of freedom are, in order, $\phi_{0,\text{opt}} = \phi_1/2$, $\hat{D}_{1,\text{opt}} = \phi_1/2$, and

$$\left(\frac{D_1}{D_0}\right)_{\text{opt}} = \frac{1}{2}(\tilde{k}\phi_1)^{1/2} \gg 1, \tag{4.26}$$

$$n_{1,\text{opt}} = (2\tilde{k}\phi_1)^{1/2} \gg 1, \tag{4.27}$$

$$\left(\frac{\Delta T_1 k_0}{q''' A_1}\right)_{mm} = \frac{3}{2\tilde{k}\phi_1}. \tag{4.28}$$

The differences among these results and their counterparts in the growth formulation [Eqs. (4.9)–(4.11)] are worth noting. Most evident is the optimized external shape of A_1, which is now exactly square $(H_1/L_1)_{\text{opt}} = 1$.

When is the first-assembly design preferable to the elemental design? This question is answered in Fig. 4.16 by a comparison of the thermal resistances of the two designs on an equal basis (A, ϕ), that is, when $A_0 = A$ and $\phi_0 = \phi$ are placed in Eq. (4.3) and $A_1 = A$ and $\phi_1 = \phi$ are placed in Eq. (4.28). The first-assembly design is preferable when $\tilde{k}\phi > 9$. The group $\tilde{k}\phi$, which accounts for the properties of the composite material, plays a critical role in the optimization of the design. As this group increases, i.e., as the k_p inserts become more conductive and/or more voluminous, there comes a point at which the internal architecture makes a *transition* from one structure (A_0) to a more complex one (A_1). The transition is toward lower resistance by means of higher complexity (e.g., turbulence, Sections 7.1–7.3).

Second-Construct Design. The next question is whether a second complexity transition is possible, perhaps at a larger $\tilde{k}\phi$ value. Consider for this purpose the second-construct configuration shown in the third frame of Fig. 4.15. The new central blade (D_2, L_2) collects the heat currents contributed by n_2 blades of type (D_1, L_1). In other words, the A_2 structure is spread over a space divided into n_2 compartments such that each compartment $(A_1 = A_2/n_2)$ has the external and the internal shapes optimized in this section. The overall size A_2 is fixed, and the number n_2 is considerably greater than 1.

The overall resistance from A_2 to the sink has two terms: the temperature drop along the D_2 blade, which by analogy with the start of the first-construct design is equal to $q''' H_2 L_2^2/(2k_p D_2)$, and the temperature drop sustained by the corner compartment A_1, which can be deduced from Eq. (4.28). The result is

$$\left(\frac{\Delta T_2 k_0}{q''' A_2}\right) = \frac{3}{2n_2 \tilde{k}\phi_1} + \frac{n_2^{1/2}}{4\tilde{k}\hat{D}_2}, \tag{4.29}$$

where $\hat{D}_2 = D_2/A_2^{1/2}$. Minimizing expression (4.29) with respect to n_2, we obtain

$$n_{2,\text{opt}} = (12\hat{D}_2/\phi_1)^{2/3}, \tag{4.30}$$

$$\left(\frac{\Delta T_2 k_0}{q''' A_2}\right)_m = \frac{3^{4/3}}{2^{7/3}\tilde{k}\phi_1^{1/3}\hat{D}_2^{2/3}}. \tag{4.31}$$

The second degree of freedom is with respect to the allocation of k_p material, i.e., the trade-off between $\hat{D}_1$ and ϕ_1 subject to the $\phi_2 = A_{p2}/A_2$ constraint, which can be written as

$$\phi_2 = (2/3)^{1/3}\phi_1^{1/3}\hat{D}_2^{2/3} + \phi_1. \tag{4.32}$$

We want to minimize resistance (4.31) with respect to ϕ_1 subject to constraint (4.32). This means that we must maximize the product $\phi_1^{1/3}\hat{D}_2^{2/3}$, which in view of the ϕ_2 constraint is equal to $(3/2)^{1/3}(\phi_2 - \phi_1)$. In conclusion, we must use the smallest ϕ_1 value possible. This corresponds to the largest n_2 that can be built [see Eq. (4.30)],

$$n_2 = 6\left(\frac{\phi_2}{\phi_1} - 1\right), \tag{4.33}$$

the largest $\hat{D}_2 = \phi_2 n_2^{3/2}/[2(n_2 + 6)]$, or the most slender external shape, $(H_2/L_2)_{\text{opt}} = 4/n_2$. This recommendation is also made by the form of the thermal resistance expression obtained when these results are combined with Eq. (4.31):

$$\left(\frac{\Delta T_2 k_0}{q''' A_2}\right)_m = \frac{3(n_2 + 6)}{4n_2 \tilde{k}\phi_2}. \tag{4.34}$$

In conclusion, in the second-construct design the number of components (n_2) is left as a parameter constrained from above by manufacturing considerations. In the limit $n_2 \gg 1$ resistance (4.34) approaches $3/(4\tilde{k}\phi_2)$. Figure 4.16 shows this limit on an equal basis with the other designs that are being compared (i.e., by setting

$A_2 = A$ and $\phi_2 = \phi$). Once again, we see that the performance improves when the complexity of the design increases. The second-construct design performs better than the first-construct design when $n_2 > 6$.

Third-Construct Design. It is instructive to consider at least one more design, that is, one more step toward higher complexity. In the third-construct design, shown in the fourth frame of Fig. 4.15, the $A = A_3$ space is covered by n_3 ($\gg 1$) components of variable size A_2 ($= A_3/n_2$). The external and the internal shapes (dimensionless aspect ratios) of these components have been optimized, as shown in the material covered already in this section. The end-to-end temperature difference sustained by the D_3 stem of the k_p tree is equal to $q''' H_3 L_3^2/(2k_p D_3)$. Adding this to the temperature difference sustained by the corner component, we obtain the overall resistance of the third-construct design:

$$\left(\frac{\Delta T_3 k_0}{q''' A_3}\right) = \frac{3a}{4n_3 \tilde{k}\phi_2} + \frac{n_3^{1/2}}{2n_2^{1/2}\tilde{k}\hat{D}_3}, \tag{4.35}$$

where $\hat{D}_3 = D_3/A_3^{1/2}$ and $a = (n_2 + 6)/n_2$. In this design an optimal number of components exists, and we obtain it by minimizing expression (4.35) with respect to n_3:

$$n_{3,\text{opt}} = (3b\hat{D}_3/\phi_2)^{2/3}, \tag{4.36}$$

$$\left(\frac{\Delta T_3 k_0}{q''' A_3}\right)_m = \frac{3^{4/3}a^{1/3}}{4n_2^{1/3}\tilde{k}\phi_2^{1/3}\hat{D}_3^{2/3}}, \tag{4.37}$$

where $b = n_2^{1/2}a$. Equation (4.37) shows that we can minimize the resistance further by maximizing the product $\phi_2^{1/3}\hat{D}_2^{2/3}$. This second optimization is constrained by the total amount of k_p material that is allocated to the third-construct design, namely, $\phi_3 = A_{p3}/A_3$, or

$$\phi_3 = \hat{D}_3(n_3/n_2)^{1/2} + \phi_2. \tag{4.38}$$

The result of this second optimization is $\phi_{2,\text{opt}} = \phi_3/2$, which can be combined with Eqs. (4.36) and (4.37) to obtain, in order,

$$\hat{D}_{3,\text{opt}} = \frac{n_2^{3/8}\phi_3}{2(3b)^{1/4}}, \tag{4.39}$$

$$\left(\frac{\Delta T_3 k_0}{q''' A_3}\right)_{mm} = \frac{3^{3/2}(n_2 + 6)^{1/2}}{2n_2\tilde{k}\phi_3}, \tag{4.40}$$

$$\left(\frac{H_3}{L_3}\right)_{\text{opt}} = \frac{n_2}{3^{1/2}(n_2 + 6)^{1/2}}, \tag{4.41}$$

$$\left(\frac{D_3}{D_2}\right)_{\text{opt}} = \frac{2(n_2 + 6)^{3/4}}{3^{1/4}n_2}, \tag{4.42}$$

$$n_{3,\text{opt}} = [3(n_2 + 6)]^{1/2}. \tag{4.43}$$

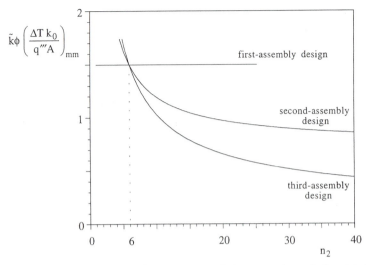

FIGURE 4.17. The minimized thermal resistances of the second- and the third-construct designs [9].

Several geometric features are worth noting. First, the number of constituents at the second-construct level (n_2) continues to play the role of parameter. The ratio $(D_3/D_2)_{opt}$ is of the order of 1 when n_2 is of the order of 10. The ratio $(H_3/L_3)_{opt}$ is equal to 1.44 when $n_2 = 10$. The number $n_{3,opt}$ is equal to 6 when $n_2 = 6$. An important result is Eq. (4.40), which shows that the overall resistance continues to decrease as n_2 increases. This trend could be added to Fig. 4.16, but an even better comparison is provided on an equal basis by Fig. 4.17. In this figure we see the effect of n_2 on the best performances of the first, second, and third constructs. The performance improves in every case as the complexity (n_2) increases, provided n_2 is greater than 6. In other words, for the same combination of materials ($\tilde{k}\phi$), it makes sense to increase the complexity of the internal architecture of the design. The limit to this sequence of complexity increase will be dictated by manufacturing considerations.

4.10 Design with Unrestricted Elemental Features

The sequence of four designs optimized in the preceding section can be continued toward constructs of higher order. An important feature that deserves emphasis is that in each design the constituents inherited the optimized shapes (aspect ratios) of the design of preceding order. For example, in the first-construct design of Fig. 4.15 (second frame) we assumed that the shape of each constituent (elemental volume A_0) matches the optimal value derived in Eq. (4.2). This choice is beneficial, because by reducing the number of degrees of freedom it simplifies the optimization work, but does it really lead to the best possible design?

In this section we examine this question by considering again the optimization of the first-construct design. We write that the peak temperature difference ΔT_1

is equal to the temperature drop along the D_1 blade [namely, $q''' H_1 L_1^2/(2k_p D_1)$] plus the peak temperature sustained by the upper-right element of size $H_0 L_0$. This second contribution is provided by Eq. (4.1). Unlike in Section 4.9, here we do not assume that H_0/L_0 is given by Eq. (4.2). After some algebra, we find that the overall resistance of the first-construct design is given by the dimensionless expression

$$\frac{\Delta T_1 k_0}{q''' A_1} = \frac{1}{2\tilde{k}\hat{D}_1}\left(\frac{L_1}{H_1}\right)^{1/2} + \frac{1}{2n_1^2}\frac{L_1}{H_1} + \frac{n_1^2}{8\tilde{k}\phi_0}\frac{H_1}{L_1}. \tag{4.44}$$

This expression shows that this time the architecture is described by four dimensionless numbers: the overall aspect ratio H_1/L_1, the thickness of the central blade $\hat{D}_1$, the number of constituents n_1, and the thickness of the elemental blade ϕ_0 (or D_0/H_0). Three of these parameters are free to vary, because the amount of k_p material is constrained by Eq. (4.25), in which ϕ_1 is a constant. In this formulation the number of degrees of freedom exceeds by one the corresponding number in the optimization conducted in Section 4.9.

The minimization of thermal resistance (4.44) begins with eliminating $\hat{D}_1$ in favor of ϕ_0, by use of constraint (4.25). Taking the derivative of this resistance with respect to ϕ_0 and setting it equal to zero, we obtain

$$3y^{7/4} - y^{3/4} = \left(\frac{H_1}{L_1}\right)^{3/2}, \tag{4.45}$$

where $y = \phi_{0,\text{opt}}/(\phi_1 - \phi_{0,\text{opt}})$. Equation (4.45), i.e., the selection of $\phi_{0,\text{opt}}$, completes the first resistance-minimization step.

The second minimization begins with eliminating from resistance formula (4.44) the ratio H_1/L_1 in favor of y, by use of Eq. (4.45). After some algebra, the once-minimized resistance assumes the form

$$\left(\frac{\Delta T_1 k_0}{q''' A_1}\right)_m = \frac{F(y, \varepsilon)}{n_1^2} \tag{4.46}$$

where $\varepsilon = n_1^2/(2\tilde{k}\phi_1)$, and

$$F = \varepsilon \frac{1+y}{y^{1/2}}\left[\frac{4}{(3y-1)^{1/3}} + (3y-1)^{2/3}\right] + \frac{1}{2y^{1/2}(3y-1)^{2/3}}. \tag{4.47}$$

The function F can be minimized with respect to y: The results for y_{opt} and $F_{\text{min}} = F(y_{\text{opt}}, \varepsilon)$, Ref. 9, complete the second resistance-minimization step.

We obtain the twice-minimized resistance by substituting $F = F_{\text{min}}$ into Eq. (4.46), namely, $(\Delta T_1 k_0/q''' A_1)_{mm} = F_{\text{min}}/n_1^2$. This function depends on n_1 and the product $\tilde{k}\phi_1$, as is shown by the upper curves in Fig. 4.18. The main effect is due to $\tilde{k}\phi_1$, especially as n_1 increases: In the limit $n_1 \gg 1$, the overall thermal resistance $(\Delta T_1 k_0/q''' A_1)_{mm}$ approaches $5/\tilde{k}\phi_1$. This trend is comparable with that of Eqs. (4.11) and (4.28), which are also shown in Fig. 4.18. The lowest resistance estimate happens to be due to the growth method of Eq. (4.11), which is by far the simplest analytical method that we have considered. The constructal design method, Eq. (4.28), leads to a higher estimate, and, finally, the design with

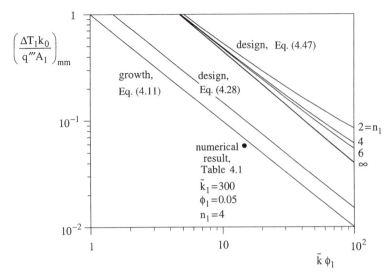

FIGURE 4.18. The minimized thermal resistance of the first-construct design with unrestricted elemental features [9].

the largest number of degrees of freedom – this section – leads to an even higher estimate.

Which of the three methods is the most accurate? To shed light on this issue we solved numerically the steady-state conduction problem in the heterogeneous domain A_1 shown in the second frame of Fig. 4.15. The heat generation rate ($q''' A_1$, fixed) was distributed over only the k_0 domain, not over the D_1 and D_0 fibers of k_p material. The length of the D_1 fiber is less than L_1, specifically, $L_1 - (H_0 - D_0)/2$. The geometry has 3 degrees of freedom: H_1/L_1, D_1/D_0, and n_1. We solved the conduction problem in many geometrical configurations and developed the overall resistance ($\Delta T_1 k_0 / q''' A_1$) as a function of H_1/L_1, D_1/D_0, and n_1. We were able to minimize this function with respect to H_1/L_1 and D_1/D_0, but not with respect to n_1. The numerical work was based on the finite-element method [9].

Table 4.2 shows that the twice-minimized resistance decreases monotonically as n_1 increases and that this effect becomes weak when n_1 is greater than 4 or 6. This behavior agrees with the analytical solution developed in the first part of this section and plotted at the top of Fig. 4.18. Coincidental are the numerical values reported for $(\Delta T_1 k_0 / q''' A_1)_{mm}$ in Table 4.2, which are very close to those predicted by the growth method: Note that Eq. (4.11) with $\tilde{k} = 300$ and $\phi_1 = 0.05$ yields $(\Delta T_1 k_0 / q''' A_1)_{mm} = 0.067$. In sum, the three analytical methods anticipate the proper order of magnitude of the minimized resistance and the effect of the group $\tilde{k} \phi_1$. The design analysis developed in this section also anticipates correctly the effect of n_1, which is weak.

It is worth commenting on the geometric features that accompany the twice-minimized resistance based on Eq. (4.47). The optimal external aspect ratio $(H_1/L_1)_{opt}$ emerges as a function of only $\tilde{k} \phi_1 / n_1^2$. For example, when $\tilde{k} = 300$, $\phi_1 = 0.05$ and $n_1 = 4$, the optimal shape is $(H_1/L_1)_{opt} = 1.86$, which agrees approximately with the numerical results shown in Table 4.2.

TABLE 4.2. **Results of the Geometric Optimization of the First-Assembly Design, when the Overall Thermal Resistance is Determined by Direct Numerical Simulation** $(\tilde{k} = 300, \phi_1 = 0.05)$ [9]

n_1	$(\Delta T_1 k_0 / q''' A_1)_{mm}$	$(H_1/L_1)_{opt}$	$(D_1/D_0)_{opt}$	$\phi_{0,opt}/\phi_1$
2	0.0818	3.80	3.2	0.7
4	0.0616	2.56	5	0.587
6	0.0575	2.37	6.8	0.564
8	0.0561	2.33	9	0.548

The corresponding internal ratio $(D_1/D_0)_{opt}$ is a function of $\tilde{k}\phi_1/n_1^2$ and n_1. For the numerical example used in the preceding paragraph, the result is $(D_1/D_0)_{opt} = 2.1$, which also verifies the order of magnitude of the numerical solution reported in Table 4.2. The volume fraction ratio $\phi_{0,opt}/\phi_1$ is close to 0.5 when $\tilde{k} = 300$, $\phi_1 = 0.05$, and $n_1 = 4$ and increases toward 1 as $\tilde{k}\phi_1/n_1^2$ increases. The result $\phi_{0,opt}/\phi_1 \cong 0.5$ agrees with the corresponding analytical results of the growth method [listed under Eq. (4.8)] and the design method [listed above Eq. (4.26)].

4.11 Constructal Heat Trees Are Robust

In this chapter we saw that in addition to the constructal *growth* method of minimizing the thermal resistance between one point and a volume, there is the *design* approach in which the volume is fixed. In both approaches, the overall volume-to-point resistance decreases as the internal complexity of the high-conductivity path increases. In the growth method the number of constituents in each new, higher-level assembly can be optimized. In the two design approaches illustrated in Sections 4.9 and 4.10, some of the number of constituents cannot be optimized and are carried along, as parameters, on the road to higher complexity (e.g., n_2 in Fig. 4.17 and n_1 in Fig. 4.18). An important characteristic of the design method is that the effect of the number of constituents becomes weaker as that number increases (e.g., Fig. 4.18).

What is even more important is that the growth and design approaches lead to designs that are comparable not only in overall performance but also in geometric appearance. Review, for example, the results for the overall volume-to-point resistance, as we have done in Fig. 4.18. Being the simplest, the growth method provides a useful shortcut to and an approximation of the architecture that would result at the end of the application of the fixed-volume design method.

In Fig. 4.18, the numerical value for the twice-minimized resistance of the first-assembly design with unrestricted elemental features (Table 4.2) is smaller than the corresponding analytical estimate based on Eq. (4.47). This discrepancy is explained by the fact that in the analysis of the design of Section 4.10 the elemental volumes were assumed to have adiabatic boundaries. When n_1 of these constituents were grouped together, the first assembly inherited $(n_1 - 1)/2$ internal adiabatic partitions, which were situated equidistantly between the D_0-thin blades. Internal partitions inhibit the free flow of heat through the volume and account for an increase in the global resistance. The first-construct design optimized numerically in Table 4.2 did not have internal adiabatic partitions and, consequently, its overall resistance was lower than in the analytical solution.

To test this explanation we repeated the numerical work reported for $n_1 = 4$ in Table 4.2; however, this time we imposed the adiabatic-wall condition on the vertical plane situated halfway between the D_0 blades of the first assembly. The results of the double minimization of the overall resistance are $(\Delta T_1 k_0 / q''' A_1)_{mm} = 0.065$, $(H_1/L_1)_{opt} = 2.56$, $(D_1/D_0)_{opt} = 5.25$, and $\phi_{0,opt}/\phi_1 = 0.576$. These results show that, as expected, the thermal resistance is higher when the internal partitions are present. The optimized geometry, however, is practically insensitive to how the vertical midplane of the k_0 medium is modeled. This is an important conclusion, because what matters in design are the optimal dimensions of the structure that connects the volume to the point. From a design standpoint, the optimized *shapes* produced analytically are *robust*.

This conclusion is illustrated in Fig. 4.19, which shows side by side the isotherm patterns in the first-assembly design optimized numerically without [Fig. 4.19(a)] and with [Fig. 4.19(b)] internal adiabatic partitions. The curves of Figs. 4.19(a) and 4.19(b) correspond to the same isotherms, i.e., the same values of dimensionless internal temperature. The plane in which the isotherms of Fig. 4.19(b) are interrupted is the plane of the two adiabatic partitions. The isotherm patterns differ somewhat in the vicinity of the midplane partitions, but the external shape $(H_1/L_1)_{opt}$ and the internal shape of the two structures are nearly identical.

We can expect the same degree of accuracy from the analytical methods presented in Fig. 4.18: fairly good predictions for the geometric shapes that place the system close to the point of minimum global resistance and only order-of-magnitude estimates of the minimized global resistance. It is only a coincidence that the two analytical methods (growth, design) predict global resistances that fall below the resistance based on Eq. (4.47). This coincidence is responsible for the better agreement between the numerical result and the two lowest curves of Fig. 4.18. It is a *useful* coincidence, because we may start any design of this type with the simplest method first (namely, the growth method) and expect from this first cut not only a fairly accurate image of the optimized architecture but also the least inaccurate global resistance estimate from the rough estimates that can be expected from all the analytical methods.

The correct path to the ultimate (exact) global optimum is clearly the numerical method that led to Table 4.2. The number of elemental details (n_1) can be increased until costs (e.g., computation) indicate that the point of diminishing returns has been reached. Furthermore, the geometry that is being optimized can be endowed with considerably more degrees of freedom in order to provide additional flexibility in decreasing the global resistance (e.g., Section 4.6).

FIGURE 4.19. The imperceptible effect of the adiabatic midplane partition on the geometry of the first assembly optimized numerically ($n_1 = 4$, $k_p/k_0 = 300$, $\phi_1 = 0.05$): (a) without partitions, (b) with partitions [9].

PROBLEMS

4.1 The elemental volume optimized in Section 4.2 may be viewed as an example of external shaping, as a mechanism of achieving global resistance minimization. Consider the potentially more promising method of practicing simultaneously external shaping and internal structuring; compare it with the lessons learned in Chap. 2. To optimize the internal structure of the elemental volume of Fig. 4.1 is to relax the assumption that the k_p blade thickness D_0 is uniform. Assume instead that D_0 has an arbitrary power-law shape,

$$D_0(x) = bx^n,$$

where b and n are two constants. Show that the D_0 shape affects only the second term on the right-hand side of Eq. (4.1): Minimize this term with respect to the shape exponent n and subject to the k_p material constraint $\phi_0 = A_{p,0}/A_0$, constant. Next, derive the corresponding global resistance that replaces Eq. (4.1) and minimize this expression with respect to the external shape H_0/L_0. Show that the global resistance minimized in this fashion (internally and externally) is

$$\left(\frac{\Delta T_0 k_0}{q''' A_0}\right)_{\min} = \frac{2^{1/2}}{3(\tilde{k}\phi_0)^{1/2}}$$

and that it is 6% smaller than the minimized resistance of the corresponding system with constant D_0 blade, Eq. (4.3).

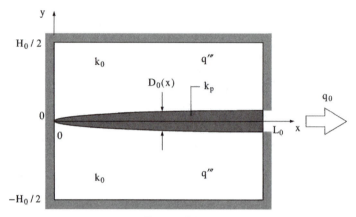

FIGURE P4.1

4.2 The elemental cylinder shown in Fig. 4.11 generates heat volumetrically and uniformly at the rate q'', in the material of low conductivity k_0. The generated heat current escapes through the origin $(0, 0)$, which also serves as root point for an axial cylindrical insert of high conductivity k_p. Place the analysis in the limit were the heat conduction through the k_0 material is oriented radially, while the conduction through the k_p insert is oriented axially. Verify later that this limit is valid when

$$\frac{k_p}{k_0} \gg 1 \qquad \frac{H_0}{L_0} \ll 1 \qquad \phi_0 = \left(\frac{D_0}{H_0}\right)^2 \ll 1.$$

Develop the analytical solution for the overall temperature difference in two parts,

$$T_{\text{hot}} - 0 = (T_{\text{hot}} - T_f) + (T_f - 0)$$

where we wrote zero for the heat sink (root point) temperature. Minimize T_{hot} with respect to the shape of the cylinder (e.g., H_0) when constrained are the total volume $V_0 = (\pi/4)H_0^2 L_0$ and the volume of k_p material, or the volume fraction ϕ_0. Show analytically that the optimal shape is represented by Eqs. (4.20). Use this solution to demonstrate that the assumed perpendicularity of the k_0 and k_p heat flow paths is valid when $k_0/k_p \ll \phi_0 \ll 1$.

REFERENCES

1. A. Bejan, Constructal-theory network of conducting paths for cooling a heat generating volume, *Int. J. Heat Mass Transfer*, Vol. 40, 1997, pp. 799–816.
2. A. Bejan and D. Tondeur, Equipartition, optimal allocation, and the constructal approach to predicting organization in nature, *Rev. Gen. Thermique*, Vol. 37, 1998, pp. 165–180.
3. J. H. Holland, *Hidden Order*, Addison-Wesley, Reading, MA, 1995.
4. A. Bejan, *Advanced Engineering Thermodynamics*, 2nd ed., Wiley, New York, 1997.
5. G. A. Ledezma, A. Bejan, and M. R. Errera, Constructal tree networks for heat transfer, *J. Appl. Phys.*, Vol. 82, 1997, pp. 89–100.

6. M. Almogbel and A. Bejan, Conduction trees with spacings at the tips, *J. Heat Transfer*, Vol. 42, 1999, pp. 3739–3756.

7. G. A. Ledezma and A. Bejan, Constructal three-dimensional trees for conduction between a volume and one point, *J. Heat Transfer*, Vol. 120, 1998, pp. 977–984.

8. N. Dan and A. Bejan, Constructal-tree networks for the time-dependent discharge of a finite-size volume to one point, *J. Appl. Phys.*, Vol. 84, 1998, pp. 3042–3050.

9. A. Bejan and N. Dan, Two constructal routes to minimal heat flow resistance via greater internal complexity, *J. Heat Transfer*, Vol. 121, 1999, pp. 6–14.

CHAPTER FIVE

FLUID TREES

5.1 Bathing a Volume: Objective and Constraints

We now turn our attention to the workings of the constructal (purpose and constraints) principle in systems with internal fluid streams. The analogy between fluid streams and heat currents is well established; in fact, it is the heart of modern heat transfer and transport phenomena theory [1]. For this reason, we expect that the optimal architectures determined for heat flow in Chap. 4 have pictorial counterparts in the realm of pure fluid flow. In this chapter we explore the many opportunities for minimizing the resistance to fluid flow between a point (source, sink) and volume or area and show how geometry – the tree network – springs out as the solution.

Fluid flows are the natural volume-to-point patterns that have been studied the most. The features of treelike patterns in fluid flow have been described quantitatively in surprisingly sharp (reproducible) terms, notably in the study of lungs, vascularized tissues, botanical trees, and rivers [2–19]. In recent years, these natural phenomena have been visualized on the computer by means of repetitive fracturing algorithms, which had to be postulated. The origin of the phenomena was left to the "explanation" that the broken pieces (or building blocks from the reverse point of view of constructal theory) are the fruits of a nondeterministic process of *self*-optimization and *self*-organization.

In this chapter we continue to show that constructal theory places a purely deterministic method behind the word self: the search for the easiest flow path when global constraints (flow rate, volume size) are imposed. For example, previous studies of the fluid access problem used as the starting point the empirical observation that tubes and streams *bifurcate*. This property is known as dichotomous branching (e.g., Figs. 5.1 and 5.2), which is a term derived from the Greek noun *dichotomia* (division into two parts). Predicting this geometric feature – the integer 2 – is an essential objective of constructal theory (see the end of Section 5.3).

To construct a theory that is both effective and transparent, it is advisable to rely on the simplest possible model that still retains the most essential feature of the network: its function. Such a model begins with Fig. 5.3, which shows that the

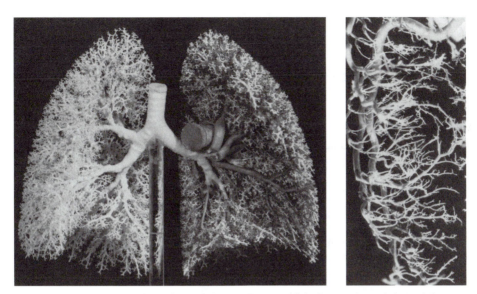

FIGURE 5.1. Dichotomous branching in the air passages of the human lung (left) and the coronary arteries of the heart muscle (right). (Courtesy of Professor Ewald R. Weibel, University of Bern. Reprinted from Ref. 5 with permission from Birkhäuser Verlag, Basel, Switzerland.)

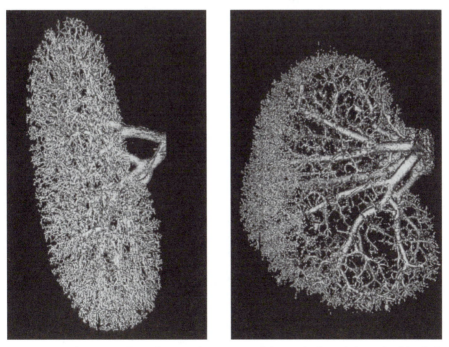

FIGURE 5.2. Dichotomous branching in the arteries of the pig kidney (left) and rabbit kidney (right). (Reprinted from Ref. 8 with permission from World Scientific Publishing, Singapore.)

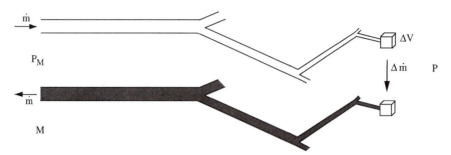

FIGURE 5.3. Network of ducts connecting one point (M) with every point residing inside a finite volume (V) [18].

function of the network is to distribute the stream $\dot{m}$ from the point M to every elemental volume ΔV.

An important consequence of the function of the network is that in some cases two superimposed networks are needed. For example, in the bronchial tree (Fig. 5.1, left) the flow is periodic (in and out), and this flow is accommodated by a single network. In the circulatory system (Figs. 5.1, right, and 5.2) two identical networks (in counterflow) are needed such that each elemental volume ΔV receives arterial blood at the rate dictated by local metabolism $\Delta \dot{m}$. The elemental volume returns the blood at the same rate to the network of veins (e.g., the dark network in Fig. 5.3). Similar pairs of networks in counterflow are encountered in trees, roots, and leaves. A river basin needs only one network (Section 5.8) because the volume V is flat (the basin surface) and $\Delta \dot{m}$ is proportional to the rainfall per unit surface.

In sum, the function of the network is well represented by the volumetric mass flow-rate density $\dot{m}''' = \Delta \dot{m}/\Delta V$, which must be collected (integrated) by the network over the volume V and channeled as a single stream ($\dot{m}$) to the point M. This operational characteristic ($\dot{m}'''$) is assumed given and serves as constraint in the thermodynamic optimization of the network.

We focus on only one of the networks of Fig. 5.3, namely the dark one, and recognize that $\Delta \dot{m}$ is driven from the elemental volume to the origin by the pressure difference $P - P_M$. This difference varies with the position of the ΔV element relative to M: Of special interest is the maximum pressure difference, $\Delta P = (P_{max} - P_M)$, which is needed by the elemental volumes that are situated the farthest from the end of the network and serves as local constraint. In the lungs and the circulatory system, for example, ΔP is the pressure level that must be generated by the thorax and heart muscles. The time-averaged mechanical power consumed by these pumps, or the entropy generation rate of the network, is proportional to the product $\dot{m}\,\Delta P$. The total flow rate $\dot{m}$ is fixed because V and $\dot{m}'''$ are fixed. In conclusion, the thermodynamic optimization of the network is equivalent to minimizing the maximum pressure difference.

5.2 Elemental Volume

Consider the fundamental problem of linking a finite volume to one point with minimal flow resistance [17–19]. For simplicity we assume that the volume is two

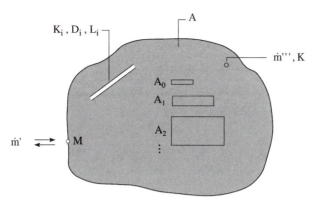

FIGURE 5.4. Two-dimensional fluid flow between a finite-size volume and one point [19].

dimensional and represented by the area A, shown in Fig. 5.4. The total mass flow rate $\dot{m}'$ (in units of kg/s m) flows through the point M and reaches (or originates from) every point that belongs to A. We also assume that the volumetric mass flow rate $\dot{m}'''$ (in units of kg/s m^3) that reaches all the points of A is distributed uniformly in space; hence $\dot{m}' = \dot{m}''' A$.

The space A is filled by a porous medium saturated with a single-phase fluid with constant properties. The flow is in the Darcy regime [20]. If the permeability of the porous medium is uniform throughout A, then the pressure field $P(x, y)$ and the flow pattern can be determined uniquely by the solution of the Poisson-type problem associated with the point-sink or point-source configuration: This classical problem is not the subject of this chapter. Instead, we consider the more general situation in which space A is occupied by a nonhomogenous porous medium composed of a material of low permeability K and a number of layers (e.g., cracks, filled or open) of much higher permeabilities $(K_1, K_2, \ldots)$. The thicknesses $(D_1, D_2, \ldots)$ and lengths $(L_1, L_2, \ldots)$ of these layers are not specified. We assume that the volume fraction occupied by the high-permeability layers is small relative to the volume represented by the K material. There are a very large number of ways in which these layers can be sized, connected, and distributed in order to collect and channel $\dot{m}'$ to the point M.

It is true that the design that minimizes the overall volume-to-point flow resistance is the one in which the most permeable material fills the entire volume. In that case the solution for the flow pattern is the radial flow (diffusion) mentioned in the preceding paragraph. In contrast, in this chapter the types and amounts of the permeable materials are constrained. In the assumed conglomerate (Fig. 5.4) the largest volume fraction is occupied by the material with the lowest permeability (K), not the highest. The constraints to which the volume-to-point flow optimization is subjected are essential characteristics of the constructal route to the optimal flow structure (e.g., tree networks). Constraints are the specified types and amounts of permeable materials and the volume whose shape is ultimately optimized for minimal volume-to-point resistance. These constraints are operational in the work described in Sections 5.2 and 5.3. Additional constraints are included in the optimization outlined in Section 5.4.

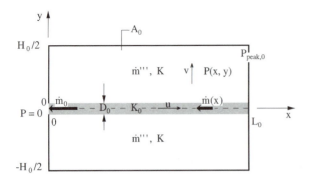

FIGURE 5.5. Elemental volume with Darcy flow through a low-permeability porous medium (K) and channel flow through a high-permeability layer (K_0) [19].

The approach we have chosen is illustrated in Fig. 5.4. We regard A as a patchwork of rectangular elements of several sizes (A_0, A_1, A_2, ...). The smallest element (A_0) contains mostly low-permeability material and only one high-permeability layer (K_0, D_0), Fig. 5.5. Each successively larger volume element (A_i) is an assembly of elements of the preceding size (A_{i-1}), which act as tributaries to the collecting layer (K_i, D_i, L_i) that defines the assembly.

In Fig. 5.5 the smallest volume $A_0 = H_0 L_0$ is fixed, but its shape H_0/L_0 may vary. The flow $\dot{m}_0' = \dot{m}''' A_0$ is collected from the K medium by a layer of much higher permeability K_0 and thickness D_0. The flow is driven toward the origin $(0, 0)$ by the pressure field $P(x, y)$. The rest of the rectangular boundary $H_0 \times L_0$ is impermeable. Because the flow rate $\dot{m}_0'$ is fixed, to minimize the global flow resistance means minimizing the peak pressure (P_{peak}) that occurs at a point inside A_0. The pressure at the origin is zero.

The analysis is greatly simplified by the assumptions that $K \ll K_0$ and $D_0 \ll H_0$, which, as we will show in Eq. (5.9), also mean that the optimized A_0 shape is such that H_0 is considerably smaller than L_0. According to these assumptions the flow through the K domain is practically parallel to the y direction [$P(x, y) \cong P(y)$ for $H_0/2 > |y| > D_0/2$], whereas the flow through the K_0 layer is aligned with the layer itself [$P(x, y) \cong P(x)$ for $|y| < D_0/2$]. Symmetry and the requirement that P_{peak} be minimum dictate that the A_0 element be oriented such that the K_0 layer is aligned with the x axis. The mass flow rate through this layer is $\dot{m}'(x)$, with $\dot{m}'(0) = \dot{m}_0'$ at the origin $(0, 0)$, and $\dot{m}'(L_0) = 0$. The K material is an isotropic porous medium bathed by Darcy flow [20]:

$$v = \frac{K}{\mu} \left(-\frac{\partial P}{\partial y} \right), \tag{5.1}$$

where v is the volume-averaged velocity in the y direction (Fig. 5.5). The actual flow is oriented in the opposite direction. We can determine the pressure field $P(x, y)$ by eliminating v between Eq. (5.1) and the local mass continuity condition,

$$\frac{\partial v}{\partial y} = \frac{\dot{m}'''}{\rho}, \tag{5.2}$$

and applying the boundary conditions $\partial P / \partial y = 0$ at $y = H_0/2$ and $P = P(x, 0)$ at

$y \cong 0$ (recall that $D_0 \ll H_0$):

$$P(x, y) = \frac{\dot{m}'''v}{2K}(H_0 y - y^2) + P(x, 0). \tag{5.3}$$

Equation (5.3) holds only for $y \gtrsim 0$. We obtain the corresponding expression for $y \lesssim 0$ by replacing H_0 with $-H_0$ in Eq. (5.3).

We obtain the pressure distribution in the K_0 layer, namely $P(x, 0)$, similarly by assuming Darcy flow along a D_0-thin path near $y = 0$:

$$u = \frac{K_0}{\mu}\left(-\frac{\partial P}{\partial x}\right), \tag{5.4}$$

where u is the average velocity in the x direction. The flow proceeds toward the origin, as shown in Fig. 5.2. The mass flow rate channeled through the K_0 material is $\dot{m}'(x) = -\rho D_0 u$. Furthermore, mass conservation requires that the mass generated in the infinitesimal volume slice ($H_0\, dx$) contribute to the $\dot{m}'(x)$ stream at this rate: $\dot{m}''' H_0\, dx = -d\dot{m}'$. Integrating this equation away from the impermeable plane $x = L_0$ (where $\dot{m} = 0$) and recalling that $\dot{m}'_0 = \dot{m}''' H_0 L_0$, we obtain

$$\dot{m}'(x) = \dot{m}''' H_0(L_0 - x) = \dot{m}'_0\left(1 - \frac{x}{L_0}\right). \tag{5.5}$$

Combining $\dot{m}'(x) = -\rho D_0 u$ with Eqs. (5.4) and (5.5), we find the pressure distribution along the x axis:

$$P(x, 0) = \frac{\dot{m}'_0 v}{D_0 K_0}\left(x - \frac{x^2}{2 L_0}\right). \tag{5.6}$$

Equations (5.3) and (5.6) provide a complete description of the $P(x, y)$ field. The peak pressure occurs in the farthest corner ($x = L_0$, $y = H_0/2$):

$$P_{\text{peak},0} = \dot{m}'_0 v\left(\frac{H_0}{8 K L_0} + \frac{L_0}{2 K_0 D_0}\right). \tag{5.7}$$

We can minimize this pressure with respect to the shape of the element (H_0/L_0) by noting that $L_0 = A_0/H_0$ and $\phi_0 = D_0/H_0 \ll 1$. The number ϕ_0 is carried in the analysis as an unspecified parameter. For example, if the D_0 layer was originally a crack caused by volumetric shrinking (e.g., cooling, drying) of the K medium, then D_0 must be proportional to the thickness H_0 of the K medium (see also Section 3.8). The resulting geometric optimum is described by

$$\tilde{H}_0 = 2^{1/2}(\tilde{K}_0\phi_0)^{-1/4}, \qquad \tilde{L}_0 = 2^{-1/2}(\tilde{K}_0\phi_0)^{1/4}, \tag{5.8}$$

$$\frac{H_0}{L_0} = 2(\tilde{K}_0\phi_0)^{-1/2}, \qquad \Delta\tilde{P}_0 = \frac{1}{2}(\tilde{K}_0\phi_0)^{-1/2}. \tag{5.9}$$

The nondimensionalization used in Eqs. (5.8) and (5.9) is based on $A_0^{1/2}$ as the length scale and K as the permeability scale:

$$(\tilde{H}_i, \tilde{L}_i) = \frac{(H_i, L_i)}{A_0^{1/2}}, \qquad \tilde{K}_i = \frac{K_i}{K}, \qquad (5.10)$$

$$\Delta \tilde{P}_i = \frac{P_{\text{peak},i}}{\dot{m}''' A_i \nu / K}, \qquad \phi_i = \frac{D_i}{H_i}. \qquad (5.11)$$

At optimum, the two terms on the right-hand side of Eq. (5.7) are equal, and this means that the principle of *equipartition* [21] (Section 4.2) once again rules the shape of the element. The shape of the A_0 element is such that the pressure drop that is due to flow through the K material is equal to the pressure drop that is due to the flow along the K_0 layer. Note also that the first of Eqs. (5.9) confirms the assumptions made about the D_0 layer at the start of this section: High permeability ($\tilde{K}_0 \gg 1$) and small volume fraction ($\phi_0 \ll 1$) mean that the optimized A_0 shape is slender, $H_0 \ll L_0$, provided that $\tilde{K}_0 \gg \phi_0^{-1}$.

5.3 First and Higher-Order Constructs

Consider next the immediately larger volume $A_1 = H_1 L_1$, Fig. 5.6, which can contain only elements of the type optimized in Section 5.2. The streams $\dot{m}'_0$ collected by the D_0-thin layers are now united into a larger stream $\dot{m}'_1$ that connects A_1 with the point $P = 0$. The $\dot{m}'_1$ stream is collected in the new layer (K_1, D_1, L_1).

When the number of A_0 elements assembled into A_1 is large, the composite material of Fig. 5.6 is analogous to the composite of Fig. 5.5, provided that the permeability K of Fig. 5.5 is replaced in Fig. 5.6 with an equivalent volume-averaged permeability (K_{e1}). We obtain the K_{e1} value by writing that the pressure drop across an A_0 element [Eq. (5.9)] is equal to the pressure drop over the distance $H_1/2$ in the K_{e1} medium: This second pressure drop can be read from Eq. (5.3), after H_0 is replaced with H_1, y with $H_1/2$, and K with K_{e1}. The result is $K_{e1} = (1/2) K_0 \phi_0$:

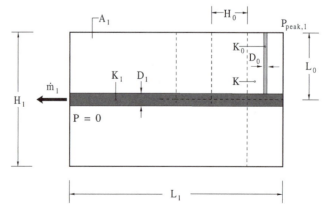

FIGURE 5.6. First construct (A_1) of elements of size A_0 and the central high-permeability channel K_1 [19].

TABLE 5.1. The optimized geometry of the elemental area A_0 and the subsequent constructs when the channel permeabilities are unrestricted

i	H_i/L_i	$\tilde{H}_i$	$\tilde{L}_i$	$n_i = A_i/A_{i-1}$	$\Delta\tilde{P}_i$
0	$2C_0^{-1/2}$	$2^{1/2}C_0^{-1/4}$	$2^{-1/2}C_0^{1/4}$	—	$\frac{1}{2}C_0^{-1/2}$
1	$(2C_0/C_1)^{1/2}$	$2^{1/2}C_0^{1/4}$	$C_0^{-1/4}C_1^{1/2}$	$(2C_1)^{1/2}$	$(2C_0C_1)^{-1/2}$
2	$(2C_1/C_2)^{1/2}$	$2C_0^{-1/4}C_1^{1/2}$	$2^{1/2}C_0^{-1/4}C_2^{1/2}$	$2(C_2/C_0)^{1/2}$	$(2C_1C_2)^{-1/2}$
$i \geq 2$	$(2C_{i-1}/C_i)^{1/2}$	$2^{i/2}C_0^{-1/4}C_{i-1}^{1/2}$	$2^{(i-1)/2}C_0^{-1/4}C_i^{1/2}$	$2(C_i/C_{i-2})^{1/2}$	$(2C_{i-1}C_i)^{-1/2}$

Note: $C_i = \tilde{K}_i\phi_i$ [19].

This value is then used in place of K_0 in an analysis that repeats the steps executed in Eqs. (5.7)–(5.9) for the A_0 optimization problem.

A more instructive alternative to this analysis begins with the observation that the peak pressure ($P_{\text{peak},1}$) in Fig. 5.6 is due to two contributions, the flow through the upper-right-corner element ($P_{\text{peak},0}$) and the flow along the (K_1, D_1) layer:

$$P_{\text{peak},1} = \dot{m}''' A_0 \frac{v}{K}\frac{1}{2}(\tilde{K}_0\phi_0)^{-1/2} + \dot{m}'_1 v\frac{L_1}{2K_1D_1}. \quad (5.12)$$

We can rearrange this expression by using the first of Eqs. (5.9) and $H_1 = 2L_0$:

$$\frac{P_{\text{peak},1}}{\dot{m}''' A_1 v/K} = \frac{1}{4\tilde{K}_0\phi_0}\frac{H_1}{L_1} + \frac{1}{2\tilde{K}_1\phi_1}\frac{L_1}{H_1}. \quad (5.13)$$

The corner pressure $P_{\text{peak},1}$ can be minimized by the selection of the H_1/L_1 shape of the A_1 rectangle. The resulting expressions for the optimized geometry (H_1/L_1, $\tilde{H}_1$, $\tilde{L}_1$) are listed in Table 5.1. The minimized peak pressure ($\Delta\tilde{P}_1$) is divided equally between the flow through the corner A_0 element and the flow along the collecting (K_1, D_1) layer. In other words, as in the case of the A_0 element of Section 5.2, the geometric optimization of the A_1 assembly is ruled by a principle of equipartition of pressure drop between the two main paths of the assembly.

The assembly and shape-optimization sequence can be repeated for larger subsystems (A_2, A_3, ...). Each new construct A_i contains a number n_i of constructs of the immediately smaller size A_{i-1}, the flow of which is collected by a new high-permeability layer (K_i, D_i, L_i). As shown in Fig. 5.6 for A_1, it is assumed that the number of constituents n_i is sensibly larger than 2. Each analysis begins with the statement that the maximum pressure difference sustained by A_i is equal to the sum of the pressure difference across the optimized constituent A_{i-1} that occupies the farthest corner of A_i, and the pressure drop along the K_i central layer:

$$P_{\text{peak},i} = P_{\text{peak},i-1} + \dot{m}'_i v\frac{L_i}{2K_iD_i}. \quad (5.14)$$

The geometric-optimization results are summarized in Table 5.1, in which we used $C_i = \tilde{K}_i\phi_i$ for the dimensionless flow conductance of each layer. The optimal shape of each rectangle $H_i \times L_i$ is ruled by the pressure-drop equipartition principle.

Beginning with the second construct, the results fall into the pattern represented by the recurrence formulas listed for $i \geq 2$. If the algorithm represented by these

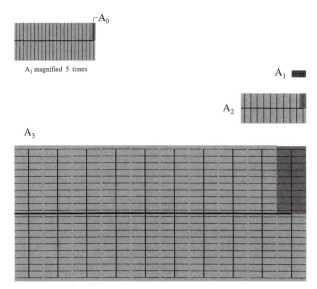

FIGURE 5.7. Tree architecture for minimal resistance to volume-to-point flow in a composite medium ($C_0 = 100$, $C_i/C_{i-1} = 10$; $i = 1, 2, 3$) [19].

formulas were to be repeated ad infinitum toward smaller A_i, then the pattern formed by the high-permeability paths (K_i, D_i) would be a fractal. In the present solution the pattern is not a fractal because the construction always has a finite number of steps (Ref. 17, p. 765; see also Ref. 22). Note that the construction begins with an element of finite size A_0 and ends when the given volume (A) is covered. Access to the infinity of points contained by the given volume is not made by making A_0 infinitely small, as in fractal patterns. Instead, the infinity of points of the given volume is reached by a flow that bathes A_0 *volumetrically*, because the permeability K of the material that fills A_0 (finite) is the lowest of all the permeabilities of the composite porous medium.

Figure 5.7 illustrates the minimal-resistance architecture recommended by the results of Table 5.1. At each level of assembly, the calculated number of constituents n_i was rounded off to the closest even number. The optimal design of the composite porous medium contains a tree network of high-permeability layers (K_0, K_1, K_2, . . .), where the interstitial spaces are filled with low-permeability material K. The actual shape of the tree depends on the relative size of the flow conductance parameters C_i. The conductance increase ratio C_i/C_{i-1} is essentially equal to the permeability ratio K_i/K_{i-1}, because the volume fraction $\phi_i \ll 1$ is expected to vary little from one construct to the next; see the comment made above Eq. (5.8). In other words, the conductance parameters C_i can be specified independently because the porous-medium characteristics of the materials that fill the high-permeability channels have not been specified. This feature is changed (i.e., constrained) in Section 5.4.

Several trends are revealed by constructions such as that of Fig. 5.7. When the conductance ratio C_i/C_{i-1} is large, the number n_i is large, the optimal shape of each assembly is slender ($H_i/L_i < 1$), and the given volume is covered "fast," i.e.,

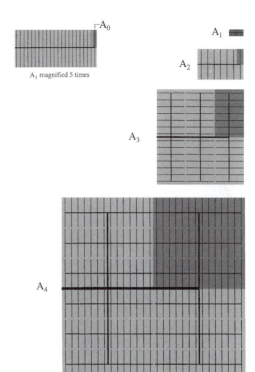

FIGURE 5.8. The effect of decreasing C_i/C_{i-1} ratios on the tree architecture for minimal resistance in a composite medium ($C_1/C_0 = 10$, $C_2/C_1 = 5$, $C_3/C_2 = 2$; $C_0 = 100$) [19].

in a few large steps of assembly and optimization. When the ratio C_i/C_{i-1} is large but decreases from one assembly to the next (Fig. 5.8), the number of constituents decreases and the shape of each new assembly becomes closer to square. Combining the limit $C_i/C_{i-1} \to 1$ with the n_i formula of Table 5.1, we see that the number 2 (i.e., dichotomy, bifurcation, pairing) emerges as a result of geometric optimization of volume-to-point flow. Note that the actual value $n_i = 2$ is not in agreement with the $n_i > 2$ assumption that was made in Fig. 5.6 and the analysis that followed. This means that when C_i/C_{i-1} is of the order of 1 the analysis must be refined, as was illustrated in Ref. 23 for heat trees (see also the end of Section 4.4).

5.4 Channels with Hagen–Poiseuille Flow

The objective of Sections 5.2 and 5.3 was to illustrate the application of the constructal method to fluid flow, i.e., to develop a general strategy for minimizing point-to-volume flow resistance. It was not to develop specific results for natural or artificial flow systems (e.g., lungs, capillary beds, river drainage basins, mass exchangers, electric windings). The simplifications adopted in the model (Fig. 5.4) do not jeopardize the fully deterministic character of the resulting structure. The algorithms of Table 5.1 are results, not assumptions or postulates.

In many natural volume-to-point flows the high-permeability paths (K_i, Fig. 5.4) are not filled with permeable materials. They are simply open spaces: cracks

TABLE 5.2. **The Optimized Geometry of the Point-to-Volume Flow Path when the Channel Permeabilities are Given by Eqs. (5.15)**

	H_i/L_i	$\tilde{H}_i$	$\tilde{L}_i$	$n_i = A_i/A_{i-1}$	$\Delta\tilde{P}_i$
0	$\dfrac{2^{5/3}3^{1/2}\hat{K}^{1/3}}{\phi_0}$	$\dfrac{2^{5/6}3^{1/6}\hat{K}^{1/6}}{\phi_0^{1/2}}$	$\dfrac{\phi_0^{1/2}}{2^{5/6}3^{1/6}\hat{K}^{1/6}}$	—	$\dfrac{3^{4/3}\hat{K}^{1/3}}{2^{7/3}\phi_0}$
1	$\dfrac{2^{5/3}\phi_0^{1/2}\hat{K}^{1/3}}{3^{1/6}\phi_1^{3/2}}$	$\dfrac{2^{1/6}\phi_0^{1/2}}{3^{1/6}\hat{K}^{1/6}}$	$\dfrac{\phi_1^{3/2}}{2^{3/2}\hat{K}^{1/2}}$	$\dfrac{\phi_1^{3/2}\phi_0^{1/2}}{2^{4/3}3^{1/6}\hat{K}^{2/3}}$	$\dfrac{3^{3/2}\hat{K}}{\phi_1^{3/2}\phi_0^{3/2}}$
2	$\dfrac{2^{7/6}\phi_0^{1/2}\hat{K}^{1/3}}{3^{1/6}\phi_2^{3/2}}$	$\dfrac{\phi_1^{3/2}}{2^{1/2}\hat{K}^{1/2}}$	$\dfrac{3^{1/6}(\phi_2\phi_1)^{3/2}}{2^{5/3}\phi_0^{1/2}\hat{K}^{5/6}}$	$\dfrac{3^{1/3}(\phi_2\phi_1)^{3/2}}{2^{5/6}\phi_0\hat{K}^{2/3}}$	$\dfrac{2^{11/6}3^{7/6}\hat{K}^{5/3}}{\phi_2^{3/2}\phi_1^3\phi_0^{1/2}}$
3	$\dfrac{2^{2/3}\phi_0^{1/2}\hat{K}^{1/3}}{3^{1/6}\phi_3^{3/2}}$	$\dfrac{3^{1/6}(\phi_2\phi_1)^{3/2}}{2^{2/3}\phi_0^{1/2}\hat{K}^{5/6}}$	$\dfrac{3^{1/3}(\phi_3\phi_2\phi_1)^{3/2}}{2^{4/3}\phi_0\hat{K}^{7/6}}$	$\dfrac{2^{1/6}3^{1/3}(\phi_3\phi_2)^{3/2}}{\phi_0\hat{K}^{2/3}}$	$\dfrac{2^{8/3}3^{5/6}\phi_0^{1/2}\hat{K}^{7/3}}{\phi_3^{3/2}\phi_2^3\phi_1^3}$
4	$\dfrac{2^{1/6}\phi_0^{1/2}\hat{K}^{1/3}}{3^{1/6}\phi_4^{3/2}}$	$\dfrac{3^{2/3}(\phi_3\phi_2\phi_1)^{3/2}}{2^{1/3}\phi_0\hat{K}^{7/6}}$	$\dfrac{3^{1/2}(\phi_4\phi_3\phi_2\phi_1)^{3/2}}{2^{1/2}\phi_0^{3/2}\hat{K}^{2/3}}$	$\dfrac{2^{7/6}3^{1/3}(\phi_4\phi_3)^{3/2}}{\phi_0\hat{K}^{2/3}}$	$\dfrac{2^{5/2}3^{1/2}\phi_0^{3/2}\hat{K}^3}{\phi_4^{3/2}\phi_3^3\phi_2^3\phi_1^3}$

Note: $\hat{K} = K/A_0$ [19].

(fissures) in two dimensions and tubes in three dimensions (Section 5.5). When the spacing D_i is sufficiently small the permeability K_i is proportional to D_i^2; see the solution for Hagen–Poiseuille flow between two parallel plates [1]:

$$K_i = \frac{1}{12}D_i^2 \qquad (i = 0, 1, 2, \ldots). \qquad (5.15)$$

In such cases we can repeat the analysis of Sections 5.2 and 5.3 step by step, noting that the high permeabilities K_i are no longer independent if the porosities ϕ_i are specified. Equations (5.15) act as constraints. The results of this geometric-optimization program are summarized in Table 5.2, where $\hat{K}$ is the dimensionless permeability of the interstitial material, $\hat{K} = K/A_0$. This analysis was performed up to the fourth construct, to see whether recurrence formulas become evident. The effects that are due to $\hat{K}$ and ϕ_i fall into a pattern beyond $i \geq 2$.

5.5 Optimization of Void-Space Distribution

In many flow systems a premium is put on the total volume and, implicitly, on the volume occupied by the network of high-permeability channels D_i. Examples of this kind (animate and inanimate) are the lung and virtually every forced-cooled winding, package of electronics, and heat exchanger. In such cases the volume of all the channels (the void space, in the class optimized in Table 5.2) is constrained and can be distributed optimally from one level of assembly to the next.

To illustrate the optimization of void-space distribution, consider a system with high-permeability paths that can be modeled as Hagen–Poiseuille flow channels with parallel and smooth walls [19]. The smallest subsystem in which there is a choice of how the void space is allocated is the first construct A_1. Here the void space ($A_{v,1} = D_1 L_1 + n_1 D_0 H_1/2$) can be expressed as an overall porosity constraint

$$\alpha_1 = \phi_1 + \phi_0, \tag{5.16}$$

where the average porosity of the A_1 assembly is $\alpha_1 = A_{v,1}/A_1$. The overall pressure drop $\Delta \tilde{P}_1$ can be minimized subject to the α_1 constraint. Because $\Delta \tilde{P}_1$ varies as $(\phi_1 \phi_0)^{-3/2}$, the result of this optimization step is $\phi_1 = \phi_0$. In other words, relative to the geometric-optimization work presented already in Table 5.2, the pressure drop associated with each first assembly is minimized once more if the internal volume of the lone D_1 channel equals the volume occupied by all the channels of size D_0.

The void-allocation optimization can be repeated at higher levels of assembly. For example, the void space distributed through A_2, namely $A_{v,2} = D_2 L_2 + n_2 A_{v,1}$, furnishes the constraint $\alpha_2 = \phi_2 + \alpha_1$, where $\alpha_2 = A_{v,2}/A_2$. The exponents of the ϕ_0, ϕ_1, and ϕ_2 factors in the $\Delta \tilde{P}_2$ expression (Table 5.2) are such that the minimum of $\Delta \tilde{P}_2$ occurs when $\phi_2 = (3/7)\,\alpha_1$ or, in view of the preceding paragraph, $\phi_2 = (6/7)\,\phi_0$. The corresponding results for the third and the fourth levels of assembly are $\phi_3 = (60/77)\,\phi_0$ and $\phi_4 = (8/11)\,\phi_0$.

Summarizing these results, we see that the optimal ϕ_i values decrease slowly as i increases, whereas their order of magnitude is always ϕ_0. An approximate way to restate this conclusion is to note that the flow resistance is minimal if ϕ_i is nearly the same from one volume (construct) size to the next. The same results can be expressed in terms of the porosities averaged over each volume size, $\alpha_i = A_{v,1}/A_i$. We obtain $\alpha_1 = 2\phi_0$, $\alpha_2 = (10/7)\alpha_1$, $\alpha_3 = (14/11)\alpha_2$, $\alpha_4 = (6/5)\alpha_3$, and $\alpha_{i+1} \to \alpha_i$ as i becomes large. In this way we see more clearly that the porosity is distributed nearly uniformly over all the volume elements.

Another way to summarize these results is to calculate the corresponding ratios between successive channel thicknesses, D_i/D_{i-1}. The results are $D_1/D_0 = 0.437\,\phi_0 \hat{K}^{-1/3}$, $D_2/D_1 = 0.648\phi_0 \hat{K}^{-1/3}$, $D_3/D_2 = 0.772\phi_0 \hat{K}^{-1/3}$, and $D_4/D_3 = 1.401\,\phi_0 \hat{K}^{-1/3}$. In conclusion, the thickness ratio has the scale $\phi_0 \hat{K}^{-1/3}$, which is a number greater than 1 because of the assumed limit $H_0/L_0 \ll 1$ and the H_0/L_0 results listed in Table 5.2. This means that in the tree network formed by D_i channels the thicknesses must be magnified by nearly the same factor if the network is to minimize flow resistance while respecting the imposed constraints (total volume and void volume). This feature agrees with earlier minimal-resistance results for natural fluid networks, for example, bronchial trees [2] and vascularized tissues [6, 9, 24, 25]. In the present case the agreement is only conceptual, because in the cited examples each channel (tube) and surrounding volume element form a three-dimensional structure, whereas in the present analysis the geometry was assumed two dimensional.

The same void-allocation results are obtained if instead of minimizing the pressure drop subject to the void-space constraint we minimize at every level of assembly the void space subject to fixed pressure drop.

We can minimize the point-to-volume flow resistance further by varying the angle between tributaries D_{i-1} and the main channel D_i of each new volume assembly i. This optimization principle is well known in physiology [6, 9, 24, 25], in which the work always begins with the assumption that a tree network of tubes *exists*. It can be shown numerically that the reductions in flow resistance obtained by optimizing the angles between channels are small relative to the reductions that are due to optimizing the shape of each volume element and assembly of elements. This effect was illustrated for heat flow in Section 4.6. In this section we fixed the angles at 90°, and focused on the optimization of volume shape. It is the optimization of shape subject to the volume constraint – the consistent use of this principle at every volume scale – that is responsible for the emergence of a tree network between the volume and the point. We focused on the optimal shapes of building blocks because our objective was to invoke a single optimization principle that can be used to explain the origin of tree-shaped networks in natural flow systems.

The preceding analyses were organized intentionally in a sequence that mimics the presentation made for heat trees (Sections 4.1–4.6). The objective of this analogy was to show that what flows, heat vs. fluid, is not the issue. What is important is that when the flow connects a volume to one point, the consistent appeal to the constructal law leads in a completely deterministic way to tree networks and their construction algorithms (e.g., Tables 4.1, 5.1, and 5.2).

The main difference between the heat trees of Chap. 4 and the fluid trees discussed so far in this chapter is that the heat trees were simpler. The conduction medium had only two materials, k_0 and k_p. The porous medium of Fig. 5.4 has an unspecified number of materials: the permeabilities K, K_0, K_1, . . ., or the spacings D_i of Eqs. (5.15). This explains why Fig. 5.7 does not look like Fig. 4.4: before we could draw Fig. 5.7 we had to assume values for the four permeability ratios, whereas for the heat trees we had to assume only one ratio, k_p/k_0.

5.6 Constructal Design: Increasing Complexity in a Volume of Fixed Size

We now take another look at the design method (Section 4.9) as an alternative to the construction and optimization cycle pursued until now in this chapter. We no longer memorize the shapes optimized at lower levels of assembly when we proceed outward and assemble the next, larger construct. Instead, we take the volume (one volume) as fixed, and work inward as we *assume* an internal structure and optimize it.

The work illustrated in this section was performed based on complete numerical simulations [26] of the Darcy flow through the multicomponent porous medium of Fig. 5.4. As an example, consider the design approach to the optimization of the first construct that has the internal structure defined in Fig. 5.6. The composite porous medium has three permeabilities: K in the interstitial material, K_0 in the more permeable layer of each elemental volume, and K_1 in the layer that collects the streams contributed by the K_0 layers. Unlike in the simplified analysis of Sections 5.2 and 5.3, here the permeabilities and the relative amount of channeling material (ϕ_1) are not restricted. The new expression for ϕ_1 is

$$\phi_1 = \frac{\hat{D}_1}{\hat{H}_1} + \frac{n_1}{2}\hat{D}_0(\hat{H}_0 - \hat{D}_1) - \hat{D}_1\left(\frac{1}{n_1\hat{H}_1} - \frac{\hat{D}_0}{2}\right). \tag{5.17}$$

We attach the coordinate system (x, y) to the $H_1 \times L_1$ frame such that the y axis is parallel to H_1, the x axis is the midplane of the K_1 layer, and the sink point is located at the origin $(0, 0)$. We determine the pressure distribution over the space covered by the first assembly by solving numerically

$$\frac{\partial^2 \hat{P}}{\partial \hat{x}^2} + \frac{\partial^2 \hat{P}}{\partial \hat{y}^2} + \hat{C} = 0, \tag{5.18}$$

where $\hat{C} = 1$ in the K domain of Fig. 5.6, $\hat{C} = K/K_0$ in the K_0 layers, and $\hat{C} = K/K_1$ in the central K_1 layer. The dimensionless variables are

$$(\hat{x}, \hat{y}, \hat{D}_0, \hat{H}_0, \hat{L}_0, \hat{D}_1, \hat{H}_1, \hat{L}_1) = (x, y, D_0, H_0, L_0, D_1, H_1, L_1)/A_1^{1/2},$$

$$\hat{P} = \frac{P - P(0, 0)}{\dot{m}''' \nu A_1 / K}, \tag{5.19}$$

where $A_1 = H_1 L_1$ is constant. The perimeter of A_1 is impermeable except for the spot of height D_1 located at the origin. Internally, there is flow over the entire A_1 space, which means that, unlike in the first construct of Section 5.3, here the elemental volumes $(H_0 L_0)$ *interact* with each other.

The numerical-optimization method is described in detail in Ref. 26. It consisted of determining the location and value of $\hat{P}_{peak}$ and minimizing this value with respect to the 3 degrees of freedom of the assembly: the shape of the assembly $(\hat{H}_1/\hat{L}_1)$, the ratio $\hat{D}_1/\hat{D}_0$ (or $\hat{D}_1$, or $\hat{D}_0$), and the number of elemental volumes n_1 assembled into A_1. There are only 3 degrees of freedom because of the two constraints: the overall size constraint $A_1 = $ constant (or $\hat{L}_1 = 1/\hat{H}_1$), and ϕ_1 constraint (5.17) written for a modified version of Fig. 5.6, in which the dead-end tip of the K_1 layer (near $\hat{x} = \hat{L}_1$) was left out.

The minimization of the overall flow resistance (or $\hat{P}_{peak}$) proceeded in three phases. In the first, the shape of the assembly $(\hat{H}_{1,opt})$ was optimized while K/K_0, K/K_1, ϕ_1, n_1, and $\hat{D}_0$ were held constant. Figure 5.9 illustrates the minimum of $\hat{P}_{peak}$ with respect to the overall shape parameter $\hat{H}_1$.

In the second phase the $\hat{H}_1$ optimization was repeated for several values of $\hat{D}_0$ until $\hat{P}_{peak,min}$ was minimized for a second time. The minimum with respect to $\hat{D}_0$ is also illustrated in Fig. 5.9: Note that the $\hat{P}_{peak}$ curves move downward as $\hat{D}_0$ decreases from 0.01 to 0.006 and then move upward as $\hat{D}_0$ decreases below 0.006. The optimization with respect to $\hat{D}_0$ is also illustrated in Fig. 5.10, this time for a smaller ϕ_1 value. Note that each point of a constant-n_1 curve is the result of optimizing the shape of the assembly $(\hat{H}_1)$.

The result of minimizing $\hat{P}_{peak,min}$ for a second time is labeled $\hat{P}_{peak,mm}$ and is shown in Fig. 5.11. This figure illustrates the third phase of the numerical-optimization strategy: The double-minimization procedure of Figs. 5.9 and 5.10 was repeated for several values of the number of elemental volumes, n_1. The group plotted on the ordinate of Fig. 5.11 measures the departure of the numerical results from the theoretical limit derived for the same configuration [26]:

$$\hat{P}_{peak,mm} = \frac{1}{\phi_1}\left(\frac{3K^2}{2K_0 K_1}\right)^{1/2}. \tag{5.20}$$

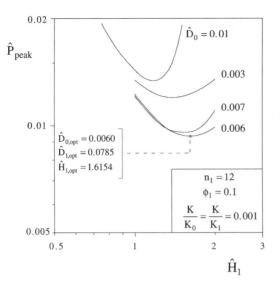

FIGURE 5.9. The numerical minimization of the flow resistance of the first construct with respect to the overall shape and the elemental layer thickness [26].

Figure 5.11 confirms the correctness of Eq. (5.20) in the limit of small ϕ_1, K/K_0, and K/K_1. The effect of increasing n_1 is weak. In the range of material properties covered by the numerical results of Fig. 5.11, we find that $\hat{P}_{peak,mn}$ decreases slowly as n_1 increases. This finding does not agree with the constructal approach (Table 5.1), which indicates the existence of an optimal n_1 value. The reason is that in Section 5.3 the number n_1 was not a degree of freedom: There were only 2 degrees of freedom, because each elemental volume ($H_0 L_0$) already had its shape selected in accordance with the geometric optimization shown in Section 5.2. Another difference is that in the analysis of Section 5.3 the elemental volumes communicated through only one end of their (K_0, D_0) layers, whereas in the numerical optimization [26] the fluid flowed across any portion of the perimeter of each elemental rectangle $H_0 \times L_0$.

In spite of these differences, the optimal n_1 value deduced from Table 5.1 agrees in an order-of-magnitude sense with the smallest n_1 values for which $\hat{P}_{peak,mn}$

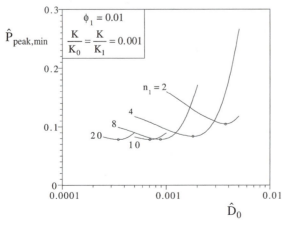

FIGURE 5.10. The second minimization of the global flow resistance when the elemental layer thickness is varied [26].

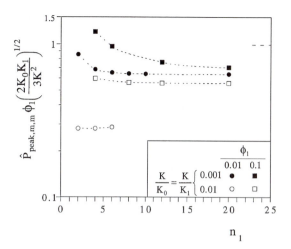

FIGURE 5.11. The effect of the number of elemental volumes n_1 on the twice-minimized flow resistance of the first construct [26].

reaches its ultimate, third minimum. For example, if we substitute $\phi_1 = 0.01$ and $K/K_1 = 0.001$ into the formula $n_1 = (2C_1)^{1/2} = (2\phi_1 K_1/K)^{1/2}$ of Table 5.1, we obtain $n_1 \cong 4.5$, which is in agreement with the n_1 range, in which the third minimum of $\hat{P}_{peak}$ is established in Fig. 5.11. If in the same calculation ϕ_1 is increased to 0.1, the calculated n_1 is 14, which also agrees with the location of the bend in the appropriate curve in Fig. 5.11. This agreement suggests a possible shortcut to the three-way optimization that was conducted numerically en route to Fig. 5.11. We might obtain results similar to those of Fig. 5.11 by assuming from the start that each elemental volume has the optimal shape determined numerically at the elemental level. In this way the number of degrees of freedom in the architecture of the first construct drops from 3 to 2. We examine this abbreviated method later in this section.

The numerical results obtained for the optimal shape of the first construct are reported in Fig. 5.12. The group plotted on the ordinate is recommended by the theoretical result developed for the same geometry in Ref. 26, $\hat{H}_{1,opt} = (8K_0/3K_1)^{1/4}$.

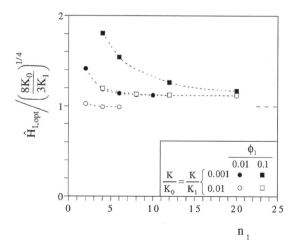

FIGURE 5.12. The effect of n_1 and material properties on the optimal shape of the first construct [26].

TABLE 5.3. **Example of an Optimized First Construct**

	Nearly Optimal	Optimal
$\hat{P}_{peak,mm}$	0.00987	0.00934
$\hat{H}_{1,opt} = (H_1/L_1)_{opt}^{1/2}$	1.32	1.65
$\hat{D}_{0,opt}$	0.0067	0.0060
$\hat{D}_{1,opt}$	0.0712	0.0785
$(\hat{D}_1/\hat{D}_0)_{opt}$	10.6	13.1

Note: Middle column, the use of elemental volumes optimized individually; right column, results of the three-way optimization described in the first part of Section 5.6 ($\phi_1 = 0.1$, $K/K_0 = K/K_1 = 0.001$, $n_1 = 12$) [26].

The numerical results show that only minor departures from the theoretical estimate occur when the parameters of the porous composite (ϕ_1, K/K_0, K/K_1) vary. The robustness of the external shape adds to the view that we might abbreviate the numerical optimization by adopting at the elemental level of the first construct the optimal elemental shape determined numerically. Along this route, the relations between the dimensionless elemental features and the dimensionless elemental features based on the fixed size of the first construct are

$$\tilde{D}_0 = n_1^{1/2}\hat{D}_0, \tag{5.21}$$

$$\hat{H}_{1,opt} = 2/\left(n_1^{1/2}\,\tilde{H}_{0,opt}\right). \tag{5.22}$$

The calculation starts with selecting $\hat{D}_0$, calculating the corresponding $\tilde{D}_0$, and obtaining $\tilde{H}_{0,opt}$ by minimization of the global flow resistance of the element [26]. The optimal shape of the first assembly is calculated next with Eq. (5.22). The numerical optimization continues with the remaining 2 degrees of freedom, $\hat{D}_0$ and n_1, as was described in the first part of this section.

The results of this abbreviated ("nearly optimal") approach are reported in Table 5.3 for a first construct that was also optimized with the full (three-way, "optimal") method. The computational time required by the abbreviated approach was 12 times shorter than the full method. Table 5.3 shows that the more direct method (the middle column) produces a design that geometrically "looks like" the optimal design (the right column). The most important message is conveyed by the top line: the performance of the nearly optimal configuration approaches within 5.7% of the performance of the optimal configuration.

In conclusion, if the minimization of the flow resistance is the real objective and if computational costs are a constraint, then the nearly optimal design is practically as good as the truly optimal design. The same trend is revealed by the second construct optimized based on the "design" method [26]. This conclusion – the nearly optimal design options that come close to the performance of the optimal design – is a common occurrence in engineering design in general [27].

5.7 Three-Dimensional Fluid Trees

Constructal theory was also applied to a general three-dimensional flow that links a volume with one point (source or sink) [17, 18]. This also happens to be the class of flows that received a lot of attention in the past, in search of *models** for the respiratory and circulatory systems. This is why it is important to stress the basic difference between the earlier analyses and the *theory*[†] presented in this section: The following results are deduced, one by one, from the constructal law. This process of deduction has a definite direction: from small subsystems toward larger subsystems (assemblies, or constructs). The smallest scale of the network is finite and known (predictable): This scale serves as the starting point in the step-by-step optimization and organization procedure.

To stress our purely theoretical course, we outline all the analytical steps [17, 18] even though some of these steps were executed in the two-dimensional analysis that opened this chapter. The role of smallest volume is played by the parallelepiped $V_1 = H_1 L_1 t$, shown in Fig. 5.13. We call this volume a first construct because of the pores (sacks) that we recognize as necessary features of its internal structure. The volume V_1 is fixed because the thickness t and the area $A_1 = H_1 L_1$ are fixed. What is variable is the shape of the element, which is represented by the aspect ratio H_1/L_1. The volume V_1 is visited uniformly by the mass flow rate $\dot{m}_1 = \dot{m}''' V_1$. At this first level only one tube (diameter D_1) is used to collect the $\dot{m}_1$ stream and lead it to one point on the boundary. In Fig. 5.13, that point is the origin $(0, 0)$. Symmetry and the requirement that ΔP be minimum suggest that the tube should be placed along the x axis. The mass flow rate through this tube is $\dot{m}(x)$, with $\dot{m}(0) = \dot{m}_1$ at the origin $(0, 0)$, and $\dot{m}(L_1) = 0$. Except for the point of origin, the surfaces of the elemental volume V_1 are impermeable. The thickness t is assumed to be sufficiently small, $t < (H_1, L_1)$, such that the pressure field that drives the flow is essentially two-dimensional, $P(x, y)$.

The rest of the flow path between the origin $(0, 0)$ and any other point (x, y) is located in the material situated above and below the x axis. To account for the flow that originates from an arbitrary point of the finite volume V_1, we model this material as an anisotropic porous medium in which Darcy flow is oriented purely in the y direction (when $y \neq 0$):

$$v = \frac{K}{\mu} \left(-\frac{\partial P}{\partial y} \right). \tag{5.23}$$

In this equation, v and K are the volume-averaged velocity and, respectively, the permeability in the y direction. The fluid may flow in the x direction along only the x axis. This anisotropic model is clearly a simplification, which later will restrict our geometric constructions to drawing only 90° angles between successive assemblies. At this early stage in the theory, however, this is a modeling approximation that is reasonable, especially if K is small.

---

* Model is a small copy or imitation of an *existing* (observed) object.
† Theory is a purely mental viewing (contemplation, speculative idea, plan, design, scheme, principle) of how something *might be* done.

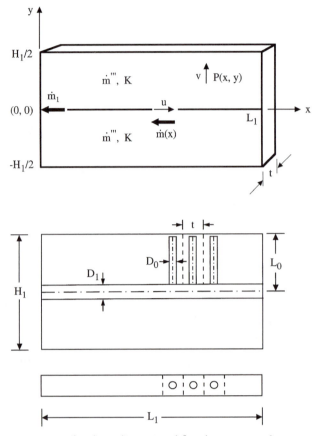

FIGURE 5.13. First construct for three-dimensional flow between a volume and one point [18].

We can determine the pressure field $P(x, y)$ by eliminating v between Eq. (5.23) and the local mass continuity condition,

$$\frac{\partial v}{\partial y} = \frac{\dot{m}'''}{\rho},\qquad(5.24)$$

and applying the boundary conditions $\partial P / \partial y = 0$ at $y = H_1/2$ and $P = P(x, 0)$ at $y = 0$:

$$P(x, y) = \frac{\dot{m}''' v}{2K}(H_1 y - y^2) + P(x, 0).\qquad(5.25)$$

We can determine the pressure distribution along the x axis after making similar assumptions about the fluid mechanics of the stream $\dot{m}_1$ that eventually exits through the origin. Let us assume that this stream is in Hagen–Poiseuille flow through a round* tube of length L_1 and diameter D_1. First, we use the mean velocity in the x

---

* The constructal theory of optimizing the shape of the stream cross section is outlined in Chap. 6. The roundness of the duct cross section is *demanded* by the constructal law: It is not an assumption or an empirical observation.

direction [1],

$$u = \frac{(D_1/2)^2}{8\mu}\left(-\frac{\partial P}{\partial x}\right)_{y=0},$$ (5.26)

to estimate the local mass flow rate $\dot{m}(x)$, which points toward the origin:

$$\dot{m}(x) = \frac{\pi D_1^4}{128\nu}\left(\frac{\partial P}{\partial x}\right)_{y=0}.$$ (5.27)

Mass conservation requires that the mass generated in the infinitesimal volume slice ($H_1 t dx$) contribute to the $\dot{m}(x)$ stream: $\dot{m}''' H_1 t dx = -d\dot{m}$. Integrating this equation away from the impermeable plane $x = L_1$ (where $\dot{m} = 0$) and recalling that $\dot{m}_1 = \dot{m}''' H_1 L_1 t$, we obtain

$$\dot{m}(x) = \dot{m}''' H_1 t(L_1 - x) = \dot{m}_1\left(1 - \frac{x}{L_1}\right).$$ (5.28)

Finally, Eqs. (5.27) and (5.28) yield the pressure distribution along the x axis:

$$P(x,0) = P_0 + 128\frac{\dot{m}''' \nu H_1 t}{\pi D_1^4}\left(L_1 x - \frac{x^2}{2}\right).$$ (5.29)

This result can be combined with Eq. (5.25) to determine the pressure distribution over the rectangular domain $H_1 \times L_1$, where $P(0,0) = P_0$. The resulting expression shows that the maximum pressure occurs in the two farthest corners, $P_{max} = P(L, \pm H/2)$, namely,

$$\Delta P_1 = P_{max} - P_0 = \dot{m}''' \nu\left(\frac{H_1^2}{8K} + \frac{64 H_1 t L_1^2}{\pi D_1^4}\right).$$ (5.30)

We can rearrange this result by using $A_1 = H_1 L_1$ to show explicitly the trade-off character of the shape of A_1:

$$\frac{\Delta P_1}{\dot{m}''' \nu} = \frac{H_1^2}{8K} + \frac{64 t A_1^2}{\pi D_1^4 H_1}.$$ (5.31)

This expression has a minimum with respect to H_1: The optimal shape is represented by

$$H_1 = \left(\frac{256}{\pi}tK\right)^{1/3}\frac{A_1^{2/3}}{D_1^{4/3}}.$$ (5.32)

The permeability K may be kept as a known constant and carried through the analysis: see, for example, the permeability of soil in the river basin constructed in Section 6.4. Because most of the optimization studies of fluid networks have dealt with the respiratory and vascular systems, it is important that we say more about K to show that it is indeed known (predictable).

The volumetric permeability of the V_1 element is due to a number of "pores" of diameter D_0 and length L_0, which are aligned with y (Fig. 5.13). In Chap. 10

we will see that in the lung the D_0 scale is dictated (fixed) by the diffusion of mass during the breathing interval t_b,

$$D_0 \sim (t_b D)^{1/2}, \tag{5.33}$$

where D is the mass diffusivity of oxygen or carbon dioxide through the "solid" (e.g., blood-saturated medium) that surrounds the pore. Substituting the appropriate orders of magnitude into approximation (5.33), namely $t_b \sim 1$ s and $D \sim 10^{-9}$ m²/s (e.g., Ref. 28, p. 589), we obtain $D_0 \sim 10^2$ μm. Averaged over time, the periodic (in and out) flow through the pore sustains a quasi-steady mass concentration field in the solid material that surrounds the pore. This is in fact the reason why a pore must exist and why an amount of solid material must be assigned to it. The solid material is to be "contaminated" by the species that are present in the pore. From the classical solutions of steady diffusion [28] we know that the pore geometry that *maximizes** the ratio of the contaminated solid material divided by the pore volume is the spherical configuration. In the present case (e.g., Fig. 5.13) the pores cannot be spherical inclusions because they must communicate at one end with the collecting tube (D_1) placed on the x axis. The only option then is for each pore to be shaped as a finger, which means that the pore length L_0 is a small multiple of D_0:

$$L_0 = \lambda D_0. \tag{5.34}$$

Another aspect of the classical solution for steady diffusion around an embedded sphere is that the scale of the thickness of the spherical annulus of contaminated solid material is dictated by the scale of the inclusion (D_0). This means that the finger-shaped pore and the surrounding contaminated solid occupy a finger-shaped volume with a thickness t that is a small multiple of D_0:

$$t = \delta D_0. \tag{5.35}$$

The precise values of the factors λ and δ are not important, although more exact estimates of these geometric features could be made by the numerical solution of the "finger" steady diffusion problem. The important conclusion is that both λ and δ are numbers of the order of 1 (comparable, say, with 2 or 3) such that the ratio λ/δ is a *number comparable with* 1.

The lower part of Fig. 5.13 summarizes these conclusions. Now the vertical permeability constant (K) of the material can be estimated as the equivalent permeability of a group of parallel tubes of diameter D_0 and spacing t (e.g., Ref. 1, p. 570):

$$K = \frac{\pi D_0^4}{128 \, t^2}. \tag{5.36}$$

---

* This maximization step is another demand made by the constructal law: In this case, again, the access to every material point of the volume is maximized geometrically through the generation of the optimal internal structure.

This formula can be combined with Eqs. (5.35) and (5.32) to obtain

$$H_1 = \left(\frac{2}{\delta}\right)^{1/3} A_1^{2/3} D_0 D_1^{-4/3}. \tag{5.37}$$

Noting in Fig. 5.13 that $H_1 = 2L_0$ and recalling that $L_1 = A_1/H_1$, we arrive at

$$L_1 = \frac{8}{9} \frac{\lambda^{3/2}}{\delta^{1/2}} D_0. \tag{5.38}$$

The geometric implications of these results become clear if we calculate the now-optimized aspect ratio,

$$\frac{H_1}{L_1} = \frac{9}{4}\left(\frac{\delta}{\lambda}\right)^{1/2} \sim 2. \tag{5.39}$$

Surprisingly, this ratio is a constant comparable with 2 when δ/λ is a number comparable with 1. The number of tubes of size (D_0, L_0) contained in this first assembly is also represented closely by the number 2 (the smallest integer greater than 1):

$$n_1 = 2\frac{L_1}{t} = \frac{16}{9}\left(\frac{\lambda}{\delta}\right)^{3/2} \sim 2. \tag{5.40}$$

In conclusion, the optimization of the rectangular shape A_1 leads to a first construct that contains only two of the smallest elements; see Fig. 5.14. It is important to keep in mind that the road traveled from Fig. 5.13 to Fig. 5.14 was one on which the number n_1 was free to vary, and the conclusion that n_1 must be the smallest even number available is *theoretical*. It was the optimization of the A_1 shape that

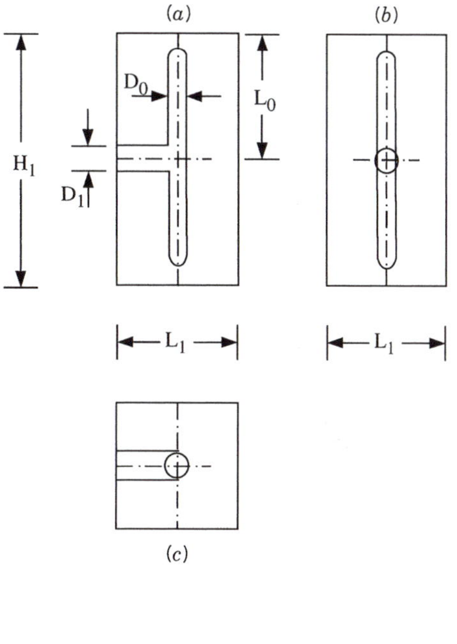

FIGURE 5.14. The optimized first construct for three-dimensional flow between a volume and one point [18].

recommended that two D_0 tubes must come together and form one tube of diameter D_1. In other words, with the present geometric construction read in the reverse, the fact that in Fig. 5.14 the D_1 tube bifurcates is of purely theoretical origin.

In Fig. 5.13 the D_1 tube ran from one end of the assembly to the other end (length L_1) because the number of vertical D_0 tubes was arbitrary, presumably large. This is no longer the case when the first assembly is optimized: In Fig. 5.14 the length of the D_1 tube is $L_1/2$ because its function is to collect the $\dot{m}_1$ stream from the center of the assembly and lead it along the shortest path out of the assembly. In place of Eq. (5.31), the pressure-drop formula can be written specifically for Fig. 5.14:

$$\Delta P_1 = \dot{m}_1 \frac{128\nu}{\pi D_1^4} \frac{L_1}{2} + \frac{1}{2} \frac{\dot{m}_1}{2} \frac{128\nu}{\pi D_0^4} L_0, \tag{5.41}$$

where the two terms account for the D_1 and the D_0 portions of the flow path. The $1/2$ factor in front of the second term accounts for the fact that one end of the D_0 tube is closed.

Equation (5.41) can be rewritten to show the geometric parameters of the flow resistance encountered by $\dot{m}_1$:

$$\frac{\pi}{64} \frac{\Delta P_1}{\dot{m}_1 \nu} = \frac{L_1}{D_1^4} + \frac{L_0}{2D_0^4}. \tag{5.42}$$

When the total volume occupied by the ducts is constrained,

$$V_{p1} = \frac{\pi}{4} D_1^2 \frac{L_1}{2} + 2 \frac{\pi}{4} D_0^2 L_0, \tag{5.43}$$

there is an optimal way to distribute this volume such that the flow resistance is minimized. Noting that $L_1 = L_0$, we find that the results of minimizing expression (5.42) subject to constraint (5.43) become

$$\frac{D_1}{D_0} = 2^{1/2}, \qquad V_{p1} = \frac{3}{4} \pi L_0 D_0^2, \tag{5.44}$$

$$\Delta P_1 = \frac{48}{\pi} \dot{m}_1 \nu \frac{L_0}{D_0^4}. \tag{5.45}$$

Worth noting is the internal structuring that emerges: The first construct is optimized when the collecting duct (D_1) is $2^{1/2}$ times thicker than its tributaries.

The optimization of subsequent constructs of larger size follows the same steps as the procedure used for the first construct. The method begins with taking a large and unspecified number n_2 of first assemblies and then arranging them on both sides of a new collecting tube of diameter D_2. In this way we arrive at Fig. 5.15, which – it is worth noting – is the same as the generic model analyzed earlier in Fig. 5.13. It can be shown [18] that the best number of first constructs is $n_2 = 2$: This number has been used to draw Fig. 5.16. We see that the shape-optimized second assembly contains two first assemblies, which form the larger stream $(\dot{m}_2 = \dot{m}_1 + \dot{m}_1)$ right in the center of the second assembly. The $\dot{m}_2$ stream is guided out of the second assembly through a new duct of diameter D_2 and length $L_2/2\,(=L_0/2)$.

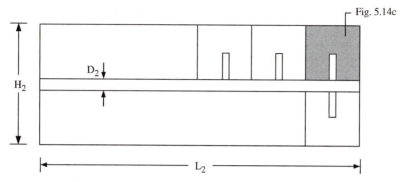

FIGURE 5.15. Second construct containing an unspecified number of the optimized first constructs shown in Fig. 5.14 [18].

The second step in the optimization of the second assembly consists of minimizing the total flow resistance overcome by $\dot{m}_2$,

$$\Delta P_2 = \dot{m}_2 \frac{128}{\pi D_2^4} \frac{L_2}{2} + \Delta P_1, \tag{5.46}$$

subject to the total pore volume constraint

$$V_{p2} = \frac{\pi}{4} D_2^2 \frac{L_2}{2} + 2 V_{p1}. \tag{5.47}$$

In these expressions, ΔP_1 and V_{p1} are provided by Eqs. (5.44) and (5.45). The optimal distribution of pore volume is described by the results listed on line $i = 2$ in Table 5.4. For example, in going from the first construct to the second construct the diameter of the collecting duct increases by the factor $D_2/D_1 = 2^{1/3}$.

The next shape-optimization step leads to a third construct that contains only two second constructs, as shown in Fig. 5.17. The new stream ($\dot{m}_3 = \dot{m}_2 + \dot{m}_2$) is formed in the center of the cube of side H_3 ($= 2L_0$) and is led to the outside through a new duct of diameter D_3 and length $L_3/2$ ($= L_0$). The pore volume distribution step of the optimization method consists of determining the overall fluid-resistance

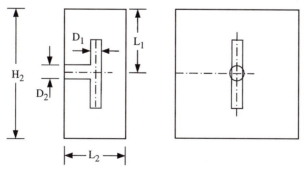

FIGURE 5.16. The optimized second construct for three-dimensional flow between a volume and one point [18].

TABLE 5.4. The Optimized Geometry of the Three-Dimensional Volume-to-Point Tree (Fig. 5.18) [17, 18]

Construct i	$\dfrac{\bar{L}_i}{\bar{L}_{i-1}}$	$\dfrac{D_i}{D_{i-1}}$	$R_i = \dfrac{\Delta P_i D_0^4}{\dot{m}_i \nu L_0}$	$\dfrac{V_{pi}}{D_0^2 L_0}$
1	$\dfrac{1}{2}$	$2^{1/2}$	$\dfrac{16}{\pi}(3)$	$\dfrac{\pi}{4}(3)$
2	1	$2^{1/3}$	$\dfrac{8}{\pi}(3 + 2^{-1/3})$	$\dfrac{\pi}{2}(3 + 2^{-1/3})$
3	2	$2^{1/3}$	$\dfrac{4}{\pi}(3 + 2^{-1/3} + 2^{1/3})$	$\pi(3 + 2^{-1/3} + 2^{1/3})$
4	1	$2^{1/3}$	$\dfrac{2}{\pi}(4 + 2^{-1/3} + 2^{1/3})$	$2\pi(4 + 2^{-1/3} + 2^{1/3})$
5	1	$2^{1/3}$	$\dfrac{1}{\pi}(4 + 2 \times 2^{-1/3} + 2^{1/3})$	$4\pi(4 + 2 \times 2^{-1/3} + 2^{1/3})$
6	2	$2^{1/3}$	$\dfrac{1}{2\pi}(4 + 2 \times 2^{-1/3} + 2 \times 2^{1/3})$	$8\pi(4 + 2 \times 2^{-1/3} + 2 \times 2^{1/3})$
7	1	$2^{1/3}$	$\dfrac{1}{4\pi}(5 + 2 \times 2^{-1/3} + 2 \times 2^{1/3})$	$16\pi(5 + 2 \times 2^{-1/3} + 2 \times 2^{1/3})$
8	1	$2^{1/3}$	$\dfrac{1}{8\pi}(5 + 3 \times 2^{-1/3} + 2 \times 2^{1/3})$	$32\pi(5 + 3 \times 2^{-1/3} + 2 \times 2^{1/3})$
9	2	$2^{1/3}$	$\dfrac{1}{16\pi}(5 + 3 \times 2^{-1/3} + 3 \times 2^{1/3})$	$64\pi(5 + 3 \times 2^{-1/3} + 3 \times 2^{1/3})$

function,

$$\Delta P_3 = \dot{m}_3 \frac{128\nu}{\pi D_3^4} \frac{L_3}{2} + \Delta P_2, \tag{5.48}$$

and minimizing it subject to the pore volume constraint,

$$V_{p3} = \frac{\pi}{4} D_3^2 \frac{L_3}{2} + 2V_{p2}, \tag{5.49}$$

where ΔP_2 and V_{p2} are given by Eqs. (5.46) and (5.47). The results of this analysis are summarized on line $i = 3$ in Table 5.4.

In particular, we note the emergence of the proportionality $D_3/D_2 = 2^{1/3}$, which means that, beginning with the second construct, the diameter of each new

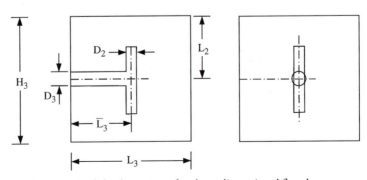

FIGURE 5.17. The optimized third construct for three-dimensional flow between a volume and one point [18].

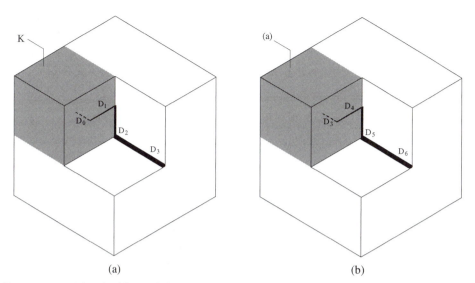

FIGURE 5.18. The doubling of the outer dimension in going from (a) the optimized third construct to (b) the optimized sixth construct [18].

collecting duct increases by a factor of $2^{1/3}$. We have deduced much more than Murray's law (the $2^{1/3}$ factor [24, 25, 29]): We have deduced that Murray's law must break down at the smallest volume scale. We have also deduced the geometric layout (lengths L_i) and numbers of constituents (n_i). In sum, we have deduced everything (the complete geometry) from nothing more than one principle – the constructal law.

The optimization and construction procedure can be repeated until the assembly size matches the scale of the available volume (V). The optimized structure is summarized in Table 5.4 and Fig. 5.18. The column $\bar{L}_i / \bar{L}_{i-1}$ in Table 5.4 represents the factor by which the length of the new collecting duct ($\bar{L}_i$) increases relative to the preceding duct ($\bar{L}_{i-1}$). The duct length ($\bar{L}_i$) is to be distinguished from the horizontal dimension of the respective assembly (L_i); in Fig. 5.17, for example, $\bar{L}_3 = L_3/2$.

The duct-length increase factor $\bar{L}_1 / \bar{L}_{i-1}$ exhibits a cyclical pattern for each sequence of three assembly sizes, provided that $i \geq 2$. This three-assembly cycle is more evident when $i \geq 4$, especially as we examine the evolution of the numerical coefficients obtained for ΔP_i and V_{pi}, where V_{pi} is the void volume of construct i. Each of the preceding (smaller) assembly sizes left its mark on the form of these coefficients. When the assembly order (i) is a multiple of 3, the numerical results of Table 5.3 can be extended with the formulas

$$\frac{\Delta P_i}{\dot{m}_i} \frac{D_0^4}{\nu L_0} = \frac{1}{\pi 2^{i-5}} \left[2 + \frac{i}{3} \left(1 + 2^{-1/3} + 2^{1/3} \right) \right], \tag{5.50}$$

$$\frac{V_{pi}}{D_0^2 L_0} = \pi 2^{i-3} \left[2 + \frac{i}{3} \left(1 + 2^{-1/3} + 2^{1/3} \right) \right], \tag{5.51}$$

where $i = 3, 6, 9, \ldots$.

The outer linear dimension of the construct of the order of ($i + 3$) is the double of the outer dimension of the construct of order (i). This factor of 2 increase also applies to the diameters of the largest (collecting) ducts of the two assemblies. Figure 5.18 illustrates the doubling of the size, from the third construct (a) to the sixth construct (b). We have deduced in this way not only Cohn's [6] assumed (postulated) dichotomous three-dimensional structure, but also the requirement that the rules of this structure must break down at the smallest volume scale.

It is absolutely essential to note that the internal details do not double their sizes in going from construct (i) to construct ($i + 3$). In other words, construct ($i + 3$) is not the same as magnifying by a factor of 2 every feature of construct (i). The reason is that the fluid-flow structure constructed theoretically [17, 18] has a definite (finite, known) beginning: the smallest scale (K or D_0) and the optimized first assembly. The geometry and finite size of this beginning distinguish the theoretical (principle-generated) construction from the deconstruction algorithms used in fractal geometry. In the latter, the algorithm (from large to small) is postulated and repeated ad infinitum [30], all the way down to the scale of size zero. It is only because of the infinite series of steps that the fractals-generated image of a certain size could, in principle, be obtained by simply magnifying an image of a smaller size.

The geometric optimization of tube flow through constrained volumes can be studied in much simpler settings such as the planar T-shaped structure [31] considered in Problem 5.1. A very simple analysis shows that when the flow is laminar the optimal ratio between successive tube diameters is $2^{1/3}$ (Murray's law). The optimal ratio between successive tube lengths is also $2^{1/3}$.

In the corresponding problem with fully developed turbulent flow (Problem 6.1) [31] the optimal ratio of diameters is $2^{3/7}$ and the optimal ratio of lengths is $2^{1/7}$. In both laminar and turbulent flow, the optimal architecture is characterized by the equipartition of pressure drop: the total pressure difference between the inlets and the outlet of the T construct (Figs. P5.1 and P6.1) is divided in half by the junction.

5.8 Scaling Laws of Living Trees

During 1995–1996, when I was drawing Figs. 5.13–5.18, I was concerned strictly with deducing geometric form from the constructal principle. I was not interested in specific biological examples. This is why in Table 5.4 I compiled only the numerical values that describe the optimized geometry and the associated flow resistance. More can be read off this table and, as shown in this section, it is quite relevant to placing on a purely theoretical foundation the existence of some of the empirical allometric laws of known living tree structures.

With reference to the construct level i listed in Table 5.4, the volume of each construct is

$$V_i = 2^i L_0^3 \qquad (i = 1, 2, 3, \ldots). \tag{5.52}$$

The void volume occupied by all the ducts in each construct is listed in Table 5.4; for $i = 3, 6, 9, \ldots$, it is also listed analytically in Eq. (5.51). Dividing V_{pi} by V_i, we record the porosity of each construct,

$$\phi_i = \frac{1}{8\lambda^2}\left[2 + \frac{i}{3}\left(1 + 2^{-1/3} + 2^{1/3}\right)\right], \qquad (i = 3, 6, 9, \ldots), \qquad (5.53)$$

where, according to the discussion of Eq. (5.34), λ is a number of the order of 1 but greater than 1, for example, 3.

In biology, in which the observer looks at a complete construct of high order, the most immediate measurements are global and external quantities such as the volume (V_i), the total mass (M_i), the cross-sectional area of the root of the tree (S_i), and the total internal surface of all the tubes in the construct (A_i). In constructal theory, these quantities follow from the reported dimensions of the optimized structure.

The mass $M_i = \rho V_i(1 - \phi_i)$ can be expressed in dimensionless form as

$$\hat{M}_i = \frac{M_i}{\rho L_0^3} = 2^i(1 - \phi_i), \qquad (5.54)$$

where ρ is the density of the tissue. The root cross-sectional area, $S_i = (\pi/4)D_i^2$, is the size of the largest tube that connects the tree to the external source or sink,

$$\hat{S}_i = \frac{S_i}{L_0^2} = \frac{\pi}{\lambda^2} 2^{(2i-5)/3}. \qquad (5.55)$$

The calculation of the internal area of each construct A_i consists of adding the internal surfaces of all the tubes that are present. The calculation proceeds toward larger constructs and yields

$$\hat{A}_i = \frac{A_i}{L_0^2} = \frac{\pi}{\lambda} a_i. \qquad (5.56)$$

where a_i is a numerical value that could be added as a new column to Table 5.4. Representative values are

i	a_i	i	a_i
2	6.31	9	1.28×10^3
3	14.86	12	1.07×10^4
4	34.5	15	8.72×10^4
5	68.6	18	7.04×10^5
6	146.3	21	7.43×10^6

Interesting relations emerge between these quantities when they are plotted against each other, as in Figs. 5.19 and 5.20. The curves were drawn through the points represented by each level of assembly i. In this presentation, the level of the construct can be absent (hidden from view), in the same way that it is absent from the measurements reported by the biologist who examines complete organs from large and small animals. The level i was plotted on top of the curves for clarity, to show their origin.

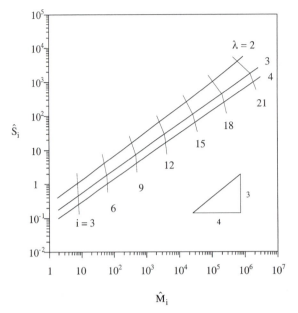

FIGURE 5.19. The relation between the cross-sectional area of the root of the tree and the total mass of the construct.

Figures 5.19 and 5.20 are summaries of the view from the outside, i.e., from the point of view of the observer who is unaware of the principle that generated the organ. From this direction, they are quite intriguing because they suggest the existence of power-law relations among mass, contact area, and root cross section. In constructal theory, power laws (straight lines on logarithmic plots) are

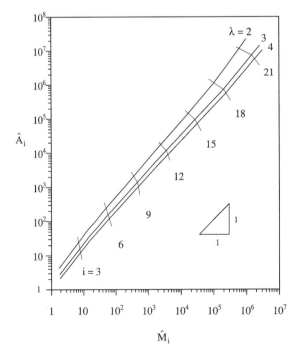

FIGURE 5.20. The relation between the total internal surface of the tubes and the total mass of the construct.

approximate curves drawn through the discrete points represented by each i. The solid lines drawn on Figs. 5.19 and 5.20 connect discrete points, and are not even straight.

In particular, the root cross-sectional area S_i increases almost in proportion with M_i raised to the power 0.7. An order-of-magnitude summary for the position of the bundle of curves of Fig. 5.19 is

$$\hat{S}_i \sim O(10^{-1})\hat{M}_i^{0.7}. \tag{5.57}$$

This trend agrees well with the allometric law averaged over organs of many sizes, where the exponent of M was approximated as 3/4.

Figure 5.20 shows that the total exposed surface of the tree of ducts is almost proportional to the mass of the construct. The bundle of curves is represented by

$$\hat{A}_i \sim O(1)\hat{M}_i^{1.03}. \tag{5.58}$$

As in Fig. 5.19, the effect of the aspect ratio of the smallest tube ($\lambda = L_0/D_0$) is small. We may say that macroscopic relations (5.57) and (5.58) are robust with respect to the architectural detail represented by λ.

More robustness with respect to λ is exhibited in Fig. 5.21, which shows the relation between volume and mass. The volume and the mass are almost proportional; however, the average density M_i / V_i decreases slightly as the order of the construct (or M_i, V_i) increases. Smaller and simpler constructs (small i) are more dense than higher-order constructs (large i). For practical purposes, Fig. 5.21 is

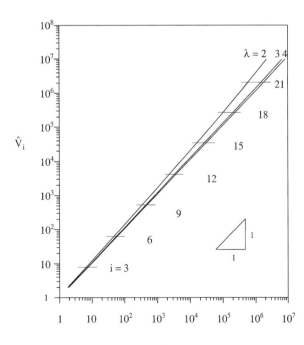

FIGURE 5.21. The relation between the volume and the mass of the construct.

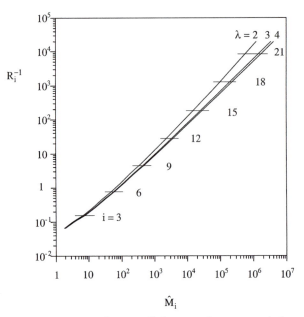

FIGURE 5.22. The relation between the overall flow conductance and the total mass of the construct.

summarized by

$$\hat{V}_i \cong \hat{M}_i. \tag{5.59}$$

These theoretical predictions agree with the known allometric laws not only qualitatively, but also quantitatively. For example, the accepted allometric law for the respiratory contact surface of mammals is [32, 33]

$$\frac{A}{1\,\mathrm{m}^2} \cong 3.31 \left(\frac{M}{1\,\mathrm{kg}}\right)^{0.98}, \tag{5.60}$$

where it has been indicated that A and M are expressed in m^2 and kg, respectively. In the body-mass range covered by mammals ($10^{-3}-10^3$ kg), this curve fit is approximated adequately by

$$\frac{A}{\mathrm{m}^2} \sim 3\frac{M}{\mathrm{kg}}. \tag{5.61}$$

On the other hand, theoretical relation (5.58) is approximated adequately by $\hat{A}_i \sim \hat{M}_i$, which means that $A_i \sim M_i/(\rho \lambda D_0)$. Substituting $\rho \sim 10^3$ kg/m^3 for the density of tissue and $D_0 \sim 10^2$ μm for the smallest (alveolus) length scale, we obtain

$$A_i \sim \frac{10}{\lambda}\frac{M_i}{\mathrm{kg}}\mathrm{m}^2. \tag{5.62}$$

Because λ is of the order of 3, the theoretical formula (5.62) is essentially the same as the empirical correlation (5.61).

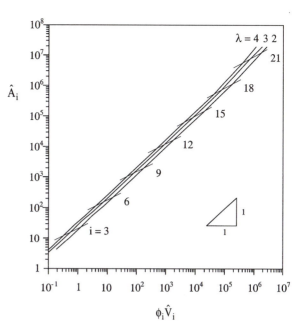

FIGURE 5.23. The relation between the total internal surface of the tubes and the total volume of the tree of tubes.

The flow conductance of each construct increases monotonically with the size, as shown in Fig. 5.22. The conductance R_i^{-1} is the inverse of the resistance reported in Table 5.4. The flow conductance is roughly proportional to $\hat{M}_i^a$, where the exponent a is approximately 0.9.

Another interesting relation emerges if we calculate the total void volume, i.e., the volume of the tree of ducts. This is represented by the product $\phi_i \hat{V}_i$. Figure 5.23 was obtained by the elimination of $\hat{M}_i$ between Figs. 5.20 and 5.21 and by use of Eq. (5.53). We see that the total area of contact $\hat{A}_i$ is almost proportional to $(\phi_i \hat{V}_i)^b$, where the exponent b is comparable with 0.9. This relation is relatively insensitive to the value of λ.

We return to the allometric laws of living trees in Sections 8.10 and 10.6, where we rely once more on constructal theory to predict the relations among animal body size, metabolic rate, and breathing and heart-beating frequencies.

PROBLEMS

5.1 An incompressible fluid flows through a tube of length L_1 and diameter D_1 and continues through two identical tubes of length L_2 and diameter D_2 [31]. Assume that the flow through each tube is laminar and fully developed (Hagen–Poiseuille). Show first that the resistance of each tube is proportional to the geometric group $(L/D^4)_{1,2}$.

Minimize the flow resistance of the entire assembly subject to fixed total tube volume. Note that the tube volume is proportional to the geometric group $L_1 D_1^2 + 2L_2 D_2^2$. Show that the global flow resistance is minimal when the tube diameters obey the proportionality $D_2/D_1 = 2^{-1/3}$. Derive an expression for the minimized resistance.

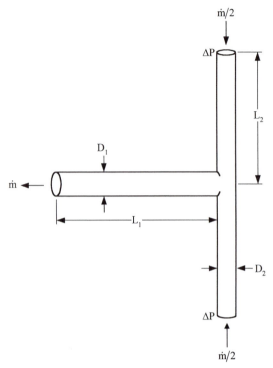

FIGURE P5.1

Minimize the global flow resistance one more time by selecting the tube lengths (L_1, L_2). Now the area (the two-dimensional territory) that houses the three tubes is constrained. If the L_2 tubes are perpendicular to the L_1, the tube area constraint is $A = 2L_2 L_1$. Show that the second resistance minimum is reached when $L_2/L_1 = 2^{-1/3}$.

What can you say about the shapes of the tributary tubes (D_2/L_2) and the collecting tube (D_1/L_1) when their total resistance is minimum subject to fixed total tube volume and fixed total two-dimensional territory?

5.2 Optimize the geometry of the Y-shaped construct of three tubes [31] shown in Fig. P5.2. The flow through each tube is laminar and fully developed. The total volume of the tubes is fixed. The only difference between this configuration and the configuration of the preceding problem is that this time the territory that houses the construct is a disk of fixed radius r. The layout of the construct is defined by the angles α and β shown on the figure.

Review the statement of Problem 5.1, and recognize that the optimal ratio of tube diameters is $D_2/D_1 = 2^{-1/3}$, while the overall flow resistance minimized with respect to D_2/D_1 is proportional to $(L_1 + 2^{1/3}L_2)^3$. These results are independent of the assumed lengths (L_1, L_2) and their relative positions, therefore they hold in the present configuration as well. Minimize the flow resistance with respect to α and β, and report the optimal Y shape and the corresponding lengths ratio L_2/L_1.

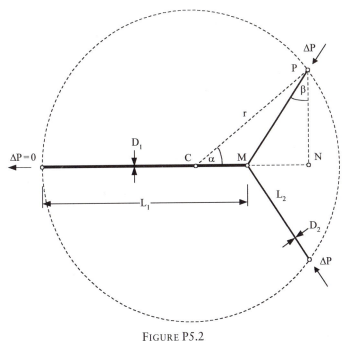

FIGURE P5.2

REFERENCES

1. A. Bejan, *Convection Heat Transfer*, 2nd ed., Wiley, New York, 1995.
2. E. R. Weibel, *Morphometry of the Human Lung*, Academic, New York, 1963.
3. K. Horsfield and G. Cumming, Morphology of the bronchial tree in man, *J. Appl. Physiol.*, Vol. 24, 1968, pp. 373–383.
4. K. Horsfield, G. Dart, D. E. Olson, G. F. Filley, and G. Cumming, Models of the human bronchial tree, *J. Appl. Physiol.*, Vol. 31, 1971, pp. 207–217.
5. E. R. Weibel, Design of biological organisms and fractal geometry, in T. F. Nonnenmacher, G. A. Losa, and E. R. Weibel, eds., *Fractals in Biology and Medicine*, Birkhäuser Verlag, Basel, Switzerland, 1994, pp. 69–85.
6. D. L. Cohn, Optimal systems: I. The vascular system, *Bull. Math. Biophys.*, Vol. 16, 1954, pp. 59–74.
7. G. S. Krenz, J. H. Linehan, and C. A. Dawson, A fractal continuum model of the pulmonary arterial tree, *J. Appl. Physiol.*, Vol. 72, 1992, pp. 2225–2237.
8. M. Sernetz, M. Justen, and F. Jestczemski, Dispersive fractal characterization of kidney arteries by three-dimensional mass-radius-analysis, *Fractals*, Vol. 3, 1995, pp. 879–891.
9. N. MacDonald, *Trees and Networks in Biological Models*, Wiley, Chichester, UK, 1983.
10. T. A. McMahon, Tree structures: deducing the principle of mechanical design, *J. Theor. Biol.*, Vol. 59, 1976, pp. 443–466.
11. A. L. Bloom, *Geomorphology*, Prentice-Hall, Englewood Cliffs, NJ, 1978, p. 204.
12. L. B. Leopold, M. G. Wolman, and J. P. Miller, *Fluvial Processes in Geomorphology*, Freeman, San Francisco, 1964.
13. A. E. Scheidegger, *Theoretical Geomorphology*, 2nd ed., Springer–Verlag, Berlin, 1970.
14. R. J. Chorley, S. A. Schumm, and D. E. Sugden, *Geomorphology*, Methuen, London, 1984.

15. M. Morisawa, *Rivers. Form and Process*, Longman, London, 1985.

16. S. A. Schumm, M. P. Mosley, and W. E. Weaver, *Experimental Fluvial Geomorphology*, Wiley, New York, 1987.

17. A. Bejan, *Advanced Engineering Thermodynamics*, 2nd ed., Wiley, New York, 1997.

18. A. Bejan, Constructal tree network for fluid flow between a finite-size volume and one source or sink, *Rev. Gen. Therm.*, Vol. 36, 1997, pp. 592–604.

19. A. Bejan and M. R. Errera, Deterministic tree networks for fluid flow: geometry for minimal flow resistance between a volume and one point, *Fractals*, Vol. 5, 1997, pp. 685–695.

20. D. A. Nield and A. Bejan, *Convection in Porous Media*, 2nd ed., Springer, New York, 1999.

21. A. Bejan and D. Tondeur, Equipartition, optimal allocation, and the constructal approach to predicting organization in nature, *Rev. Gen. Therm.*, Vol. 37, 1998, pp. 165–180.

22. D. Avnir, O. Biham, D. Lidar, and O. Malcai, Is the geometry of nature fractal? *Science*, Vol. 279, 1998, pp. 39–40.

23. A. Bejan, Constructal-theory network of conducting paths for cooling a heat generating volume, *Int. J. Heat Mass Transfer*, Vol. 40, 1997, pp. 799–816.

24. D'A. W. Thompson, *On Growth and Form*, Cambridge Univ. Press, Cambridge, UK, 1942.

25. C. D. Murray, The physiological principle of minimal work, in the vascular system, and the cost of blood-volume, *Proc. Acad. Nat. Sci.*, Vol. 12, 1926, pp. 207–214.

26. M. R. Errera and A. Bejan, Tree networks for flows in composite porous media, *J. Porous Media*, Vol. 2, 1999, pp. 1–18.

27. A. Bejan, G. Tsatsaronis, and M. Moran, *Thermodynamic Design and Optimization*, Wiley, New York, 1996, p. 4.

28. A. Bejan, *Heat Transfer*, Wiley, New York, 1993.

29. T. A. Wilson, Design of the bronchial tree, *Nature (London)*, Vol. 213, 1967, pp. 668–669.

30. B. B. Mandelbrot, *The Fractal Geometry of Nature*, Freeman, New York, 1982.

31. A. Bejan, L. A. O. Rocha, and S. Lorente, Thermodynamic optimization of geometry: T- and Y-shaped constructs of fluid streams, *International Journal of Thermal Sciences*, Vol. 40, 2001, to appear.

32. E. R. Weibel, Morphometric estimation of pulmonary diffusion capacity. V. Comparative morphometry of alveolar lungs, *Resp. Physiol.*, Vol. 14, 1972, pp. 26–43.

33. K. Schmidt-Nielsen, *Scaling (Why is Animal Size So Important?)*, Cambridge Univ. Press, Cambridge, UK, 1984, p. 112.

CHAPTER SIX

DUCTS AND RIVERS

6.1 Geometric Puzzles

Next to the plants and trees that surround us, rivers provide the most numerous and stunning natural images that exhibit the features anticipated by constructal theory. Rivers are an extremely important subject even without the large-scale similarities that challenge us to come up with a theory. First, rivers are the primary mechanism responsible for shaping the Earth's surface. Second, rivers are the most common high Reynolds number flows known to man. Fluid mechanics in general, and turbulence research in particular, will never be *theory* if they cannot predict the tightly correlated (reproducible) geometry and performance of river flow.

The flow of a river strikes us with images, so simple and so repetitive, that cry for an explanation:

1. The river channel tends to acquire a sinusoidal shape (it meanders, buckles) with a wavelength that is proportional to the channel width.
2. The width of the channel is proportional to the maximum depth.
3. The river drainage basin and the river delta (Fig. 6.1) have the plane tree features that are now anticipated based on the area-to-point flow access optimization (Chap. 5).

These similarity types have been documented extensively and very successfully in geophysics: The correlations can be found in all the modern treatises [1–6]. In this section we run only briefly through points 1–3 to show that all these geometrical features can be anticipated now based on pure theory.

(a) Meanders. The river meander puzzle (1) is how I started to write about large-scale organization in nature [7–9], when I proposed a purely geometric theory for turbulent mixing regions (Sections 7.3 and 7.4). The river channel of cross-sectional area A, mean longitudinal velocity V, and density ρ can be viewed as a finite-size column in end-to-end compression. The "compressive" force acting axially is $\rho A V^2$. When the Reynolds number $V A^{1/2}/\nu$ is large such that the bulk flow may be modeled as inviscid, the river column is elastic (irreversibility free)

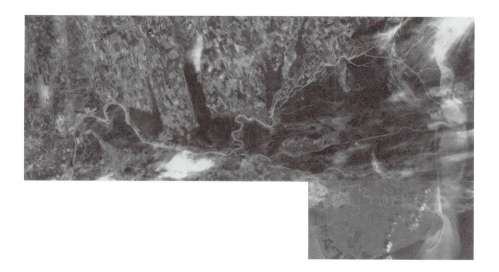

FIGURE 6.1. Danube delta and meanders (Landsat photo), and drainage basin near the mouth of the Colorado River, San Felipe, Mexico (U.S. Navy photo).

and develops a resistive bending moment. The Euler-type infinitesimal buckling analysis of the river column leads to the following conclusion [7, 8]: The natural shape of the river must be sinusoidal, and the buckling wavelength λ_B must be approximately twice the river width W:

$$\lambda_B \sim 2\,W. \qquad (6.1)$$

The buckling theory is in very good agreement with field observations on the *incipient* formation of river meanders in nearly straight channels [9], as shown in Table 6.1. More useful is the fact that the geometric feature $\lambda_B/W \sim 2$ provides a tape measure and a stopwatch for constructing theoretically other turbulent flows, not just rivers. The method is reviewed in Refs. 8 and 16.

TABLE 6.1. **Summary of Physical Observations Regarding the Wavelength/Width Ratio of Meandering Rivers [9]**

Wavelength/Width	Reference
2.22	Theory of buckling in nearly straight rivers [9]
2–3	Stream plate simulations [10]
2–3	Field studies in straight rivers [1, 11, 12]
3.24 ± 0.64	Laboratory channel during development [13]
3–5	Channel experiments during development [14]
5–7	Channel experiments during late stages of development [14]
6.5–11	Natural rivers (long history) [15]

Note: Note that the age of the meanders increases downward through this table.

Most of the applications of the buckling theory of turbulent flow have come in the field of convective heat transfer, where they have been reviewed [16]. For example, the theory anticipates the transition to turbulence in all flow configurations, the size of the smallest eddy, the $y^+ \sim 10$ thickness of the viscous sublayer, the vortex shedding frequency, the linear growth of all turbulent mixing regions, the Colburn analogy between turbulent heat and momentum transfer, and the frequency of pulsating fire plumes.

(b) Drainage Basins and Deltas. The second puzzle – the proportionality between river width and maximum depth – forms the subject of the next section. It is an astonishingly simple geometric feature that must be taken into account when treating the third puzzle: the dendritic (i.e., treelike) construction of drainage basins and deltas. The start of the analysis has the ingredients given in Fig. 5.13, where the volume is flat ($H_1 L_1 t$) and the thickness t can be assumed to represent the depth (d) of the first river. The width of the first river channel is W_1 instead of D_1. The permeability K is a constant dictated by the characteristics of the material through which the volumetric Darcy flow covers the elemental surface $H_1 L_1$.

The excess pressure field $P(x, y)$ is another way of accounting for gravity – that is, the slopes that develop over the elemental area $H_1 L_1$ in order to drive the flow. The first river width is also a characteristic of the terrain: The order of magnitude of W_1 is dictated by the size of the grains that can be displaced and carried away by the elemental flow. We return to this idea in Section 6.4.

It is worth repeating that constructal theory starts from the smallest subsystem of known finite size, where the volumetric (diffusive) flow coexists with the first stream flow (Fig. 6.2; see also Fig. 1.2). This observation is especially important in the field of river morphology, because in that field the elemental system has not been recognized. Special care must be taken to see such small "first rivers" in nature. Furthermore, the elemental system is not a feature of the drainage basin models that are being assumed, simulated, and optimized numerically on the computer [17–21], after a general network, such as that of Fig. 6.3, is assumed and the many links (ducts, channels) are moved around until an overall power dissipation figure is minimized.

FIGURE 6.2. The smallest river channels in the beach sand of Dana Point, California. The scale is indicated by the footprints of seagulls.

Strong support for constructal theory is provided by the end results of computer optimizations of drainage basins and vascularized tissues [17–21]: They reveal patterns that are strikingly similar to those that occur naturally. What is also relevant is the authors' speculation that some form of global optimization principle underlies the existence of natural fractal structures. Constructal theory unveils that very principle and structure.

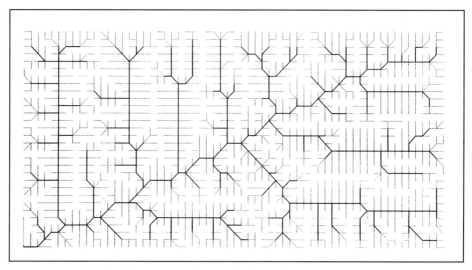

FIGURE 6.3. Example of near-optimal network of river channels optimized numerically ([17]; copyright by the American Geophysical Union).

To illustrate the coexistence of the two flow regimes (volumetric plus duct) at the elemental level, I experimented with coffee sediment, as I showed in Fig. 1.2 The water flows volumetrically through the layer of coffee grounds that covers the surface. At the same time, each elemental area develops its first river, along which the sacrificed grains of sediment are swept downstream. The merit of this antiestablishment experimental method (other than the zero cost) is that it works every time. Those who still believe in fractal rivers should look for the smaller and smaller rivers ad infinitum in Figs. 1.2 and 6.2, or in their own coffee cups. They will not find them.

The *infinite* sequence of fracturing steps is the defining statement of fractal geometry; the non-Euclidian dimension (Hausdorff) exists strictly in this limit, at infinity. When the sequence is cut off (quite arbitrarily) and made finite, the incomplete (read: Euclidean) images printed on paper look like patterns we see in nature. This coincidence does not mean that natural patterns are fractal! The contrary is true: Everything shown to us by nature and everything done under the table by the fractal algorithm manipulator support the view that the real image (e.g., sheet of paper or sand sculpture) is Euclidean. This is even why the image can be distinguished (i.e., seen), because otherwise we would be seeing nothing but blurred images and shades of gray. Constructal theory is supported by the sharp Euclidian images that we see, because the theory starts from the smallest (finite) scale, continues by predicting a finite sequence of constructs, and displays its predictions in two or three dimensions. Along the way, the finite sequence of constructs also explains why the incomplete fractal sequences assumed by the mathematician happen to look like natural patterns.

The preceding paragraph is the footnote that I wrote on the spur of the moment as I was finishing my 1997 book [22]. The physics claim that the geometry of nature is fractal [23] was called into question independently one year later, in the pages of *Science* [24]. Enlightening in this respect are two earlier voices: Shenker's essay [25] and Kadanoff's column [26], in which the emptiness of the physics claims made on behalf of descriptive fractal geometry was also exposed.

The drainage basin in reverse is the river delta; in this case the flow direction is from one point to a finite area. The water flow rate is fixed, because it is dictated by the river that arrives near the shoreline. This stream distributes itself through channels with relatively low resistance, and it finally seeps volumetrically into the ground with high flow resistance. Superimposed on this point-to-area flow is the related problem of distributing (again, with least flow resistance) a fraction of the original stream to the coastal perimeter of the delta area.

Figure 6.4 shows a remarkably simple simulation of the delta flow. The two-dimensional flow is created by the injection of dyed water into the $75\text{-}\mu\text{m}$-thin gap between two 10×10 cm^2 glass plates [27]. This apparatus is known as a Hele–Shaw cell. The gap is initially filled with glycerine. The inner surfaces of the glass plates are roughened artificially with random dots (dimples), each with a diameter and a dot-to-dot distance of the order of $250\ \mu\text{m}$ and a dot depth of the order of $22\ \mu\text{m}$. The Hele–Shaw cell is held horizontally, and the water is injected through a central hole in the top plate. The dyed water displaces the more viscous liquid

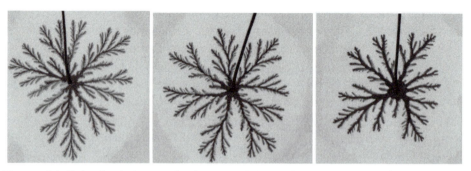

FIGURE 6.4. Delta simulations made when water is injected into a narrow parallel-plates space occupied by glycerine ([27]; reprinted with permission from Springer-Verlag, Berlin).

(glycerin) and creates the now-familiar pattern predicted by the solution to the area access problem.

A river delta looks like only a portion – a segment – of the disks shown in Fig. 6.4, because the surface of the delta is sloped toward the sea. I have two reasons for exhibiting Fig. 6.4 in this section. First, this simple experiment shows that the water chooses to organize itself into streams while invading an area characterized volumetrically by high resistance (rough plates, viscous glycerin). This geometric arrangement is completely analogous not only to delta flows, but to every other point-to-area or point-to-volume dendrite found in nature. Note that streams (dendrites) *never occur* if the roles are reversed – that is, if the injected fluid is more viscous than (or as viscous as) the displaced fluid (e.g., Fig. 6.9 in Section 6.4). Similar flows can be generated by pouring water from a cup onto a rough horizontal surface (e.g., cement, pavement, asphalt).

My second reason for using Fig. 6.4 as an illustration of optimal point-to-area access is that in the physics literature this flow is widely recognized under a different name (another class): two-dimensional *fingers*. This class has been studied exhaustively in physics, in which the focus has been on stability aspects that may control the shape (tips) of the fingers of the expanding flow. The fingers formed during the growth of bacterial colonies [28] are an analogous phenomenon, for reasons that will be explored further in Chapter 11. As with the other naturally organized phenomena discussed in this book, the contribution of constructal theory is that it brings under the same tent many examples (e.g., fingering) that have received prolonged study as *isolated* phenomena.

6.2 Optimal River Channel Cross Sections

In this section we examine from the point of view of constructal theory the second similarity puzzle (2) of river flow. Why is the width of a river a certain multiple of its depth? Why should such a simple geometric proportionality exist?

These questions appear difficult when we think of the many complications of river flow and channel development. Turbulent flow, secondary flow (cross circulation), bed erosion, sediment transport, time, and the geological characteristics of the terrain all play important roles, which have been amply recognized in the literature. These complications are responsible for significant variations in channel cross section from one river to another or along the same river. They can also be invoked when we are observing that the ratio between the river width W and the maximum depth d is not a precise constant: There is considerable scatter in the W/d data.

The theoretical approach that has been offered as an explanation for the shape of the channel cross section recognizes from the beginning the complicated, developing nature of the river bed (e.g., Refs. 1 and 2). An equilibrium was envisaged between the forces acting on a ground particle at rest on the bottom of the channel. As noted by Scheidegger [2], this equilibrium condition is insufficient and must be complemented by questionable assumptions concerning the distribution of shear stress (or bottom velocity) with depth. After numerical integration, this approach yields a bottom profile that is roughly sinusoidal, in which the river width W and the maximum depth d are two input (undetermined) constants. In sum, the river bed equilibrium theories do not predict the observed proportionality between W and d.

In spite of the many complicated phenomena that have been recognized in both river surveys and theories, one simple geometric fact stands out: Large (wide) rivers are deeper than tiny rivers. This begs for a theory that should be as simple as the scaling law itself:

$$W \sim d. \tag{6.2}$$

The cross section of an open channel is characterized by an upper straight segment (the free surface), which is shear free, and the rest of the perimeter (the bottom), which is labeled p and is characterized by shear (Table 6.2). Assume that the channel is straight. The cross-sectional area A and the mass flow rate $\dot{m}$ are known constants. The area A is the result of the slow development of the river bed, and, in this sense, it describes the age of the river bed. By fixing A, we focus our inquiry on the present, or, in view of the slow development, on the channel as a system in quasi-steady state.

The mass flow rate is fixed by the global constraint represented by the rainfall on the drainage surface situated upstream of A. Several important conclusions follow from the A and $\dot{m}$ constraints. First, the mean flow velocity is fixed, $V = \dot{m}/\rho A$, where ρ is the water density. Next the longitudinal shear stress along the bottom is also fixed, $\tau = \frac{1}{2} C_f \rho V^2$, because in turbulent flow over a very rough surface the average skin friction coefficient C_f is practically constant [16]. Here we are using the nomenclature of duct fluid mechanics (e.g., heat exchangers), not river hydraulics, because soon we will make reference to actual duct flows (e.g., blood vessels).

The total force per unit of channel length in the flow direction is $p\tau$. The flow resistance $p\tau/\dot{m}$ decreases in proportion with the bottom perimeter p. This conclusion can also be expressed in thermodynamic terms. The rate of work destruction $p\tau V$ at the longitudinal station where A is located is also proportional to p. The same holds for the rate of entropy generation in A, namely $p\tau V/T_0$, where T_0 is the ambient temperature.

TABLE 6.2. **Optimized Cross-Sectional Shapes of Open Channels**

Shape	Optimal Shape	$(W/d)_{\text{opt}}$	$p_{\min}/A^{1/2}$
Rectangle	A	2	2.828
Triangle	A	2	2.828
Parabola	A	2.056	2.561
Circle	A	2	2.507

We arrived at a very simple geometric problem that accounts for the thermo-dynamic optimization (minimum flow resistance, work destruction, or entropy generation) of the river flow through the cross section A:

> Find the cross-sectional shape that has the minimum ground perimeter p, which, along with a straight (free-surface) segment, encloses the fixed area A.

Table 6.2 shows four steps in the search for the optimal cross-sectional shape. In each step, the shape of the bottom had to be assumed. In the first example, the bottom shape is rectangular, and its perimeter ($p = W + 2d$) can be minimized subject to the area constraint ($A = Wd$). The result is the optimal proportionality

$$W = 2d. \tag{6.3}$$

This is the first instance in which the natural scaling law (6.2) is derived based on pure theory. The next three shapes (triangular, parabolic, circular) reveal the same scaling law. The minimized bottom perimeter decreases slightly from one example to the next.

Table 6.2 also shows the route to the mathematically optimal channel cross section. That optimal image has two parts, or 2 degrees of freedom: (a) the shape, or bottom profile, which had to be assumed in each of the examples of Table 6.2, and (b) the slenderness ratio W/d, which was optimized analytically in each example.

(a) The optimal shape of the bottom can be determined based on variational calculus. We assume an arbitrary function $y(x)$ for the bottom profile, where y is the depth corresponding to the coordinate x measured along the free surface (in the plane of A: see the bottom sketch in Table 6.2). The objective is to minimize the total length of the bottom perimeter:

$$p = \int_{-W/2}^{W/2} \left[1 + \left(\frac{dy}{dx} \right)^2 \right]^{1/2} dx, \tag{6.4}$$

where, for the time being, the width W is considered fixed. The optimal function $y(x)$ that minimizes the integral of Eq. (6.4) subject to the cross-sectional area constraint,

$$A = \int_{-W/2}^{W/2} y \, dx, \tag{6.5}$$

is readily obtained when the associated Euler equation is solved [29]. The conclusion is that the optimal $y(x)$ is the arc of a circle, or that A is the segment of a circle such that W is the chord. There is a family of such solutions, from the very thin slice to the full circle, depending on how W compares with the length scale $A^{1/2}$. Three examples are shown in the upper part of Fig. 6.5.

(b) The optimal slenderness ratio W/d of the optimal shape is determined next by minimization of the ground contact perimeter subject to fixed A. The

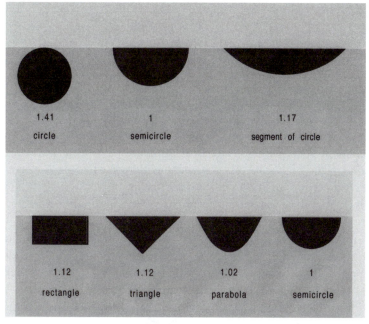

FIGURE 6.5. Channels with the same cross-sectional area, different shapes, and different wetted perimeters. The top row shows three examples with bottoms shaped as arcs of circles. The number under each drawing represents the flow resistance relative to the resistance of the channel with semicircular cross section. The four cross sections shown on the bottom row have had their depth/width ratios optimized for minimum resistance. Note that the resistance does not vary appreciably from one shape to the next. Channel cross-sectional designs are robust (see Plate VII).

result of this operation is the half-circle [$W/d = 2$ and $p_{min} = (2\pi A)^{1/2}$], which is shown as the fourth case in Table 6.2.

In summary, the scaling law between river width and maximum depth can be anticipated in nakedly simple terms by minimization of the mechanical power destroyed in the entire cross-sectional area A. In this way, the optimal geometry of the channel is associated with a *global* optimization subject to imposed, global constraints (A, $\dot{m}$). In this sense, the present theory is in agreement not only with all the natural examples covered in this book, but also with the conclusions reached in the global optimization of assumed lattices of channels in drainage basin models [18, 19]. Furthermore, the proportionality $W \sim d$ is supported by recently optimized numerical models of drainage basins [17–21], in which it was shown that in every optimized channel both W and d are proportional to the square root of the flow rate.

The examples aligned in Table 6.2 reveal two additional features of the optimal channel geometry. First, in the optimal shape (the half-circle) the river banks extend vertically downward into the water and are likely to crumble under the influence of erosion (drag on particles) and gravity. This will decrease the slopes of the river bed near the free surface and, depending on the bed material, will increase the slenderness ratio W/d somewhat. The important point is that there remains plenty of room for the equilibrium theories proposed in the past [1, 2]; in fact, their territory remains intact: The equilibrium theories begin where constructal theory leaves off.

The second feature revealed by Table 6.2 and the lower part of Fig. 6.5 is that in practical, engineering terms the four designs have nearly the same performance (the same $p_{min}/A^{1/2}$ value). The reader will note that this is a recurring theme in nature and constructal theory (e.g., Chaps. 4 and 5).

The optimized rectangular and triangular channels perform identically, whereas the parabolic and semicircular channels agree to within 2.2%. The two most extreme examples given in Table 6.2 are separated by only 12%, even though the rectangular shape is strikingly "rough" in comparison with the parabolic and the round shapes. This high level of agreement with regard to performance is very important. It accounts for the significant scatter in the data on river bottom profiles, that is, global thermodynamic performance is what matters, not local shape. Again, this is in agreement with the new work on drainage basins, in which the computer-optimized (randomly generated) network looks like the many, never-identical networks seen in the field. Yes, there is uncertainty in reproducing the many shapes that we see in nature, but this can be attributed to other random factors (Section 6.5). What is important is that there is very little uncertainty in anticipating global characteristics such as optimal performance and optimal scaling laws (the ratio W/d in this case).

This situation is completely analogous to the highly complex optimization processes that are performed routinely in the field of engineering design. Almost as a rule, in the end it turns out that the chosen configuration of the optimized system performs *nearly* the same as the truly optimal configuration, even though, visually, the nearly optimal and the optimal configurations may be different (Ref. 30, p. 4).

6.3 Optimal Duct Cross Sections

Striking support for the above conclusion is provided by the equally wider class of internal ducts found in organisms – for example, the blood vessels and the pulmonary airways. The fact that every duct is round or nearly round has long been attributed to the minimization of flow resistance or power destruction (e.g., Ref. 31). This principle, however, deserves a second look in view of the near-optimal shapes found for river cross sections.

Two features distinguish the internal duct from the open channel optimized in this section. First, there is no shear-free segment on the perimeter ($W = 0$). If the flow is highly turbulent and in the fully rough regime, then the solution is the same as that of paragraph (a) in the preceding section, in the $W/A^{1/2} \to 0$ limit: The optimal shape is the complete circle. This is why the natural caves formed by subterranean rivers and coastal wave action tend to have round cross sections. The earthworm is round for the same global optimization reason.

The second feature that may be different in an internal duct is the flow regime. In laminar flow the thermodynamic optimization is not as simple as the minimization of p subject to fixed A. The pressure drop ΔP per unit of duct length ΔL is [16]

$$\frac{\Delta P}{\Delta L} = \frac{f}{D_h} 2\rho V^2, \tag{6.6}$$

where D_h is the hydraulic diameter, $D_h = 4A/p$, and f is the friction factor. An interesting and relatively unknown geometric property of D_h is this: In a regular polygonal cross section, D_h is equal to the diameter of the circle inscribed to the polygon; see Table 3.1 in Ref. 16.

When the laminar flow is fully developed, the friction factor is inversely proportional to the Reynolds number ($\mathrm{Re} = D_h V/\nu$):

$$f = \frac{C}{\mathrm{Re}}, \tag{6.7}$$

where the factor C depends solely on the shape of the cross section [16]. Table 6.3 shows several regular polygonal cross sections. The flow resistance per unit of duct

TABLE 6.3. **The Laminar Flow Resistances of Ducts with Regular Polygonal Cross Sections with n Sides**

n	C	$p/A^{1/2}$	p^2C/A
3	40/3	4.559	277.1
4	14.23	4	227.6
5	14.74	3.812	214.1
6	14.054	3.722	208.6
8	15.412	3.641	204.3
10	15.60	3.605	202.7
∞	16	$2\pi^{1/2}$	201.1

Note: The C data are from Ref. 16.

length can be expressed as

$$\frac{\Delta P/\Delta L}{\dot{m}} = \frac{\nu}{8\,A^3}\,p^2 C. \qquad (6.8)$$

The flow resistance decreases in proportion with the geometric group $p^2 C$, when the only variable is the shape of the perimeter. Alternatively, $p^2 C$ is also proportional to the total mechanical power destroyed over the system A or the total rate of entropy generation in A. The table shows that p decreases as the shape becomes rounder; however, this effect is tempered by C, which increases in the same direction.

Even though the round shape is the best, the nearly round shapes perform almost as well. For example, the relative $p^2 C$ change from the hexagon to the circle is only 3.7%. Yet the hexagonal ducts have the great advantage in that they can be packed in parallel into bundles to fill a volume completely. Square ducts also have this packing advantage, and their flow resistance is only 9.1% greater than that of hexagonal ducts.

In conclusion, the nearly round cross sections are practically as effective as the perfectly round cross section. Even if the duct is imperfect – that is, with random features such as flat spots – its performance is already as good as it can be. Furthermore, the optimal performance ($p^2 C$) can be predicted quite accurately when the global constraints (A, $\dot{m}$) are specified. These findings reinforce the constructal theory conclusions reached throughout this book.

6.4 Deterministic River Drainage Basins

The deterministic power of constructal theory is an invitation to new theoretical work on natural flow structures that have evaded determinism in the past. This section is dedicated to such a phenomenon: the tree shape of a river drainage network. The deterministic approach employed in this section is based on the proposition that a naturally occurring flow structure – its geometrical form – is the end result of a process of geometric optimization. The objective of the optimization process is to construct the path (or assembly of paths) that provides minimal resistance to flow or, in an isolated system, to maximize the rate of approach to internal (volumetric) equilibrium.

We saw that the tree structure is generated by two or more flow regimes that work hand in glove toward minimizing the volume-to-point resistance. In Chap. 4 the two regimes were (a) volumetric conduction through a low-conductivity material (the "glove") that filled the "elemental" system, and (b) channeled conduction through a blade of high-conductivity material (the "hand") that collected the volumetric flow and led it to a point on the system boundary. In Chap. 5 the roles of (a) and (b) were played by low-permeability and high-permeability media.

Reexamine Fig. 1.2 to see the coexisting flow regimes (diffusive and channeled). The water flows volumetrically through the layer of coffee grounds that covers

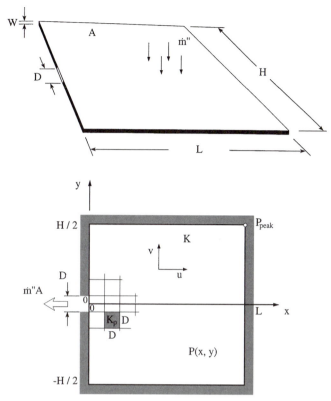

FIGURE 6.6. Model of area-to-point flow in a porous medium with Darcy flow and blocks that can be dislodged and swept downstream [32].

the surface. Each elemental area develops its first river, along which the sacrificed portions of sediment are swept downstream. The smallest area scale – the elemental system – is known because it is dictated by the size of the dislodged parts. This smallest and finite scale is also evident in Fig. 7.18: The smallest channels for organized flow have a thickness comparable with the length scale of the eroded grains of rice.

The coffee sediment demonstration of Fig. 1.2 suggested to us the two-dimensional, area-to-point flow model [32] defined in Fig. 6.6. The surface area $A = HL$ and its shape H/L are fixed. The area is coated with a homogeneous porous layer of permeability K. The small thickness of the K layer, i.e., the dimension perpendicular to the plane $H \times L$, is W, where $W \ll (H, L)$.

An incompressible Newtonian fluid is pumped through one of the A faces of the $A \times W$ parallelepiped, such that the mass flow rate per unit area is uniform, $\dot{m}''$ (kg/m^2 s). The other A face and most of the perimeter of the $H \times L$ rectangle are impermeable. The collected stream ($\dot{m}'' A$) escapes through a small port of size $D \times W$ placed over the origin of the (x, y) system. The fluid is driven to that outlet by the pressure field $P(x, y)$ that develops over A. The pressure field accounts for the effect of slope and gravity in a real drainage basin, and the uniform flow rate $\dot{m}''$ accounts for the rainfall.

We can determine the flow field after making an assumption about the flow regime, i.e., the relation between flow rate and pressure difference. The simplest assumption is that the flow through the K medium is in the Darcy regime [33]:

$$u = -\frac{K}{\mu}\frac{\partial P}{\partial x}, \qquad v = -\frac{K}{\mu}\frac{\partial P}{\partial y}. \tag{6.9}$$

In these equations, (u, v) are the volume-averaged velocity components and μ is the viscosity. The conservation of mass in every infinitesimal volume element $\Delta x\, \Delta y\, W$ requires that [16]

$$\frac{\partial u}{\partial x} + \frac{\partial v}{\partial y} - \frac{\dot{m}''}{\rho W} = 0, \tag{6.10}$$

where ρ is the density of the fluid. Combining Eqs. (6.9) and (6.10), we obtain a Poisson equation for the pressure field,

$$\frac{\partial^2 P}{\partial x^2} + \frac{\partial^2 P}{\partial y^2} + \frac{\dot{m}'' \nu}{WK} = 0, \tag{6.11}$$

where $\nu = \mu/\rho$ is the kinematic viscosity. The boundary conditions for solving Eq. (6.11) are impermeable boundaries ($u = 0$ or $v = 0$) all around the domain, except over the sink located at $x = 0$ and $-D/2 < y < D/2$, where $P = 0$.

The resistance to this area-to-point flow, which soon will become the object of geometric minimization, is the ratio between the maximal pressure difference (P_{peak}) and the total flow rate ($\dot{m}''\, A$). The location of the point of maximal pressure is not the issue, although in the model described until now its identity is clear (see Fig. 6.6, bottom). It is important to calculate P_{peak} and to reduce it at every possible turn by making appropriate changes in the internal structure of the $A \times W$ system. Determinism results from invoking a single principle and using it consistently.

Changes are possible because finite-size portions (blocks) of the system can be dislodged and ejected through the sink. Let us assume that the removable blocks are of the same size and shape (square, $D \times D \times W$). The critical force (in the plane of A) that is needed to dislodge one block is τD^2, where τ is the yield shear stress averaged over the base area D^2. The yield stress and the length scale D are assumed known. They provide a useful estimate for the order of magnitude of the pressure difference that can be sustained by the block. At the moment when one block is dislodged, the critical force τD^2 is balanced by the net force induced by the local pressure difference across the block ΔP, namely $\Delta P D W$. The balance $\tau D^2 \sim \Delta P D W$ suggests the pressure-difference scale $\Delta P \sim \tau D/W$, which along with D can be used for the purpose of nondimensionalizing Eq. (6.11):

$$\frac{\partial^2 \tilde{P}}{\partial \tilde{x}^2} + \frac{\partial^2 \tilde{P}}{\partial \tilde{y}^2} + M = 0, \tag{6.12}$$

$$(\tilde{x}, \tilde{y}) = \frac{(x, y)}{D}, \qquad \tilde{P} = \frac{P}{\tau D/W}, \tag{6.13}$$

$$M = \dot{m}'' \frac{\nu D}{\tau K}. \tag{6.14}$$

Based on what criterion do we search for the blocks that are dislodged? Let s be the direction of the resultant of all the pressure forces that act on the block perimeter $D \times D$. The block does not break away as long as $(\partial \overline{P}/\partial s)W < \tau$, which in dimensionless terms is

$$\left(\overline{\frac{\partial \tilde{P}}{\partial \tilde{s}}} \right) < 1. \qquad (6.15)$$

The pressure gradient $\overline{(\partial \tilde{P}/\partial \tilde{s})}$ is averaged over the square base area of one block.

When condition (6.15) is violated, the block is removed and its place is taken by a channel that is considerably more permeable – more conductive for fluid flow – than the original medium (K). The simplest way to implement this change is to assume that the space vacated by the block is also a porous medium with Darcy flow except that the new permeability (K_p) of this medium is sensibly greater, $K_p > K$. This happens to be the correct assumption when the flow is slow enough (and W is small enough) so that the flow regime in the vacated space is Hagen–Poiseuille between parallel plates. The equivalent K_p value for such a flow is $W^2/12$; see Eqs. (5.15). The main reason for introducing the K_p assumption in this model is to simplify the calculation of the pressure distribution over the area occupied by the dislodged block. Over that area, the pressure is governed by an equation that is the same as Eq. (6.12) except that M is replaced with MK/K_p. The pressure varies continuously between the original material K and the vacated domain K_p. When we account for mass continuity across the interface between the K and the K_p domains, the ratio K/K_p emerges as a dimensionless parameter of the system.

The pressure $\tilde{P}$ and the block-averaged gradient [condition (6.15)] increase in proportion with the imposed mass flow rate (M), because Eq. (6.12) is linear. When M exceeds a critical value M_c, condition (6.15) is violated and the first block is dislodged. For example, the critical flow-rate parameter is $M_c = 0.00088932$: This value does not depend on K/K_0 because in the beginning the entire system is occupied by K material. The first block that breaks away is the one that has the outlet port as one of it four sides. The peak pressure is located in the farthest corners of the A domain and experiences a drop when the first block is removed at constant flow rate $(M = M_c)$; when $K/K_0 = 0.1$, this drop occurs from $\tilde{P}_{\text{peak}} = 3.631$ to 2.934. This is in fact the purpose of the change in the internal structure of the area-to-point flow system, i.e., the physics principle that we invoke: The resistance to fluid flow is decreased through geometric changes in the internal architecture of the system.

To generate higher pressure gradients that may lead to the removal of a second block, we must increase the flow-rate parameter M above the first M_c, by a small amount. The removable block is one of the blocks that borders the newly created K_p domain. The peak pressure rises as M increases, and then drops partially as the second block is removed. This process can be repeated in steps marked by the removal of each additional block. In each step, we restart the process by increasing M from zero to the new critical value M_c. During this sequence the peak pressure decreases, and the overall area-to-point flow resistance $(\tilde{P}_{\text{peak}}/M_c)$ decreases monotonically.

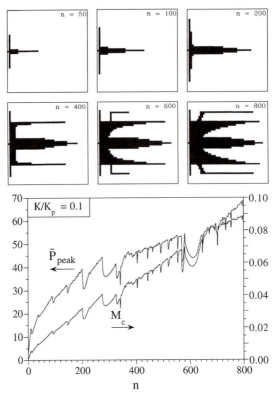

FIGURE 6.7. The evolution of the structure of the system of Fig. 6.6 when $K/K_p = 0.1$: the growth of the K_p domain and the variations in the critical flow rate and peak pressure [32].

The key result is that the removal of certain blocks of K material and their replacement with K_p material generate macroscopic internal *structure*. The generalizing mechanism is the minimization of flow resistance, and the resulting structure is deterministic: Every time we repeat this process we obtain exactly the same sequence of images.

For illustration, consider the case $K/K_p = 0.1$, shown in Fig. 6.7. The number n represents the number of blocks that have been removed. The domain A is square and contains a total of 2601 building blocks of base size $D \times D$; in other words, $H = L = 51D$. The pressure field equations, Eqs. (6.12)–(6.14), were solved with the finite-element method described in Ref. 32. In most cases the variation of $\tilde{P}$ over the edges of the blocks located near the interfaces between the two porous domains was quite smooth, almost linear. The average pressure on each face was obtained by numerical integration, which was followed by the calculation of the resulting force and its projection onto the direction in which the block would be free to move. The entire numerical procedure was automated: The cycle formed by calculations and by the necessary changes in the internal structure (K, K_p) was repeated without any interference from the outside. This is to say that the curves plotted in Fig. 6.7 are unique (reproducible).

Figure 6.7 also shows the evolution of the critical flow rate and peak pressure. The curves appear ragged because of an interesting feature of the erosion model: every time that a new block is removed, the pressure gradients redistribute themselves and blocks that used to be "safe" are now ready to be dislodged even without an increase in M. The fact that the plotted M_c values drop from time to time is due to restarting the search for M_c from $M = 0$ at each step n.

The shape of the high-permeability domain K_p that expands into the low-permeability material K is that of a tree. New branches grow in order to channel the flow collected by the low-permeability K portions. The growth of the first branches is stunted by the fixed boundaries (size, shape) of the A domain. The older branches become thicker; however, their early shape (slenderness) is similar to the shape of the new branches.

The slenderness of the K_p channels and the interstitial K regions is dictated by the K/K_p ratio, that is, by the degree of dissimilarity between the two flow paths. Highly dissimilar flow regimes ($K/K_p \ll 1$) lead to slender channels (and slender K interstices) when the overall area-to-point resistance is minimized. This behavior is completely analogous to that of the tree networks obtained by means of thermal-resistance minimization in heterogeneous heat conductors (Chap. 4) or by means of fluid-flow-resistance minimization in heterogeneous porous media (Chap. 5). In this section the effect of K/K_p is illustrated in Figs. 6.8 and 6.9: The branches (or tributaries) become thicker and fewer as K_p approaches K and disappear entirely when K_p is practically the same as K (Fig. 6.9).

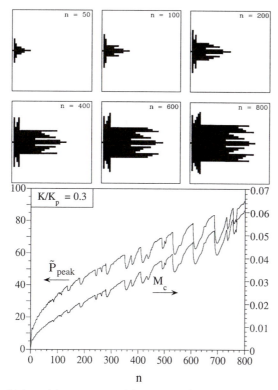

FIGURE 6.8. The evolution of the structure of the system of Fig. 6.6 when the two permeabilities are comparable ($K/K_p = 0.3$): the growth of thicker tributaries in the K_p domain and the variations in critical flow rate and peak pressure [32].

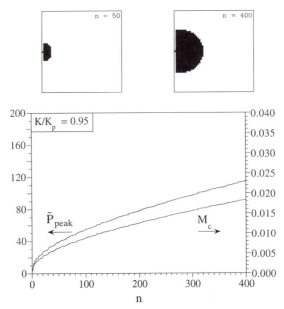

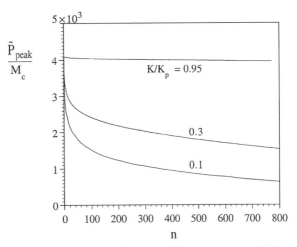

FIGURE 6.9. The disappearance of distinct channels when the two permeabilities are nearly equal ($K/K_p = 0.95$): the radially symmetric growth of the K_p domain and the monotonic rise in critical flow rate and peak pressure [32].

Taken together, Figs. 6.7–6.9 stress the observation made that the availability of two dissimilar flow regimes ($K_p \neq K$) is a necessary precondition for the formation of deterministic structures through flow-resistance minimization. The "glove" is the high-resistance regime (K), and the "hand" is the low-resistance regime (K_p): Both regimes work toward minimizing the overall resistance. Even though the $M_c(n)$ and $\tilde{P}_{\text{peak}}(n)$ functions show fluctuations that can be related to temporary changes in the internal structure, the overall flow resistance $\tilde{P}_{\text{peak}}/M_c$ decreases *monotonically* as each additional block is removed (Fig. 6.10). The decrease is

FIGURE 6.10. The monotonic decrease of the overall flow resistance during the processes documented in Figs. 6.7–6.9 [32].

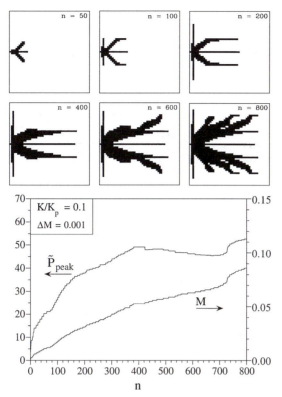

FIGURE 6.11. The evolution of the structure of the system of Fig. 6.6 when the flow rate M is increased in steps $\Delta M = 10^{-3}(K/K_p = 0.1)$ [32].

more accentuated when the vacated space is more permeable to flow than the original material.

The raggedness of the $\tilde{P}_{\text{peak}}(n)$ curves disappears when the flow-rate parameter M is increased monotonically from one step to the next (e.g., Fig. 6.11). In this new sequence, each step begins with the removal of the first block that can be dislodged by the flow rate M. Following the removal of the first block, the M value is held fixed, the pressure field is recalculated and condition (6.15) is applied again to the blocks that border the newly shaped K_p domain. The additional blocks that violate condition (6.15) are removed. To start the next step, the M value is increased by a small amount ΔM. The $M(n)$ curves shown in Figs. 6.11 and 6.12 are "stepped" because of the assumed size of ΔM and the finite number (Δn) of blocks that are removed during each step. Although the monotonic $M(n)$ curves obtained in this manner are not the same as the critical flow-rate curves $M_c(n)$ plotted in Figs. 6.7–6.9, they too are deterministic.

Figures 6.11 and 6.12 correspond to a composite porous material with $K/K_p = 0.1$, which is the same material from which the river basin of Fig. 6.7 was constructed. It is instructive to compare the shapes of the high-conductivity domains shown in these figures. The hand-in-glove structure is visible in all three figures; however, the finer details of the K_p domain depend on how the flow rate

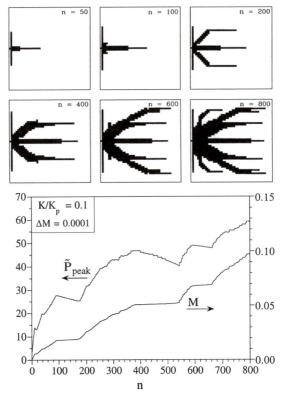

FIGURE 6.12. The increase of the flow rate in smaller steps, $\Delta M = 10^{-4}$, and the evolution of the structure of the system of Fig. 6.6 ($K/K_p = 0.1$) [32].

M is varied in time. To say that M varies in time is correct, because the removal of each new block (Fig. 6.7) or a small number of new blocks (Figs. 6.11 and 6.12) proceeds in a definite direction that can be aligned with time.

The main difference between the patterns of Fig. 6.7 and those of Figs. 6.11 and 6.12 is visible relatively early in the erosion process: Diagonal fingers form when the flow rate is increased monotonically. The size of the ΔM step also has an effect on the structure of the K_p domain. When we compare the $n = 800$ frames of Figs. 6.11 and 6.12 we see that the fingers are more regular – thinner and more numerous – when ΔM is smaller.

What is also interesting is the agreement between the patterns shown for $n \leq 100$ in Figs. 6.7 and 6.12: The agreement is due to the fact that the early portion of the $M(n)$ curve of Fig. 6.12 is practically the same as the critical curve $M_c(n)$ of Fig. 6.7. For this agreement to become visible, it is necessary to use a small enough ΔM such that the smoothness of the $M(n)$ curve matches the one-block smoothness of the $M_c(n)$ curve.

In conclusion, the details of the internal structure of the system depend on the external "forcing" that drives it, in our case the function $M(n)$. This sensitivity is illustrated by Figs. 6.7, 6.11, and 6.12, which are three responses exhibited by the same square system ($K/K_p = 0.1$). The structure is deterministic, because it is known when the function $M(n)$ is known.

6.5 River Basins with Randomly Distributed Resistance to Erosion

There are still major differences between natural drainage structures (e.g., Fig. 1.2) and the deterministic structures produced in Section 6.4. One obvious difference is the lack of axial symmetry in the trees of Fig. 1.2. How do we reconcile the lack of symmetry and unpredictability of the finer details of a natural pattern with the deterministic resistance-minimization mechanism that led us to the tree networks of Figs. 6.7–6.12?

One way is to recognize that the developing internal structure depends on two entirely different concepts: the generating mechanism, which is deterministic, and the properties of the natural flow medium, which are not known accurately and at every point [32]. To illustrate how the resistance-minimization mechanism can lead to irregular tree networks, let us assume that the resistance $(\partial \tilde{P}/\partial \tilde{s})$ that characterizes each removable block is distributed randomly over the basin area. This characteristic of river beds is well known in the field of river morphology.

In Fig. 6.13 we assume for $(\partial \tilde{P}/\partial \tilde{s})$ a normal distribution with the mean equal to 1, a standard deviation equal to 0.66, and a variance equal to 0.00425. The gray-scale code attached to the square mosaic shows that the darker blocks are easier to dislodge than the lighter ones. For the erosion process we chose the system ($K/K_p = 0.1$) and the $M(n)$ function of Fig. 6.11, in which M increased monotonically in steps of 0.001. The evolution of the drainage system is shown in Fig. 6.14. The emerging tree network is considerably less regular and symmetric than in Fig. 6.11, and reminds us more of Fig. 1.2 (with the colors inverted). The unpredictability of this pattern, however, is due to the unknown spatial distribution of system properties, not to the resistance-minimization principle. The principle is known.

6.6 River Basins with Optimized External Shape

In the study of the first tree structures (Figs. 6.7, 6.8, 6.11, and 6.12) we saw that the K_p branches propagate into the K medium faster and deeper when K_p is considerably larger than K. Simulations performed [32] for $K/K_p = 0.01$ and $K/K_p = 0.001$ strengthen this observation. The propagation of a single branch into a subdomain containing K medium can also be interpreted as the optimization of the subdomain shape so that its own area-to-point resistance is minimal. According to this interpretation, each subdomain must be slender (with the K_p branch as its axis) when the ratio K/K_p is small.

This alternative view suggests both a test and an additional way to minimize area-to-point resistance in the flow system studied until now. The new feature that we add to the model of Fig. 6.6 is the variable shape of the overall domain, which

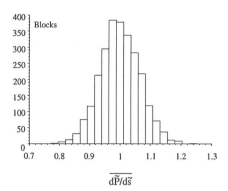

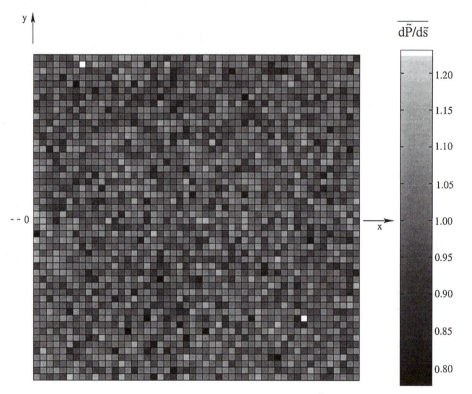

FIGURE 6.13. Basin with random resistance to erosion, when $\partial \tilde{P}/\partial \tilde{s}$ follows a normal distribution [32].

is represented by the aspect ratio H/L. The total size of the domain continues to be constrained, $HL = $ constant. We continue to rely on the block size D employed until now; therefore the dimensionless size constraint is $\tilde{H}\tilde{L} = 51^2 = 2601$. The flows described in Section 6.4 constitute a special case (the shape-constrained case $H/L = 1$) of the more general class of flows considered in this section.

The primary objective is to demonstrate that the overall shape of a collecting basin can be optimized for minimal flow resistance, as we have found for simpler systems in Section 5.2. Consider the composite medium $K/K_p = 0.1$ and assume

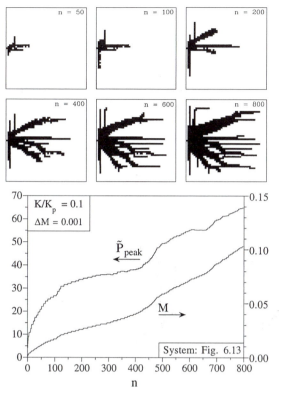

FIGURE 6.14. The development of structure in the random-resistance system of Fig. 6.13, when $K/K_p = 0.1$ and M is increased in steps of 0.001 [32].

that M increases monotonically in steps of size $\Delta M = 0.001$. The numerical procedure is the same as in the development of Fig. 6.11, except that the shape H/L is optimized after each block (or set of blocks) is removed; see condition (6.15). During each such step the shape H/L is varied from 4 to 1/4. The numerical results show that an optimal shape exists such that $\tilde{P}_{peak}$ is minimal. That shape is recorded as $(H/L)_{opt}$, and its evolution is displayed in Fig. 6.15.

The case documented in Fig. 6.15 ($K/K_p = 0.1$, $\Delta M = 0.001$) can be compared with Fig. 6.11, in which the same system was optimized for minimum resistance in a constrained outer shape ($H/L = 1$). The dendritic patterns are very similar, and their evolutions in time (n) are comparable. The main difference is that the optimal shape of the $H \times L$ domain is closely represented by the number $(H/L)_{opt} = 2$ and that this aspect ratio is relatively insensitive to the stage (n) in the evolution of the system. Recall that similar invariants were encountered in the minimization of global resistance in heterogeneous conductive media (Section 4.4) and heterogeneous Darcy-flow porous media (Sections 5.2–5.6). A direct consequence of making H/L a degree of freedom is that the system with optimized external shape (Fig. 6.15) has a flow resistance $\tilde{P}_{peak}/M$ that is consistently smaller (by 10%–20%) than the resistance in the square domain (Fig. 6.11) at the same stage (n).

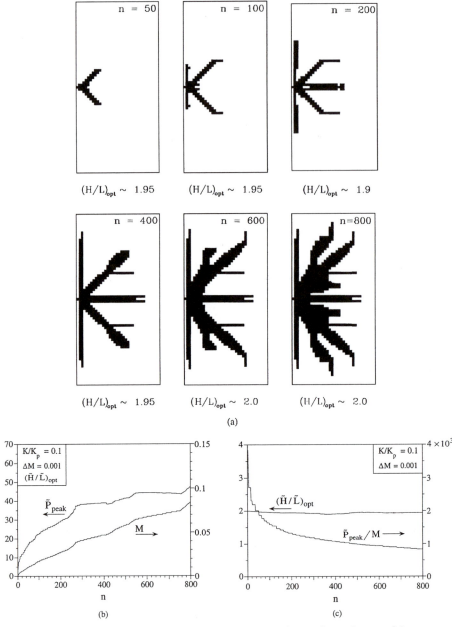

FIGURE 6.15. The optimization of the external shape H/L during the evolution of the structure of the system of Fig. 6.6 when M increases in steps of 0.001 ($K/K_p = 0.1$) [32].

These observations are reinforced by the evolution documented in Fig. 6.16 ($K/K_p = 0.01$, $\Delta M = 0.001$), which corresponds to a system with considerably less resistance through the dendritic channels than that in Fig. 6.15. The channels develop much faster than those in Fig. 6.15, and they grow mainly along the diagonals. When they reach the boundaries ($n \cong 200$) the peak pressure difference and the flow rate experience jumps to almost twice their magnitude.

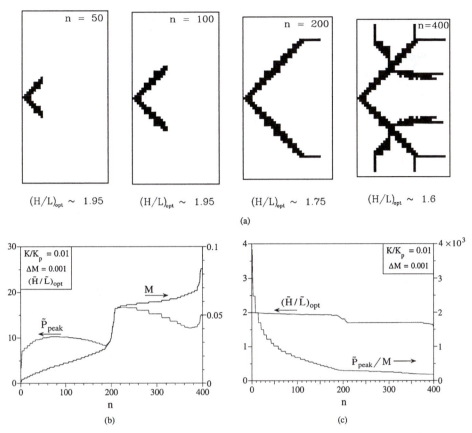

FIGURE 6.16. Channels with considerably less resistance ($K/K_p = 0.01$) and their effect on the evolution of the internal structure when the external shape is optimized and M increases in steps of 0.001 [32].

The pattern continues to develop beyond $n \cong 200$ as if the process that preceded $n \cong 200$ is repeated twice, simultaneously, around the upper and lower diagonals. When the new channels reach the outer boundaries of the domain ($n \cong 400$) the peak pressure difference and flow rate again jump to higher levels. The optimized aspect ratio $(H/L)_{opt}$ of the domain is almost equal to 2 before $n \cong 200$ and drops relatively abruptly to 1.75 in the range $200 \leq n \leq 400$. In spite of these abrupt changes, the global resistance $\tilde{P}_{peak}/M$ decreases monotonically as n increases.

Figure 6.17 is an alternative view of the structure of the system optimized in Fig. 6.16 at $n = 400$. The streaklines that originate from the center of each of the $D \times D$ building blocks of the system are plotted in a centrosymmetric montage consisting of the frame of Fig. 6.16 and its mirror image. The right half of Fig. 6.17 should be compared with the $n = 400$ frame of Fig. 6.16 because it shows, in addition to the K_p paths, the flow through the low-permeability domain K. The streaklines allow us to discover that the K flow divides itself into clearly delineated subsystems that are serviced by a single, central K_p path. This feature was also encountered in constructal tree networks of Chaps. 4 and 5.

FIGURE 6.17. Streaklines at time $n = 400$ inside the system optimized in Fig. 6.16 [32] (see Plate VIII).

6.7 Constructal Fluid Trees are Robust

In Sections 6.4–6.6 we used a simple two-dimensional erosion model with Darcy flow to demonstrate that dendritic patterns of low-resistance channels can be anticipated based on a single principle that is invoked at every step. The generating principle was the same as that in the preceding material: the constructal principle, i.e., the constrained minimization of global resistance in area-to-point flow, when the internal and external geometry of the system can be changed. The model of Section 6.4 allowed us to contemplate choices between paths with local high resistance (K) and paths with local low resistance (K_p). The internal structure emerged as a deterministic feature – a result – when the choices were made consistently toward the minimization of global resistance.

Here and in Ref. 32 many cases were simulated numerically and exhibited a larger than usual number of flow patterns. This was done intentionally, even though the large number of images (which are not identical) tends to obscure the resistance-minimization principle that generated them all. The intent was to show that the dendritic patterns are generated by a global optimization principle of the type suspected by Kadanoff [26] and Nottale [34]. We were able to rationalize the unpredictability (lack of repeatability) of the fine details of a natural internal flow structure on the basis of the unknown local properties of the heterogeneous medium (e.g., Figs. 1.2 and 6.14). The local details of the structure are sensitive to the local variations in the properties of the medium.

Even when the properties are uniform, the details of the internal flow structure are sensitive to how the stream is forced to flow through the system. In the cases documented here, the external forcing was represented by the total flow-rate parameter M. The flow rate increased from zero to a critical value at every step (e.g., Fig. 6.7), or increased monotonically from step to step (e.g., Fig. 6.11). In the latter, the structural details were also sensitive to the step size ΔM, as shown by the patterns of Figs. 6.11 and 6.12.

The overall flow resistance is remarkably insensitive to how the flow is forced, after it has been minimized at every step. For example, it can be verified [32] that the minimized-resistance curves corresponding to the monotonically increasing M programs of Figs. 6.11 and 6.12 are practically the same as the curve reported for $K/K_p = 0.1$ in Fig. 6.10, where M was increased from 0 to M_c at every step. This high degree of repeatability (or robustness) with regard to optimized global performance was also encountered in heat conduction and fluid trees through heterogeneous media (Chaps. 4 and 5).

In conclusion, if minimum global resistance is the principle that counts, then the finest details of the resulting flow path are not important: They may vary according to unknown, incidental local factors that can be labeled chance. The larger picture – the optimized global performance, the flow structure type (e.g., channels versus no channels) – can be anticipated in purely deterministic fashion; that is, if the constructal principle is recognized as law [22].

This conclusion finds additional support in the extreme experiment documented in Fig. 6.18. The system is the same as that in Fig. 6.6, with the constrained external square shape. This time we make no assumptions concerning the erosion characteristics of the medium and the manner in which the flow rate varies between changes

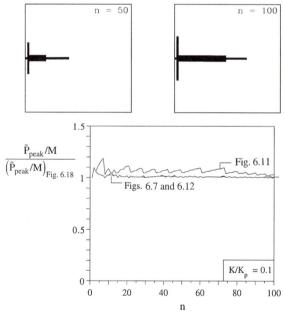

FIGURE 6.18. The evolution of the system of Fig. 6.6 when M is held constant and, at each step, the removed block is the one that leads to the largest drop in the overall resistance [32].

in structure. The flow-rate parameter M is held constant. At every step (n) we examine every block that touches the low-resistance (K_p) domain and calculate the new flow resistance that results when each block is removed alone. We select the block whose removal leads to the largest drop in the overall resistance [32].

The start of this process is simple. The first block to be removed is the one that touches the origin in Fig. 6.6. Next, we have a choice among the three blocks that border the newly created K_p space of size $D \times D$: The removal of the block centered on the x axis yields the largest drop in flow resistance. Numerically, it is necessary to determine the flow in the entire domain for every single block that is a candidate for removal. At advanced stages in the development of the structure (large n) this sequence becomes prohibitively slow: In Fig. 6.18 the sequence progressed only as far as $n = 100$, because, at this last step alone, the flow had to be simulated 42 times.

Two features of Fig. 6.18 are worth noting. First, the internal flow structure is almost the same as that in Fig. 6.7, where the K/K_p ratio was the same. Second, the global resistance of the structure optimized in Fig. 6.18 is, at every step, practically the same as in the corresponding systems optimized in Figs. 6.7, 6.11, and 6.12. Once again, we see that the resistance-minimization principle delivers the optimized global performance and a broad-brush description of the flow structure.

6.8 Rivers of People

We began our study of tree patterns for fluid flow by simulating the generation of a river basin in coffee sediment (Fig. 1.2). In this chapter we ended with computer simulations of dendritic channels formed in an erodable porous medium. Erosion creates the much-needed second flow regime – the path of lower resistance – that enters in a marriage with the original high-resistance flow and generates flow structure.

River basins are natural structures. The way in which we deduced river structures from constructal principle can be repeated when we contemplate other natural flows. An example from the animate world is the movement of individuals in a crowd that pushes toward a single opening – the entrance to the theater or the stadium or the exit from a cinema. We all know that in a tight crowd each individual's walk toward the gate is not in a straight line. The flow of people is not in a quasi-radial pattern, as in diffusion from an area to a point sink.

The walk is not straight because it is much easier to step into the space just vacated by your neighbor, preferably the neighbor who is in front of you or almost in front. You find the path of least resistance by following the individual or individuals who are already moving. Almost imperceptible rivers of people appear through the crowd that aims for the same goal. The crowd is the river basin. The space vacated by the person near or ahead of you is the eroded river bed – the path of low resistance. The conga lines formed by those who move faster through the crowd are the river branches.

Most of the people fill the spaces between these streams. They march in place, shifting their weight from one foot to the other, waiting for an opening, hoping to be carried away by a stream. They are unwitting analogs of the disorganized molecular motion that accounts for diffusion – the high-resistance mode that filled the interstices of the fluid trees discussed in this book.

What we do in a tight crowd in front of the theater entrance we do everywhere in society. With so many of us pursuing the same goals, it is natural to look for the best way, the shortest route, and the least costly methods to reach our goals. In this process we form associations, if not rivers of people, then railroads, highways, airways, and telecommunication networks. Our own movement and the flow of goods and money describe patterns on the Earth's surface.

The economics of location or spatial (geographic) economics is a wide arena in which the constructal principle has been operating from the beginnings of mankind and other forms of social animal behavior. These applications of constructal theory are explored in Section 9.6 and Chap. 11.

PROBLEMS

6.1 Two identical open-channel streams of width D_2, length L_2, and flow rate $\dot{m}/2$ continue downhill as a single stream $(D_1, L_1, \dot{m})$ [35]. The total drop in elevation between the inlets and the outlet of the three-channel assembly is given. Also fixed are the size of the territory occupied by the assembly (the area $2L_1 L_2$) and the total volume of the three channels. Assume that the shape of each channel cross section has been optimized for minimum flow resistance. In addition, assume that each channel flow is in the fully turbulent and fully rough regime, so that the friction factor is constant (flow-rate independent) and has the same value for all the channels. Minimize the overall flow resistance of the assembly and show that the optimal architecture is characterized by $(D_1/D_2)_{\text{opt}} = 2^{3/7}$, $(L_1/L_2)_{\text{opt}} = 2^{1/7}$, and a junction elevation situated halfway between the elevations of the inlets and the outlet.

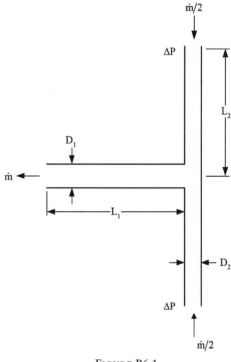

FIGURE P6.1

6.2 Consider the proposal to determine the optimal shape of a duct [36] that must carry a certain stream ($\dot{m}$) between two pressure levels (P_0 and $P_L = P_0 - \Delta P$) separated by a fixed distance (L). The objective is to minimize the overall flow resistance $\Delta P/\dot{m}$. The duct is straight with the flow pointing in the x direction; however, the cross-sectional area $A(x)$ and wetted perimeter $p(x)$ may vary with the longitudinal position. Two constraints must be taken into account. One is the total duct-volume constraint,

$$\int_0^L A(x)\,dx = V, \qquad \text{constant.}$$

The volume constraint is important in the design of compact heat exchangers. Another possible constraint refers to the total amount of duct wall material used,

$$\int_0^L p(x)\,dx = M, \qquad \text{constant.}$$

The material constraint is crucial in designs in which the unit cost of the material is high or in which the weight of the overall heat exchanger is constrained, as in aerospace applications. We can calculate the overall flow resistance by noting that regardless of whether the flow is laminar, turbulent, or fully developed, the pressure gradient is given by [16]

$$\frac{dP}{dx} = -\tau_w(x)\frac{p(x)}{A(x)},$$

where τ_w is the wall shear stress. To calculate the global resistance to flow, we substitute the friction factor $f = \tau_w/(\frac{1}{2}\rho U^2)$, the Reynolds number $\mathrm{Re}_{D_h} = D_h U/\nu$, and the mass flow rate $\dot{m} = \rho U A$, and integrate from the inlet to the outlet:

$$\frac{\Delta P}{\dot{m}} = \int_0^L f\,\mathrm{Re}_{D_h}\frac{p\nu}{2\,A^2\,D_h}\,dx.$$

Consider as an example the hydraulic entrance region to a duct of round cross section [diameter $D(x)$] through which the flow is sufficiently isothermal such that ν may be regarded as constant. Show that when the overall resistance is minimized subject to the volume constraint, the optimal diameter $D(x)$ varies as $x^{-1/12}$, which means that the duct is shaped like a trumpet. Show also that when the optimization is subjected to the wall material constraint, the optimal duct shape varies as $D \sim x^{-1/10}$. Derive the corresponding shapes for a duct with the cross section shaped like a very flat rectangle of spacing $D(x)$ and width W, such as $D(x) \ll W$.

6.3 Water is being distributed to a large number of users by using a round duct of length L and inner diameter $D(x)$, where x is the longitudinal coordinate. Each user consumes the same fraction of the initial flow rate of the water stream. The users are distributed equidistantly along the duct. The initial flow rate is $\dot{m}$, and enters the pipe through the $x = L$ end. The last user served by the stream is located near the $x = 0$ end. Assume that the users are sufficiently numerous, and their individual demands sufficiently small, so that the flow rate varies continuously along the duct, $\dot{m}(x)$. The water flow is in the fully turbulent regime, with a friction factor that is independent of flow rate (the fully rough limit). Find the optimal duct shape $D(x)$ that minimizes the total power required to pump the stream through the entire system. Assume that the resistance to water flow is due mainly to the flow through the duct. Assume also that the duct wall thickness is small in comparison with D, and that the total amount of duct wall material is fixed.

REFERENCES

1. L. B. Leopold, M. G. Wolman, and J. P. Miller, *Fluvial Processes in Geomorphology*, Freeman, San Francisco, 1964.

2. A. E. Scheidegger, *Theoretical Geomorphology*, 2nd ed., Springer-Verlag, Berlin, 1970.

3. A. L. Bloom, *Geomorphology*, Prentice-Hall, Englewood Cliffs, NJ, 1978.

4. R. J. Chorley, S. A. Schumm, and D. E. Sugden, *Geomorphology*, Methuen, London, 1984.

5. M. Morisawa, *Rivers. Form and Process*, Longman, London, 1985.

6. S. A. Schumm, M. P. Mosley, and W. E. Weaver, *Experimental Fluvial Geomorphology*, Wiley, New York, 1987.

7. A. Bejan, On the buckling property of inviscid jets and the origin of turbulence, *Lett. Heat Mass Transfer*, Vol. 8, No. 3, 1981, pp. 187–194.

8. A. Bejan, *Entropy Generation through Heat and Fluid Flow*, Wiley, New York, 1982, Chap. 4.

9. A. Bejan, Theoretical explanation for the incipient formation of meanders in straight rivers, *Geophys. Res. Lett.*, Vol. 9, 1982, pp. 831–834.

10. M. A. Gorycki, Hydraulic drag: a meander initiating mechanism, *Geol. Soc. Am. Bull.*, Vol. 84, 1973, pp. 175–186.

11. L. B. Leopold and M. G. Wolman, River channel patterns, in *Rivers and River Terraces*, G. H. Dury, ed., Praeger, New York, 1970.

12. G. H. Dury, Principles of underfit streams, U.S. Geological Survey Prof. Paper 452-A, 1964.

13. S. A. Schumm and H. R. Khan, Experimental study of channel patterns, *Geol. Soc. Am. Bull.*, Vol. 83, 1972, pp. 1755–1770.

14. E. A. Keller, Development of alluvial stream channels: a five stage model, *Geol. Soc. Am. Bull.*, Vol. 83, 1972, pp. 1531–1536.

15. L. B. Leopold and M. G. Wolman, River meanders, *Geol. Soc. Am. Bull.*, Vol. 71, 1960, pp. 769–794.

16. A. Bejan, *Convection Heat Transfer*, 2nd ed., Wiley, New York, 1995.

17. I. Rodriguez-Iturbe, A. Rinaldo, R. Rigon, R. L. Bras, A. Marani, and E. Ijjasz-Vasquez, Energy dissipation, runoff production, and the three-dimensional structure of river basins, *Water Resour. Res.*, Vol. 28, 1992, pp. 1095–1103.

18. T. Sun, P. Meakin, and T. Jossang, Minimum energy dissipation model for river basin geometry, *Phys. Rev. E*, Vol. 49, 1994, pp. 4865–4872.

19. A. Maritan, F. Colaiori, A. Flammini, M. Cieplak, and J. R. Banavar, Universality classes of optimal channel networks, *Science*, Vol. 272, 1996, pp. 984–986.

20. I. Rodriguez-Iturbe and A. Rinaldo, *Fractal River Basins*, Cambridge Univ. Press, Cambridge, UK, 1997.

21. P. Meakin, *Fractals, Scaling and Growth Far from Equilibrium*, Cambridge Univ. Press, Cambridge, UK, 1998.

22. A. Bejan, *Advanced Engineering Thermodynamics*, 2nd ed., Wiley, New York, 1997.

23. B. B. Mandelbrot, *The Fractal Geometry of Nature*, Freeman, New York, 1982.

24. D. Avnir, O. Biham, D. Lidar, and O. Malcai, Is the geometry of nature fractal? *Science*, Vol. 279, 1998, pp. 39–40.

25. O. R. Shenker, Fractal geometry is not the geometry of nature, *Stud. Hist. Philos. Sci.*, Vol. 25, No. 6, 1994, pp. 967–981.

26. L. P. Kadanoff, Fractals: where's the physics?, *Phys. Today*, September 1986, pp. 11–12.

27. J.-D. Chen, Radial viscous fingering patterns in Hele–Shaw cells, *Exp. Fluids*, Vol. 5, 1987, pp. 363–371.

28. E. Ben-Jacob, O. Shochet, I. Cohen, and A. Tenenbaum, Cooperative strategies in formation of complex bacterial patterns, *Fractals*, Vol. 3, 1995, pp. 849–868.

29. D. S. Lemons, *Perfect Form*, Princeton Univ. Press, Princeton, NJ, 1997.

30. A. Bejan, G. Tsatsaronis, and M. Moran, *Thermal Design and Optimization*, Wiley, New York, 1996.

31. S. Vogel, *Life's Devices*, Princeton Univ. Press, Princeton, NJ, 1988.

32. M. R. Errera and A. Bejan, Deterministic tree networks for river drainage basins, *Fractals*, Vol. 6, 1998, pp. 245–261.

33. D. A. Nield and A. Bejan, *Convection in Porous Media*, 2nd ed., Springer, New York, 1999.

34. L. Nottale, *Fractal Space–Time and Microphysics*, World Scientific, Singapore, 1993.

35. A. Bejan, L. A. O. Rocha, and S. Lorente, Thermodynamic optimization of geometry: T- and Y-shaped constructs of fluid streams, *International Journal of Thermal Sciences*, Vol. 40, 2001, to appear.

36. P. Jany and A. Bejan, Ernst Schmidt's approach to fin optimization: an extension to fins with variable conductivity and the design of ducts for fluid flow, *Int. J. Heat Mass Transfer*, Vol. 31, 1988, pp. 1635–1644.

CHAPTER SEVEN

TURBULENT STRUCTURE

7.1 Two Flow Regimes: High Resistance and Low Resistance, Intertwined

In every example of flow with structure that we have discussed so far we have seen that macroscopic organization becomes visible only when the flow can exist in at least *two regimes*, not one. One regime must offer a flow resistance that is significantly different from the resistance of the other regime. In Chap. 4 the two regimes were high-resistance conduction through the k_0 material and low-resistance conduction through the k_p material. In Chaps. 5 and 6 the high-resistance regime was viscous diffusion that touched every point, and the low resistance was provided by channels of higher Darcy permeability or by open fissures and ducts with Hagen–Poiseuille flow.

What we recognize as macroscopic structure is none other than the geometric arrangement of the spatial domain of one flow regime relative to (or through) the territory occupied by the other. The structure is not visible – it is not even an issue – when only one flow regime is present. In such cases the lone regime is the one with the highest resistance. Note that any of the volume-to-point flows treated in Chaps. 4–6 can also be effected solely by molecular diffusion, namely, thermal diffusion through a single material (k_0) and Darcy flow through a single porous medium (K_0). Such flows are disorganized microscopically and *shapeless* macroscopically. They become visible (as background) only when a second flow regime emerges (streams, channels). The latter represents organized motion, or *flow with shape*.

Turbulent flows are well known for their ability of combining and distributing through space two regimes, viscous diffusion and streams (eddies). This is why the existence of turbulent flow can be reasoned on the purely theoretical basis provided by the constructal law. In this chapter we continue the theoretical line of Section 4.8 and show that the flow structure called eddy is made necessary by the constructal law: Internal structure must be generated in order to speed up the approach to the state of internal uniformity.

149

7.2 Why Do Icebergs and Logs Drift Sideways?

Flying over the North Atlantic in winter, we see icebergs drifting in the wind. Tabular icebergs are relatively elongated and always drift sideways. Their short dimension aligns itself with the direction of the wind. The same observation holds for other relief forms of the water surface: wind-generated waves, tree logs, and abandoned ships. They all drift sideways. Why?

The sea water and air layer that moves along it may be regarded as a thermodynamic system that is in a state of internal nonuniformity (Fig. 7.1). In the simplest sense, the nonuniformity is due solely to the relative motion between the two parts of the system. These parts are represented by subsystems (a) and (b) in Fig. 7.1. The sea water is assumed stationary ($U_b = 0$), so that the air speed U_a is also the relative speed between the two fluid masses.

The second law of thermodynamics assures us that after a sufficiently long time the velocity gradients will disappear and the entire system will move in the same direction with the same speed. The constructal law assures us that the system will generate structure (geometry, configuration, arrangement) internally for the purpose of speeding up its march toward the state of internal uniformity. In Fig. 7.1 the only geometric degree of freedom available to the system is the orientation of the protuberances present at the interface between the moving parts. The protuberance is a floating object of length L and diameter D. Region (a) pushes this object horizontally with the drag force F. The object transmits the same force into region (b), and in this way it effects the transfer of momentum downward, from (a) to (b). The larger the force F, the shorter the time needed for the difference between U_a and U_b to disappear.

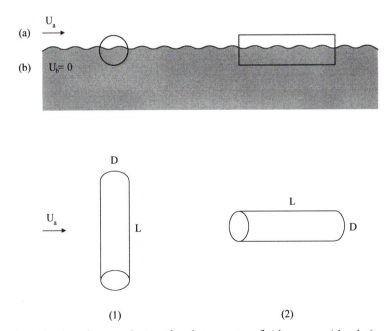

FIGURE 7.1. Floating object at the interface between two fluid masses with relative motion.

The floating object acts as a lock-and-key mechanism between (a) and (b), and the constructal law is the principle based on which this lock is designed. The floating object may have the special orientations (1) and (2) in Fig. 7.1 and many other positions in between. The orders of magnitude of the drag forces in the two extreme cases are

$$F_1 \sim LDC_D\frac{1}{2}\rho_a U_a^2, \qquad F_2 \sim D^2 C_D\frac{1}{2}\rho_a U_a^2, \tag{7.1}$$

where ρ_a is the density of the upper fluid. The drag coefficient C_D is a number of the order of 1 when the Reynolds number $(U_a D/\nu_a)$ is of the order of 10^2 or larger. It is clear that the floating object is most effective as a downward path for horizontal momentum in configuration (1) of Fig. 7.1 because

$$F_1 > F_2 \qquad (L > D). \tag{7.2}$$

We reach the same conclusion if we compare the effective shear stresses at the interface, $\tau_1 = F_1/DL$ and $\tau_2 = F_2/DL$, namely

$$\tau_1 > \tau_2 \qquad (L > D). \tag{7.3}$$

In summary, the constructal law calls for configuration (1) of Fig. 7.1, and this happens to be the configuration that nature exhibits, billions and billions of times.

7.3 The First and Smallest Eddy

In view of the success of the theoretical argument based on the constructal law and Fig. 7.1, we now invoke the same argument to predict the internal structure of an even simpler flow system. In Fig. 7.2 the two moving masses contain the same fluid, (a). Initially the upper mass moves to the right with the uniform speed U_∞, while the lower mass is stationary. Again, the second law assures us that after a sufficiently long time the two masses will acquire the same speed. The constructal law question is this: Which internal configuration will enable the system to reach internal uniformity faster?

A fluid shear layer with finite thickness (D) has the property of exhibiting two flow regimes [1]. One is the laminar regime (1), in which velocities remain nearly horizontal and momentum is being transferred downward by viscous diffusion. In

FIGURE 7.2. The two mechanisms for transferring momentum across the interface between two flow regions of the same fluid.

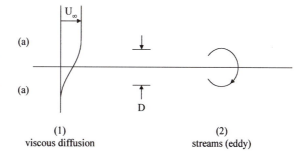

this case the effective shear stress is of the order of

$$\tau_1 \sim \mu \frac{U_\infty}{D}. \tag{7.4}$$

The other regime consists of wrinkling, mixing, and thickening the shear layer by rolling its fluid. We refer to this as regime (2). In this case the fluid flow becomes organized into a vertical counterflow (eddy) that rolls with a peripheral velocity of the order of U_∞. The mass flow rate in each stream of the counterflow is of the order of $\rho D U_\infty$. The counterflow transfers horizontal momentum in the downward direction at the rate $(\rho D U_\infty) U_\infty$. The rate of momentum transfer per unit area at the original interface, $(\rho D U_\infty) U_\infty / D \sim \tau_2$, is an effective eddy stress that connects the two parts of the system:

$$\tau_2 \sim \rho U_\infty^2. \tag{7.5}$$

Comparing approximations (7.4) and (7.5), we reason that streams are a necessary structural feature when $\tau_2 > \tau_1$, i.e., when they provide easier access to the flow of momentum between the two parts of the system. The inequality $\tau_2 > \tau_1$ and approximations (7.4) and (7.5) yield

$$\frac{U_\infty D}{\nu} \gtrsim 1 \tag{7.6}$$

as the condition for the formation of eddies at the interface. The group $U_\infty D / \nu$ is the *local Reynolds number* of the flow [1], because it is based on the local velocity scale (U_∞) and local transversal length scale (D):

$$\mathrm{Re}_l = \frac{U_\infty D}{\nu}. \tag{7.7}$$

Relation (7.6) is an important step because it accounts in purely theoretical fashion for the occurrence of the first eddy or for the occurrence of the first disturbance assumed (not predicted) by those who perform stability analyses in fluid mechanics [1]. It also accounts for the dimensionless description of the first-eddy size in terms of a transition Reynolds number. The constructal law demands eddies, but only in layers that are thick enough and (or) sheared fast enough.

The transition Reynolds number is a geometric constant that expresses the competition between configurations (1) and (2) in Fig. 7.2. The right-hand side of relation (7.6) shows O(1) only because its derivation was based on the simplest brand of scale analysis, in which we neglected all the factors of the order of 1. A more accurate estimate is obtained when such factors are retained in the analysis, as shown in the following time-minimization analysis.

Consider the time development of the shear layer [1–3] in a frame that rides at $U_\infty/2$ from left to right in Fig. 7.2. What we see is sketched in Fig. 7.3: The upper fluid moves to the right at $U_\infty/2$, and the lower fluid moves to the left with the same speed. Immediately after $t = 0$, the horizontal midplane is occupied by a one-dimensional shear layer. The instantaneous thickness of this layer (D) increases as $t^{1/2}$ and can be estimated based on the classical solution to the one-dimensional momentum diffusion problem [5], Fig. 7.4. The velocity (v) for the flow parallel

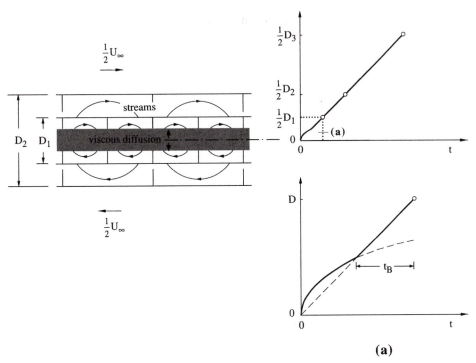

FIGURE 7.3. The expansion of the flow region in a frame that travels to the right with the velocity $U_\infty/2$ [4].

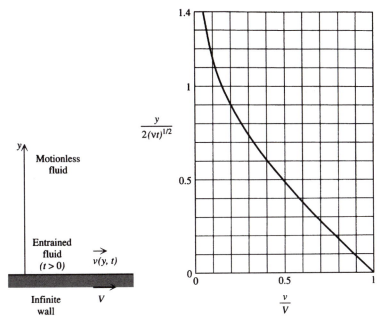

FIGURE 7.4. The velocity distribution near an infinite solid wall that moves at constant speed relative to itself in an originally motionless fluid [5].

to an infinite plane that, starting with $t = 0$, moves at constant speed (V) parallel to itself is

$$\frac{v}{V} = \text{erfc}\left[\frac{y}{2(vt)^{1/2}}\right]. \tag{7.8}$$

The "knee" of this velocity profile is located at the distance (the viscous penetration depth, or thickness) where the erfc argument is of the order of 1 (see Fig. 7.4):

$$\frac{y}{2(vt)^{1/2}} \sim 1. \tag{7.9}$$

In the notation of Figs. 7.2 and 7.3 this means that $y \sim D/2$, which in combination with approximation (7.9) yields the thickness of the layer that has been effectively mixed by viscous diffusion,

$$D \sim 4(vt)^{1/2}. \tag{7.10}$$

Approximation (7.10) has been plotted in the lower detail (a) of Fig. 7.3, which is a blow up of the upper drawing in the same figure. It shows that the viscous diffusion mechanism is effective only in the beginning when the growth rate dD/dt is large. As time increases, dD/dt decreases as $t^{-1/2}$, and the laminar growth of the mixed region slows down monotonically. This brings up the constructal question of how to optimize access: How can the mixing layer cover the space in the shortest time? Spatial growth in the frame of Fig. 7.3 means growth in the vertical direction, because the momentum transfer is unidirectional.

The alternative is the eddy formation mechanism. The transversal velocity in this case is constant, $dD/dt \sim U_\infty/2$, because $U_\infty/2$ is the eddy peripheral speed in the frame of Fig. 7.3. The spacing between two adjacent eddies is the same as the buckling wavelength [2, 3], $\lambda_B \sim 2D$, or the wavelength of neutral stability [1]. Two eddies roll into a new one of thickness $2D$ during the roll-up time:

$$t_B \sim \frac{\lambda_B}{U_\infty/2} \sim \frac{4D}{U_\infty}. \tag{7.11}$$

The rate of lateral expansion through eddy formation is $(2D - D)/t_B \sim U_\infty/4$, whereas the viscous swelling rate indicated by approximation (7.10) is $dD/dt \sim 2(v/t)^{1/2} \sim 8v/D$. The most rapid expansion occurs when the viscous diffusion regime is followed by eddy formation at the time when the viscous growth rate is just outpaced by the eddy growth rate – that is, when $DU_\infty/v \sim 32$, or, replacing approximation (7.6),

$$\frac{U_\infty D}{v} \sim 10^2. \tag{7.12}$$

This theoretical estimate of the local Reynolds number ($\text{Re}_l = U_\infty/D/v$) predicts the transition to turbulent flow in virtually all the known flows. Table 7.1 and Fig. 7.5 provide a bird's-eye view of classical observations on the laminar–turbulent transition. To calculate the corresponding Re_l value we must identify the correct local scales, U_∞ and D. Examples are discussed in Ref. 6. The table shows that the critical *local* Reynolds number of a laminar boundary-layer flow over a flat plate

TABLE 7.1. Traditional Critical Numbers for Transitions in Several Key Flows and the Corresponding Local Reynolds Number Scale [5, 6]

Flow	Traditional Critical Number	Local Reynolds Number
Boundary-layer flow over flat plate	$Re_x \sim 2 \times 10^4 - 10^6$	$Re_l \sim 94\text{-}660$
Natural convection boundary layer along vertical wall with uniform temperature ($Pr \sim 1$)	$Ra_y \sim 10^9$	$Re_l \sim 178$
Natural convection boundary layer along vertical wall with constant heat flux ($Pr \sim 1$)	$Ra_{*y} \sim 4 \times 10^{12}$	$Re_l \sim 330$
Round jet	$Re_{nozzle} \sim 30$	$Re_l \gtrsim 30$
Wake behind long cylinder in cross flow	$Re \sim 40$	$Re_l \gtrsim 40$
Pipe flow	$Re \sim 2000$	$Re_l \sim 500$
Film condensation on a vertical wall	$Re \sim 450$	$Re_l \sim 450$

is nearly the same as that of the buoyancy-driven jet along a heated vertical wall. It can be argued that the Re_l range for transition in round jet flow is actually higher than the listed nozzle Reynolds number because the laminar jet expands rapidly outside the nozzle (i.e., the D scale of the jet is larger than the nozzle diameter). The same observation applies to the transition in the wake behind a long cylinder, where the listed Reynolds number is based on the diameter of the cylinder, not on the transversal length scale of the wake.

On the high side of the transition criterion $Re_l \sim 10^2$, we note that the transition in pipe flow occurs at diameter-based Reynolds numbers of the order of 2000. The actual thickness of the centerline fiber that exhibits the sinuous motion is

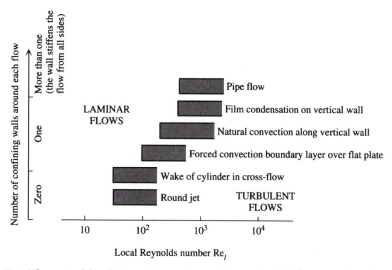

FIGURE 7.5. Theoretical local Reynolds number of the order of 10^2 as a universal transition criterion [5].

considerably smaller than the pipe diameter; therefore, the local Reynolds number is correspondingly smaller than 2000. This is why a smaller value ($Ra_l \sim 500$) is listed in the second column of the table.

Figure 7.5 reviews the local Reynolds numbers that correspond to the transitions considered in Table 7.1. One remarkable aspect of this figure is that it condenses the transition observations to a relatively narrow band of values centered around $Re_l \sim 10^2$: compare this narrow band with the 10–10^{12} range covered by the traditional critical numbers in Table 7.1.

The local Reynolds number criterion and the theory behind it have several additional consequences that can only be enumerated in these concluding paragraphs. One consequence is that the Reynolds number of the smallest eddy in the ensuing flow field is of the order of 10^2. This discovery refutes the established view in modern fluid mechanics that the smallest eddy Reynolds number is of the order of 1 [7, 8].

Another consequence is the predicted pulsating frequency of large pool fires [9],

$$f_v \sim \left(\frac{2.3 \, \text{m/s}^2}{D} \right)^{1/2}, \qquad (7.13)$$

where D is the length scale (width, diameter) of the pool (base) shape. Until 1991, correlation (7.13) was one of the "unanswered questions in fluid mechanics" [10], having been discovered and rediscovered empirically based on fire observations involving a wide variety of fuels and fire pool shapes. In addition to anticipating correlation (7.13), the theory [9] predicts correctly that the fire can pulsate only when its base dimension D is greater than approximately 2 cm.

It was also shown that Re_l criterion (7.12) predicts the existence of a viscous sublayer in turbulent flow near a wall. The thickness of this sublayer must be of the order of 10 in the classical y^+ coordinate [6]. The same theoretical starting point has been used to derive analytically Colburn's analogy between heat transfer and friction in turbulent flow near a wall, $St_x Pr^{2/3} = (1/2)C_{f,x}$, where St_x and $C_{f,x}$ are the local Stanton number and skin friction coefficient, respectively. An integral part of that analysis [6] is the proof that Colburn's analogy is valid only for fluids with Prandtl numbers of the order of 1 or greater.

In summary, the maximization of the growth rate of the mixing region accounts in a purely theoretical way for the occurrence of turbulence in a vast array of physical situations. The body of empirical evidence that supports these predictions is centuries old, voluminous, and extraordinarily well correlated and accepted.

As in the original presentation of this facet of constructal theory [1–3], we can also obtain eddy formation criterion (7.12) by equating the time scale of viscous diffusion with the time scale of eddy travel over the same distance D. The same time equality stands behind the more recent work by Mikic et al. [11] on sustained oscillations in the forced convection cooling of electronics. This time argument is one more example of the application of the principle of equipartition at the elemental level, in which the two transport mechanisms find themselves in that perfect balance that generates the elemental structure.

PLATE I. Tree and roots on the lower Danube (photograph courtesy of Teresa M. Bejan) (see Fig. 1.1).

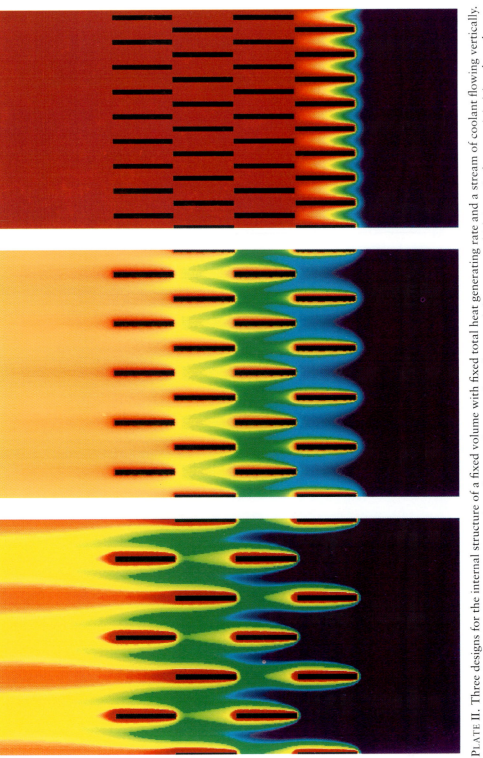

PLATE II. Three designs for the internal structure of a fixed volume with fixed total heat generating rate and a stream of coolant flowing vertically. The objective is to maximize the global thermal conductance between the volume and the stream, which is equivalent to minimizing the red areas (hot-spot temperatures) that occur inside the volume (see Fig. 3.3).

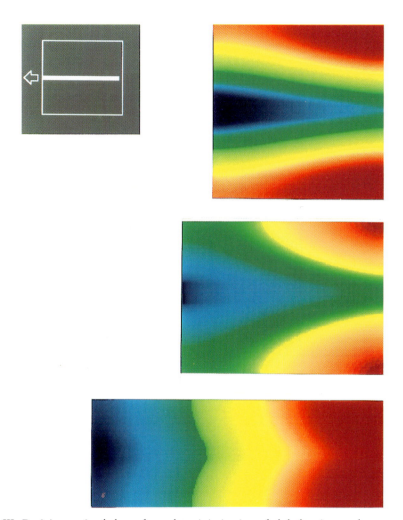

PLATE III. Deriving optimal shape from the minimization of global resistance between a volume and one point. Three competing designs are shown. The volume and the heat generation rate are fixed. The aspect ratio of the rectangular domain is variable. The resistance is proportional to the peak temperature difference, which is measured between the hot spots (red) and the heat sink (blue). The middle shape minimizes the areas covered by red and has the smallest volume–point resistance (see Fig. 4.2).

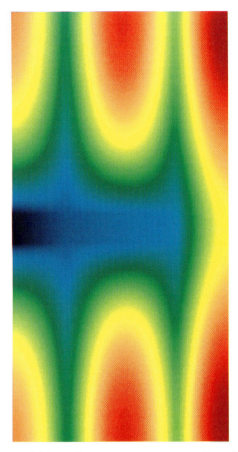

PLATE IV. The temperature field in a first construct containing four elemental volumes. Note the formation of hot spots (red) in the corners situated the farthest from the heat sink (blue) (see Fig. 4.4).

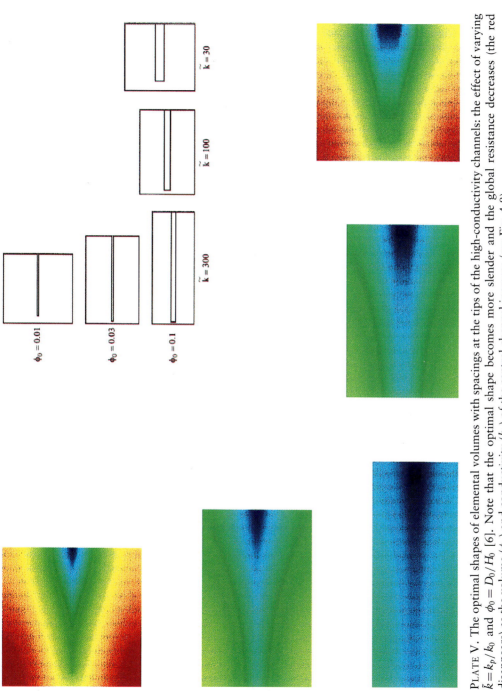

PLATE V. The optimal shapes of elemental volumes with spacings at the tips of the high-conductivity channels: the effect of varying $\tilde{k} = k_p/k_0$ and $\phi_0 = D_0/H_0$ [6]. Note that the optimal shape becomes more slender and the global resistance decreases (the red disappears) as the volume (ϕ_0) and conductivity (k_p) of the central channel increase (see Fig. 4.8).

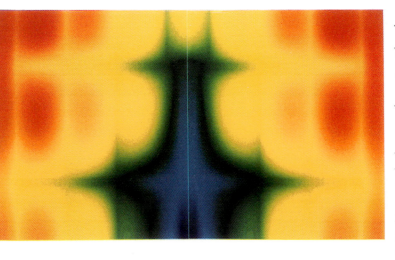

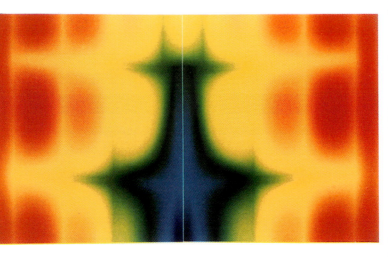

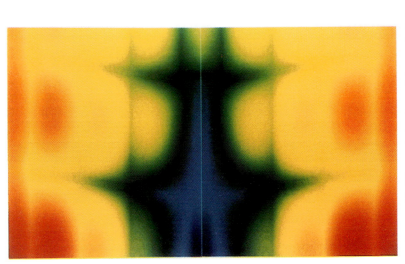

PLATE VI. The optimization of the angle between the first-construct stems and the stem of a second construct. In this example the second construct has four first constructs, and each first construct has eight elemental volumes. When the stems are perpendicular ($\alpha = 0°$, the third frame) the hot spots are concentrated in the farthest corners relative to the heat sink (blue). When the angle is too large (the first frame), the hot spots jump to the corners that are on the same side as the heat sink. In the optimal configuration (the second frame), the hot spots are spread most uniformly around the periphery of the system. More points work the hardest in this configuration. We pursue this idea further in the shapes of constant resistance constructed in Chap. 12 (see Fig. 4.10).

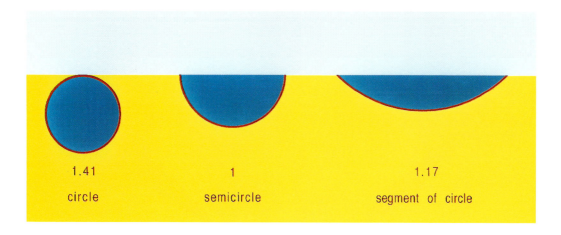

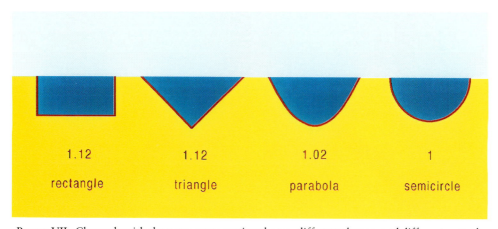

PLATE VII. Channels with the same cross-sectional area, different shapes, and different wetted perimeters. The top row shows three examples with bottoms shaped as arcs of circles. The number under each drawing represents the flow resistance relative to the resistance of the channel with semicircular cross section. The four cross sections shown on the bottom row have had their depth/width ratios optimized for minimum resistance. Note that the resistance does not vary appreciably from one shape to the next. Channel cross-sectional designs are robust (see Fig. 6.5).

PLATE VIII. Streaklines at time $n = 400$ inside the system optimized in Fig. 6.16 [32] (see Fig. 6.17).

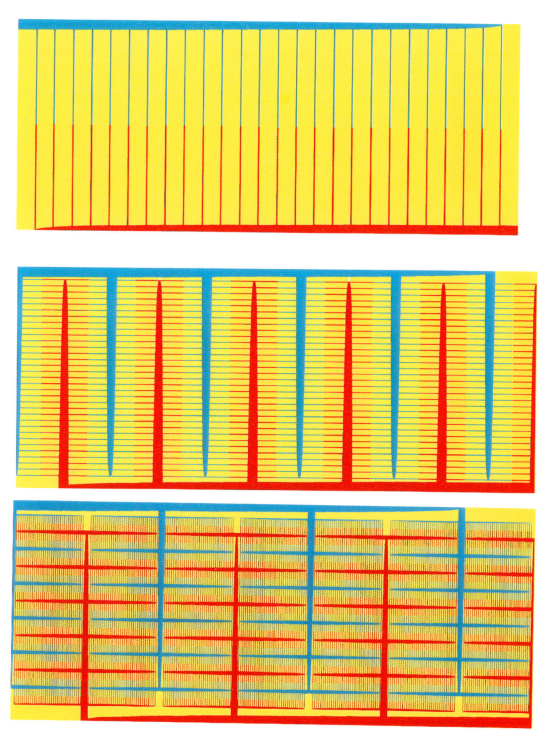

PLATE IX. Color displays of inflowing (blue) and outflowing (red) streams in the first construct (top), second construct (middle), and third construct (bottom) [19] (see Fig. 8.24).

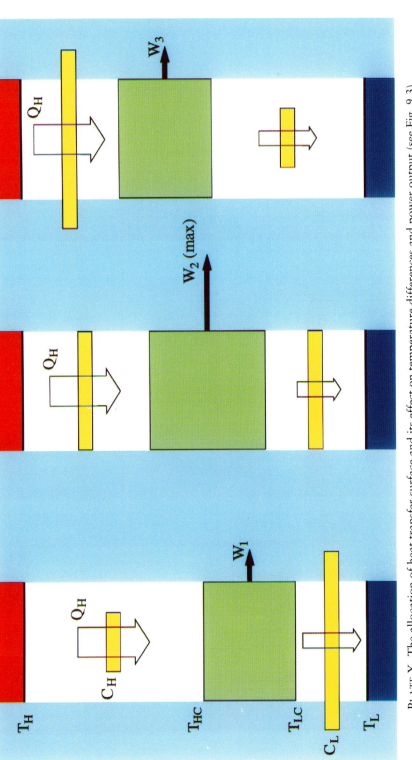

PLATE X. The allocation of heat transfer surface and its effect on temperature differences and power output (see Fig. 9.3).

PLATE XI. Flock of pelicans flying overhead at Phinda, KwaZulu-Natal, South Africa (photograph courtesy of William A. Bejan) (see Fig. 9.14).

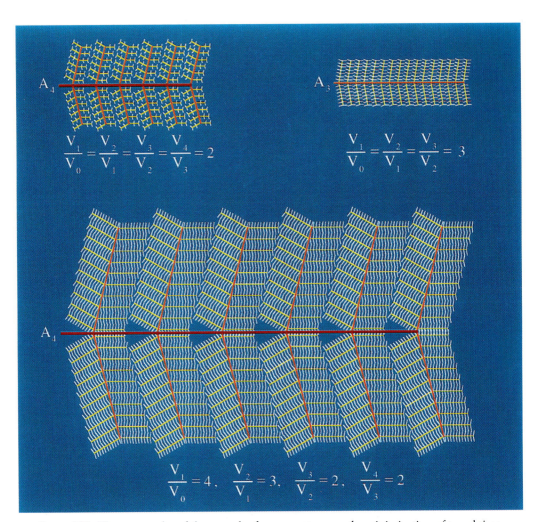

PLATE XII. Three examples of the growth of street patterns as the minimization of travel time between a finite-size area and one point. Velocities increase as the constructs become larger. Each construct has been optimized twice, for shape and angle [2] (see Fig. 11.3).

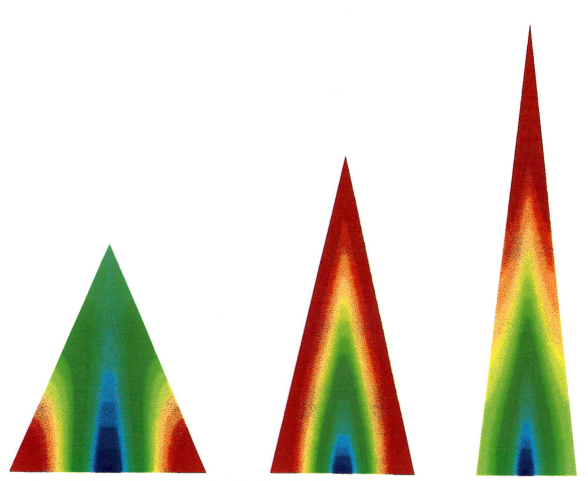

PLATE XIII. The color red shows the migration of the hot spots as the aspect ratio of the triangle changes. There is an optimal, intermediate slenderness that marks the moment when the hot spot jumps from the tip to the two base corners. In this design the entire long sides of the triangle are at the hot-spot temperature (see Fig. 12.2).

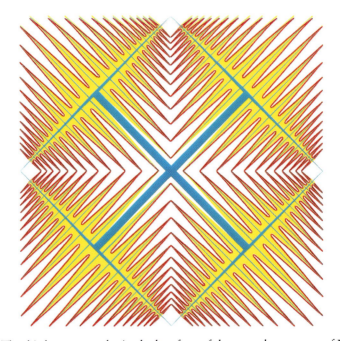

PLATE XIV. The third construct obtained when four of the second constructs of Fig. 12.5 are joined (right) [3]. Red indicates the boundaries with hot-spot temperature. Blue indicates the high-conductivity inserts. This construct should be compared with Fig. 6.17, which shows the river drainage basin derived from a resistance-minimization erosion process (see Fig. 12.8).

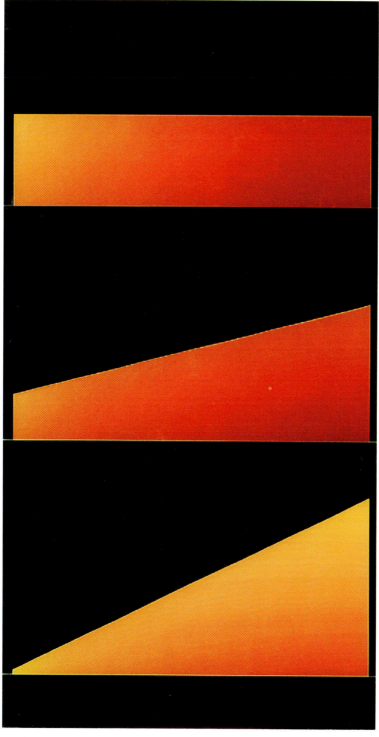

PLATE XV. The temperature field in a plate fin with adjustable crest inclination [8] (see Fig. 12.10).

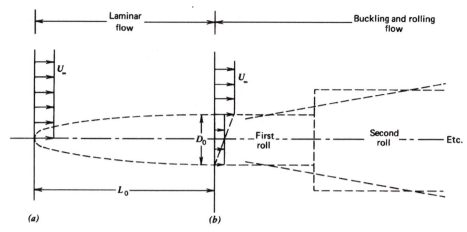

FIGURE 7.6. Laminar tip (elemental volume) in the beginning of every shear layer [2].

7.4 The Stepwise Growth of Mixing Regions

In Fig. 7.3 the size of the first eddy is labeled D_1. In the preceding analysis this length scale was $2D$. The order of magnitude of D_1 is given by local Reynolds number criterion (7.12). Beyond the first-eddy event, the flow continues to expand in steps by rolling and forming eddies. Each step leads to the doubling of the mixing region (cf. $\lambda_B \sim 2D$). Viscous diffusion does not have time to act – to compete – over larger distances such as D_2, D_3, and so on [1–3]. The outer boundaries of the mixing region are extended now by the second flow regime: streams (eddies). The analogy with the other structures deduced based on constructal theory in this book is complete.

Figures 7.6 and 7.7 display the graphic information of Fig. 7.3 in the frame of reference attached to the lower fluid region. The laminar tip of the shear layer is the elemental system that repeats itself many times in the downstream direction. This is how the flow (shearing effect) is able to cover every point of the volume called the mixing region.

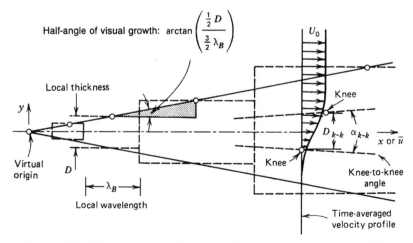

FIGURE 7.7. The constant-angle growth of turbulent mixing regions [1].

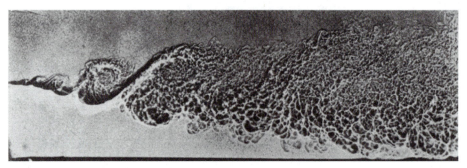

FIGURE 7.8. The discrete, stepwise growth of a turbulent shear layer (upper side: helium, 9.15 m/s; lower side: nitrogen, 0.33 m/s; pressure = 7 atm). (Reprinted with permission from G. L. Brown and A. Roshko, *J. Fluid Mechanics*, Vol. 64, 1974, pp. 775–816. Copyright © 1974 Cambridge University Press.)

The turbulent section thickens stepwise in the downstream direction, as is shown schematically in Fig. 7.7. Each step is D_i thick and $\lambda_{B,i}$ long. Furthermore, the proportionality $\lambda_B, i \sim 2D_i$ makes all the rectangular building blocks geometrically similar. By averaging in time the parade of large eddies through this sequence of steps, we anticipate that the time-averaged shear layer region must flare out linearly with a constant half-angle of the order of arctan $[(D/2)/(3\lambda_B/2)] = $ arctan $[1/(\pi 3^{1/2})] \sim 10°$. This geometric constant and the stepped structure of the theory (Fig. 7.7) are confirmed beautifully by laboratory visualizations of turbulent shear layers [12], Fig. 7.8. All the free turbulent mixing regions – jets, plumes, and shear layers – are shaped as cones or, in two dimensions, wedges [6].

The eddy doubling sequence sketched in Figs. 7.6 and 7.7 cannot continue in two dimensions indefinitely. The reason is once again the access-optimization principle that guided us up to this point. Two-dimensional eddies are curved two-dimensional jets. The infinitely flat cross sections of such jets are far from having the most efficient shape (review the optimal cross sections for ducts and river channels, Section 6.3). Consequently, what starts out as a two-dimensional curved jet is destined to become a three-dimensional flow that organizes itself into several parallel jets with nearly round cross sections (for the same reason that delta branches appear in Fig. 6.4). The most obvious examples of potentially very flat two-dimensional streams that opt for rounder cross sections are the Gulf Stream and the atmospheric jet streams. Many other illustrations are provided by jets with flat-rectangle cross sections, which always develop into jets with round mixing regions sufficiently far downstream.

7.5 The Onset of Rolls in Fluid Layers Heated from Below

The constructal approach employed throughout this book is based on the view that a naturally occurring flow – its geometric structure – is the end result of a process of internal geometric optimization, i.e., self-organization by means of constrained optimization. According to theory, the objective of this process is to construct an assembly of paths of minimal resistance for the current that must flow through the system. Natural convection in heating from below (Bénard convection) is the next testing ground chosen for the theory [13]. We show that macroscopic shape and structure can be predicted based on the geometric minimization of thermal resistance. In other words, we show that the constructal approach is deterministic with respect to the very existence of organized fluid motion (streams).

The reason for using Bénard convection as a test case for constructal theory is that this type of organized heat transfer is very well known. An added reason is to renew interest in a piece of classical work that, from today's point of view, looks even more important. The maximization of heat transfer rate (or Nusselt number) in Bénard convection was a hypothesis introduced by Malkus [14] and used in several studies of convection in pure fluids and fluid-saturated porous media. It was related by Glansdorff and Prigogine [15] to a principle of entropy generation minimization subject to imposed boundary conditions. In this section we develop the connection between conductance maximization and the optimization of geometry in a single, finite-size flow element. In this way we recognize Malkus' hypothesis as part of a general theory of geometrical shape and pattern formation in nature. This and other examples collected in this book show that constructal theory brings under the same deterministic umbrella many of the advances made by others in the past.

The geometric optimization of the flow pattern can be demonstrated in single-phase convection. We limit this discussion to two-dimensional geometries without a free surface, first, heated from below, and, later, heated from the side. The objective of this treatment is not to refine the mathematical models used by previous researchers; rather it is to demonstrate self-organization as a geometric-optimization process, and for this we rely on the simplest and most transparent models and analyses.

Why should a disorganized motion and heat transfer mechanism (chaotic, molecular thermal diffusion) change abruptly such that the disorganized entities ride together in a macroscopic motion visible as streams? Why should shapelessness (diffusion) coexist with shape and structure (streams)?

Consider the single-phase fluid layer shown in Fig. 7.9, which is characterized by the thickness H and the bottom excess temperature $\Delta T = T_h - T_c$. In

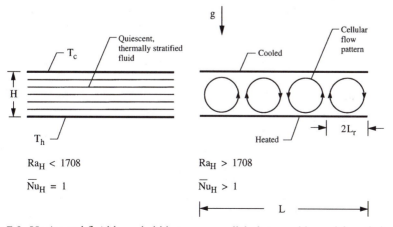

FIGURE 7.9. Horizontal fluid layer held between parallel plates and heated from below [13].

line with the access-optimization principle of constructal theory, we search for the fastest (most direct) route for heat transfer across the fluid layer. In other words, we subject heat transfer to the same principle that momentum transfer was subjected to in Sections 7.2–7.4. To start with, the classical solution for time-dependent thermal diffusion near a wall with a sudden jump in temperature (ΔT) is [cf. Eq. (7.8)]

$$\frac{T - T_c}{\Delta T} = \mathrm{erfc}\left[\frac{y}{2(\alpha t)^{1/2}}\right], \tag{7.14}$$

where T_c is the far-field temperature in the fluid. The effect of the temperature jump is felt to the distance

$$\frac{y}{2(\alpha t)^{1/2}} \sim 1, \tag{7.15}$$

which represents the knee in the temperature profile indicated by Eq. (7.14). The time needed by this heating effect to travel by thermal diffusion the distance H is

$$t_0 \sim \frac{H^2}{4\alpha}. \tag{7.16}$$

The time t_0 corresponds to the heating of the entire layer $(y \sim H)$. The factor 4 in the denominator arises from the geometry (shape) of the time-dependent temperature profile in Eq. (7.14).

Pure conduction continues to be the preferred heat transfer mechanism, and the fluid layer remains macroscopically motionless as long as H is small enough that t_0 is the shortest time of transporting heat across the layer. The alternative to conduction is convection, or the *channeling* of energy transport on the back of fluid streams, which act as conveyor belts (rolls, Fig. 7.9, right-hand side). The question is whether the convection time (t_1) around the convection cell is shorter than t_0. The convection time is $t_1 \sim 4H/v$, where v is the vertical velocity of the fluid (the peripheral velocity of the roll).

To evaluate the v and t_1 scales, we rely on the method of scale analysis [1, 6]. First, we note that the effective diameter of each roll is of the order of H, but smaller, for example, $H/2$. When the roll turns, an excess temperature of the order of $\Delta T/2$ is created between the moving stream and the average temperature of the fluid layer. This excess temperature induces buoyancy (modified gravitational acceleration) of the order of $g\beta\Delta T/2$. The total buoyancy force that drives the roll is of the order of $(g\beta\Delta T/2)\rho(H/2)^2$. When the Prandtl number is of the order of 1 or greater, the driving force is balanced by the viscous shearing force $\tau H/2$, where the shear-stress scale is $\tau \sim \mu v/(H/4)$. The force balance buoyancy $\sim$ friction yields the velocity scale $v \sim g\beta\Delta T H^2/(16v)$ and the corresponding convection time scale

$$t_1 \sim \frac{64v}{g\beta\Delta T H}. \tag{7.17}$$

To see the emergence of an opportunity to optimize the geometric features of the flow pattern, imagine that H increases. The thermal diffusion time t_0 increases in accelerated fashion, approximation (7.16), whereas the convection time t_1 (a *property* of the H-tall system, even if quiescent) decreases monotonically, approximation (7.17). Setting $t_1 \lesssim t_0$ and using approximations (7.16) and (7.17), we find that the first streams occur when $\mathrm{Ra}_H = 256 \sim O(10^2)$, where $\mathrm{Ra}_H = g\beta\Delta T H^3/(\alpha\nu)$ is the Rayleigh number.

The exact solution for this critical condition is $\mathrm{Ra}_H = 1708$ (see Ref. 6); in other words, $\mathrm{Ra}_H = O(10^3)$. The error in the result of scale analysis is understandable (and unimportant) because it can be attributed to the imprecise geometric ratios (factors of the order of 1) introduced along the argument made above in approximation (7.17). A more exact estimate can be achieved in an analysis that reflects better the scales of the flow, as we demonstrate soon in Section 7.7: see feature (b) that follows Eq. 7.30. What is important is that the predicted critical Ra_H is a *constant* considerably greater than 1. This constant expresses the compounded effect of all the geometric ratios of the roll-between-plates configuration. Had we neglected the geometric reality of how the rolls must fit or the geometric fact that 4 belongs in the denominator of approximation (7.16), we would have obtained only $\mathrm{Ra}_H \sim 1$, i.e., the correct dimensionless group Ra_H but not the fact that the critical Ra_H is a *geometric* (structural) constant. All the transition (critical) numbers of fluid mechanics are constants that reflect the geometry (shape) of the elemental system – the first roll or the first eddy (e.g., Table 7.1).

When convection occurs there are two heat transfer mechanisms, not one. Each roll characterized by $t_0 \sim t_1$ is an elemental system in the sense of constructal theory. The equipartition of time $t_0 \sim t_1$ is the analog of the equipartition of temperature drop across an optimized element of the heat generating volume of constructal theory. Conduction, or thermal diffusion, is present and does its job at every point inside the elemental volume $H \times 2L_r$, shown in Fig. 7.9. Superimposed on this volumetric heat flow is an optimal pattern of convection streets that guide the imposed heat current *faster* across H.

The usual terminology for faster in the field of heat transfer is to say that the onset of convection is followed by an increase in the overall Nusselt number [6], which is defined as $\overline{\mathrm{Nu}}_H = \bar{q}'' H/k\Delta T$. In the present discussion of heating from below, we have been fixing the heat current. If we fix the uniform heat flux q'' in Bénard convection, we see once again that the optimization of the heat flow pattern at the elemental level leads to a smaller overall ΔT and thus a larger $\overline{\mathrm{Nu}}_H$ (Fig. 7.9).

The geometric minimization of the temperature difference across H continues to manifest itself as H (or Ra_H) increases as convection becomes more intense. In this case geometric optimization means the selection of the number of rolls that fill a layer of horizontal dimension L (Fig. 7.9) or the selection of the roll aspect ratio H/L_r.

7.6 Partitioned Fluid Layer Heated from the Side

An illustration of the minimization of thermal resistance through the optimization of the roll shape is provided by the manmade configuration shown in Fig. 7.10.

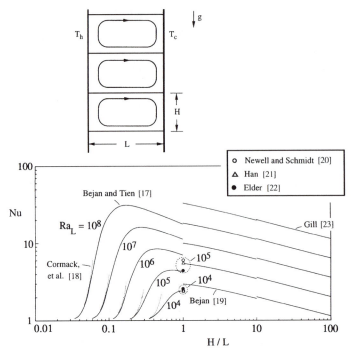

FIGURE 7.10. The optimization of roll shape and the maximization of overall thermal conductance in a cellular vertical fluid layer heated from the side [16].

The fluid layer is heated from the side ($\Delta T = T_h - T_c$), and its thickness – the horizontal dimension – is fixed. The ensuing flow is segmented into rolls by the insertion of horizontal partitions, which are impermeable and adiabatic. The partitions are equidistant, but their vertical spacing H may vary.

Figure 7.10 shows what happens to the overall thermal conductance [Nu = $q'/(k\Delta T/L)$] of the L layer when the shape of each roll (H/L) varies [16]. These changes occur at constant L, i.e., at the constant Rayleigh number based on L, $Ra_L = g\beta\Delta T L^3/(\alpha\nu)$. The thermal conductance reaches a maximum when each roll has a certain, intermediate shape (not too tall, not too shallow). The evidence that supports the maxima exhibited in Fig. 7.10 is strong and comes from seven independent studies, which are indicated on the figure. The optimal roll shape becomes more slender as the convection becomes more intense (i.e., as Ra_L increases).

The analogy between the geometric optimization of segmented vertical layers (Fig. 7.10) and the geometric maximization of thermal conductance in layers heated from below (Fig. 7.9) was not noted until recently [13]. An analogous geometric principle governs the maximization of thermal conductance across a segmented vertical layer filled with a fluid-saturated porous medium [16]. The maximum of Fig. 7.10 was rediscovered in 1999 by Frederick [24]. In another independent 1999 study, Landon and Campo [25] showed that further improvements are possible if the corners of the two-dimensional cavity are rounded. Very promising is the direction started by Lartigue et al. [26, 27], in which the cavity walls are deformable, and the effect of deformation on global heat transfer is documented.

7.7 Optimization of Flow Geometry in Layers Heated from Below

The flow geometry resulting from the optimization process in the man-made system of Fig. 7.10 shows the way to a pure and very simple theory of natural flow in a fluid layer heated from below (Fig. 7.9). The key is to regard the natural flow pattern as the result of an engineering process in which many flow patterns are evaluated until the flow with minimal thermal resistance is identified. We can describe this process analytically by intersecting the two asymptotes of the geometric configuration: many cells versus few cells. This method has been used with consistently good results in the geometric optimization of cooling arrangements for electronics, as shown in Chap. 3.

All of the physical parameters of the system of Fig. 7.9 are fixed except for L_r, or the number of cells. For simplicity, we assume that the flow is two dimensional; however, a strongly three-dimensional flow can be optimized by use of the same method [28]. Because we are free to vary L_r, we can imagine the many-cells limit shown in Fig. 7.11. Each cell is a very slender counterflow, which has the important property that it can sustain a longitudinal (i.e., vertical) temperature gradient of the same order as the imposed gradient $\Delta T/H$. The longitudinal gradient occurs because one branch of the counterflow loses heat to (or gains heat from) the other branch. This property of slender counterflows was first recognized in cryogenic engineering [2, 29] and subsequently in the modeling of heat transfer through vascularized tissues [30]. See also Sections 8.10, 9.4, and 10.6.

We are interested in more than just the orders of magnitude of the flow variables: As we showed in Section 7.5, we want to compare quantitatively the results of geometric optimization with the observed features of natural flows. This is why in addition to using scale analysis we assume a reasonable shape for the laminar temperature profile. If the temperature profile is parabolic across each branch

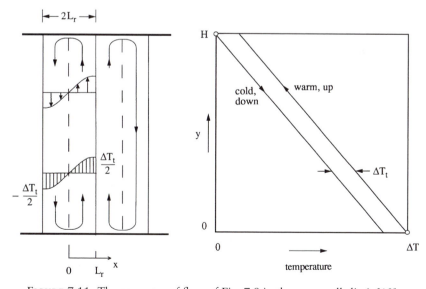

FIGURE 7.11. The geometry of flow of Fig. 7.9 in the many-cells limit [13].

(Fig. 7.11), and if the maximum temperature difference across one branch is $\Delta T_t/2$, then the average temperature difference between the two branches is $(2/3)\,\Delta T_t$.

We can estimate the mass flow rate of one branch, $\dot{m}'$, by writing the momentum equation for fully developed flow in the upflowing branch (e.g., Ref. 6, pp. 186–187):

$$\frac{d^2 v}{dx^2} = -\frac{g\beta}{\nu}(\bar{T}_{\rm up} - \bar{T}_{\rm down}). \tag{7.18}$$

In this equation $\bar{T}_{\rm up}$ and $\bar{T}_{\rm down}$ are the average temperatures of the upflowing and the downflowing branches, respectively. Noting that $(\bar{T}_{\rm up} - \bar{T}_{\rm down}) \cong (2/3)\Delta T_t$ and integrating Eq. (7.18) subject to the conditions of no-slip at $x = 0$ and zero shear at $x = L_r$, we obtain $v(x)$ and the mass flow rate as

$$\dot{m}' = \frac{2\rho g\beta L_r^3 \Delta T_t}{9\nu}. \tag{7.19}$$

The enthalpy lost by the upflowing branch is $\dot{m}' c_P (T_{\rm in} - T_{\rm out})$, where $T_{\rm in}$ and $T_{\rm out}$ are the start and the finish bulk temperatures of the branch. The right-hand side of Fig. 7.11 shows that $T_{\rm in} - T_{\rm out} = \Delta T - \Delta T_t$. This enthalpy loss is being conducted horizontally to the downflowing branch. The horizontal temperature gradient across the interface between the two branches ($x = 0$; dashed line in Fig. 7.11) is $2(\frac{1}{2}\Delta T_t/L_r)$, where the leading factor of 2 is the mark of the assumed parabolic temperature profile. Finally, the first law of thermodynamics written for the upflowing branch as a flow system of size $H \times L_r$ is

$$\dot{m}' c_P (\Delta T - \Delta T_t) = kH\frac{\Delta T_t}{L_r}. \tag{7.20}$$

Combining Eqs. (7.19) and (7.20), we find the relation between the transversal temperature difference ΔT_t and the difference imposed vertically (ΔT):

$$\frac{\Delta T_t}{\Delta T} = 1 - \frac{9}{2{\rm Ra}_H}\left(\frac{H}{L_r}\right)^4. \tag{7.21}$$

Another property of the counterflow is that it convects energy longitudinally (upward) at the rate $\dot{m}' c_P (\bar{T}_{\rm up,b} - \bar{T}_{\rm down,b})$. The present analysis is sufficiently approximate so that we may replace the bulk temperature difference with the mean temperature difference $(2/3)\Delta T_t$. The counterflow carries this energy current upward through a space of thickness $2L_r$. Consequently, the average heat flux removed by the counterflow from the bottom wall is

$$q'' = \frac{\dot{m}' c_P (2/3)\Delta T_t}{2L_r}, \tag{7.22}$$

or, after Eqs. (7.19)–(7.21) are used,

$$q'' = \frac{2kg\beta L_r^2}{27\alpha\nu}(\Delta T)^2 \left[1 - \frac{9}{2{\rm Ra}_H}\left(\frac{H}{L_r}\right)^4\right]^2. \tag{7.23}$$

This result has been sketched qualitatively in Fig. 7.12: The heat flux decreases approximately as L_r^2 as $L_r \to 0$, i.e., as the cells become more numerous.

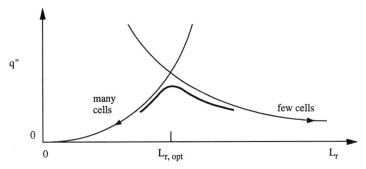

FIGURE 7.12. The intersection of the many-cells and the few-cells asymptotes, and the optimal cell thickness.

Consider now the opposite limit in which the flow is spread out, by design, and consists of a small number of upflows and downflows. This limit is illustrated in Fig. 7.13. Each vertical flow is a plume formed over a long portion of horizontal wall of length $L = 4L_r$. The thermal resistance is due to the horizontal boundary layers that line each section of length $2L_r$. Although it is possible to obtain a purely theoretical estimate for the heat transfer rate across such a boundary layer by using boundary-layer theory, in this section we rely on the most accurate estimate, which is derived from an experimental correlation.

We begin with the observation that the plume rises or sinks in a quiescent and thermally stratified fluid of average temperature difference $\Delta T/2$. This means that the effective temperature difference between the horizontal base of each plume and the fluid reservoir that surrounds it is $\theta = \Delta T/2$. The average heat flux removed by the plume is known from direct measurements [31]:

$$\frac{q''L}{\theta k} = 0.54 \, \mathrm{Ra}_{L,\theta}^{1/4}, \tag{7.24}$$

where $\mathrm{Ra}_{L,\theta}$ is the Rayleigh number based on L and θ and $\mathrm{Ra}_{L,\theta} < 10^7$. Note that according to Eq. (7.24), the heat flux is independent of the vertical dimension of the system, H. Expressed in terms of L_r and ΔT, correlation (7.24) reads as

$$q'' = 0.161 \frac{k\Delta T}{H} \left(\frac{H}{L_r} \right)^{1/4} \mathrm{Ra}_H^{1/4}. \tag{7.25}$$

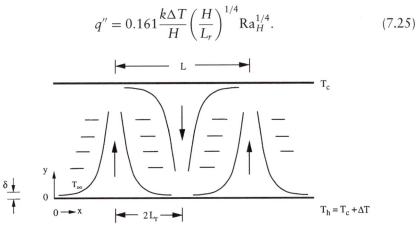

FIGURE 7.13. The geometry of the flow of Fig. 7.9 in the few-cells limit [13].

In conclusion, when the horizontal surfaces are covered sparsely with isolated plumes, the wall-averaged heat flux decreases monotonically as the spacing L_r increases. This asymptotic trend has been added to Fig. 7.12 to show that a flow geometry with maximal heat flux exists. We can locate the optimal flow structure (L_r) by intersecting asymptotes (7.23) and (7.25). Eliminating q'', we obtain

$$\mathrm{Ra}_H^{1/3}\left(\frac{L_{r,\mathrm{opt}}}{H}\right)\left[1 - \frac{9}{2\mathrm{Ra}_H}\left(\frac{H}{L_{r,\mathrm{opt}}}\right)^4\right]^{8/9} = 1.41. \tag{7.26}$$

It is convenient to define the dimensionless factor f as

$$f = \frac{L_{r,\mathrm{opt}}}{H\mathrm{Ra}_H^{-1/3}} \tag{7.27}$$

such that Eq. (7.26) provides implicitly the function $f(\mathrm{Ra}_H)$:

$$f^4 - 1.474\,f^{23/8} = \frac{9}{2}\mathrm{Ra}_H^{1/3}. \tag{7.28}$$

Equations (7.27) and (7.28) pinpoint the optimal value of the flow slenderness ratio $L_{r,\mathrm{opt}}/H$ as a function of Ra_H. The maximum heat flux that corresponds to this geometry is obtained by substitution of $L_r = L_{r,\mathrm{opt}}$ into Eq. (7.23) or Eq. (7.25):

$$\mathrm{Nu}_{\max} \le 0.161\,f^{-1/4}\mathrm{Ra}_H^{1/3}. \tag{7.29}$$

The inequality sign is a reminder that the actual peak of the $q''_{\max}(L_r)$ curve falls under the intersection of the two asymptotes (Fig. 7.12). Electronics cooling applications of the intersection-of-asymptotes method have shown repeatedly that the inequality sign accounts for a factor of approximately 1/2 between the height of the point of intersection and the peak of the actual curve. Finally, at the optimum, Eq. (7.21) reads as

$$\frac{\Delta T_t}{\Delta T} = 1.47\,f^{-9/8}. \tag{7.30}$$

Figure 7.14 shows the main features of the optimal geometry derived based on this simple analysis. The plotted $\mathrm{Nu}_{\max}$ curve is based on relation (7.29) with the equal sign, which means that the actual $\mathrm{Nu}_{\max}$ should be lower by a factor of the order of 1/2. Several features of this solution are worth noting:

(a) The $\mathrm{Nu}_{\max}(\mathrm{Ra}_H)$ curve comes close to the experimental data, e.g., the $\mathrm{P_r} \sim 1$ range of the widely accepted correlation that is due to Globe and Dropkin [32]. If the expected factor of the order of 1/2 is applied, then the $\mathrm{Nu}_{\max}$ curve falls right on top of the correlation of Globe and Dropkin, which is shown by a dashed line in Fig. 7.14 (top). At the same point, $L_{r,\mathrm{opt}}/H \cong 0.35$, which agrees with the linear stability solution $L_{r,\mathrm{opt}}/H \sim 0.5$.

(b) The $\mathrm{Nu}_{\max}(\mathrm{Ra}_H)$ curve cuts the $\mathrm{Nu}_{\max} = 1$ line at an $\mathrm{Ra}_H \cong 526$, which is a value of the order of 10^3. In other words, the predicted transition between pure diffusion and stream flow is in agreement with the time-minimization (or access-maximization) argument of Section 7.5. This improved estimate of the critical Ra_H for the onset of convection validates the comments made in the second paragraph under approximation (7.17).

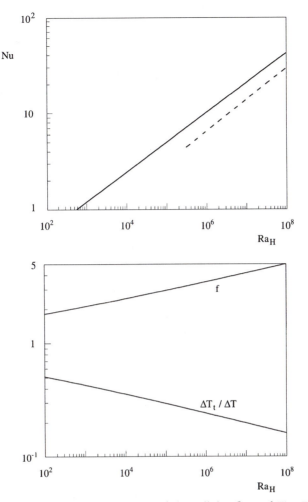

FIGURE 7.14. The heat transfer characteristics of the cellular flow of Fig. 7.9 after its L_r dimension has been optimized [13].

(c) The $Nu_{max}(Ra_H)$ curve is described approximately by $Nu_{max} \sim Ra_H^n$ where the exponent n decreases from 0.333 to 0.313 as Ra_H increases. Over most of the Ra_H range of Fig. 7.14, the n value is close to 0.313 as the f factor approaches $1.033\, Ra_H^{1/12}$. This $Nu_{max}(Ra_H)$ behavior also agrees with observations. The minor and gradual decrease of the Ra_H exponent is well known and is one point of controversy in the field of Bénard convection. In this section we derived this trend from theory.

(d) The transversal temperature difference between the vertical branches of the roll (ΔT_t) decreases as Ra_H increases, i.e., as convection intensifies. Although this trend may seem counterintuitive, it is explained by the fact that the rolls become more slender as Ra_H increases. In this limit the vertical streams find themselves in a more intimate thermal contact.

(e) The optimized slenderness ratio is given by $L_{r,opt} = Hf Ra_H^{1/3}$ and approaches $L_{r,opt} \cong 1.033\, H Ra_H^{-1/4}$ as Ra_H increases, i.e., when $f \cong 1.033 Ra_H^{1/12}$.

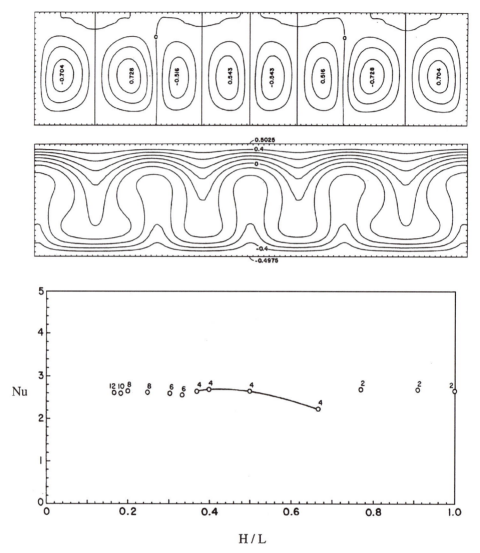

FIGURE 7.15. Convection rolls in a layer of cold water near 4 °C heated from below, and the maximization of global thermal conductance (Nu) through the optimization of roll shape (H/L) [33].

The optimization of roll shape, which in this section was pursued from principle, was encountered many times earlier in the fields of convection and fluid mechanics. One example is the numerical simulation of heat transfer and fluid flow in a two-dimensional water layer heated from below (8 °C), when its upper surface is cooled to 0 °C. These temperatures embrace the temperature associated with maximum density (4 °C), divide the system into an upper sublayer that is relatively quiescent, and a lower sublayer dominated by rolls. This flow is illustrated in the upper frame of Fig. 7.15, and the middle frame shows the corresponding pattern of isotherms. The bottom frame shows how the global thermal conductance of the layer (Nu) responds to changes in the height/length ratio (H/L) of an individual roll. The

FIGURE 7.16. Complicated boiling flow during the cooking of spaghetti (left) and the close-to-perfect organization discovered after the water is poured out (right).

number 4 listed on the solid curve represents the number of rolls found based on direct numerical simulation of the flow in a layer of fixed horizontal dimension (the upper frame shows 8 rolls). The solid curve shows that the conductance can be maximized by selection of the roll shape [33].

Organized motion (rolls) is the macroscopic flow feature even when the flow looks hopelessly complicated. This look has been the traditional impediment to theoretical progress in turbulence, as evidenced now by the structure revealed by theory in Figs. 7.6–7.8. In heating from below, it is not necessary to have a pure (clear) fluid in order to detect organization. The best counterexamples are at everyone's fingertips, in the kitchen. When we boil spaghetti (Fig. 7.16 left), we see a confusing flow dominated by boiling, foam, and steam. The mind that is driven by theory, however, begins to distinguish a radial preference in the flow of the spaghetti noodles themselves. This hunch is confirmed with great clarity on the right side of Fig. 7.16: after pouring the water out carefully, we discover that the "confusing" flow had organized the spaghetti very neatly into a toroidal shape.

7.8 Porous Layer Saturated with Fluid and Heated from Below

There is an analogy between the geometric optimization of flow structure in a space filled with fluid and the optimization of structure in a space filled with a porous medium saturated with fluid. In this section we illustrate the determinism that results from minimizing geometrically the thermal resistance across a fluid-saturated porous layer. For brevity, we rely once more on Figs. 7.9 and 7.11–7.13 and list only the analytical highlights that differ from those of Section 7.7. One difference is that in this section the two asymptotes of the flow geometry are described in exact analytical terms.

Assume that the system of Fig. 7.9 is a porous layer saturated with fluid and that, if present, the flow is two dimensional and in the Darcy regime. The height H

is fixed, and the horizontal dimensions of the layer are infinite in both directions. The fluid has nearly constant properties such that its density–temperature relation is described well by Boussinesq linearization [6]. The volume-averaged equations that govern the conservation of mass, momentum, and energy are [34]

$$\frac{\partial u}{\partial x} + \frac{\partial v}{\partial y} = 0, \tag{7.31}$$

$$\frac{\partial u}{\partial y} - \frac{\partial v}{\partial x} = -\frac{Kg\beta}{\nu}\frac{\partial T}{\partial x}, \tag{7.32}$$

$$u\frac{\partial T}{\partial x} + v\frac{\partial T}{\partial y} = \alpha_m\left(\frac{\partial^2 T}{\partial x^2} + \frac{\partial^2 T}{\partial y^2}\right), \tag{7.33}$$

where T, u, and v are the volume-averaged temperature and velocity components, and K, g, β, and ν are the permeability constant, gravitational acceleration, coefficient of volumetric thermal expansion, and kinematic viscosity respectively. The thermal diffusivity $\alpha_m = k_m/(\rho c_P)_f$ is based on the thermal conductivity of the porous matrix saturated with fluid (k_m) and the heat capacity of the fluid alone $(\rho c_P)_f$. The horizontal length scale of the flow pattern ($2L_r$), or the geometric aspect ratio of one roll, is unknown.

In the limit $L_r \to 0$ each roll is a very slender vertical counterflow, Fig. 7.11. Because of symmetry, the outer planes of this structure ($x = \pm L_r$) are adiabatic: They represent the center planes of the streams that travel over the distance H. The scale analysis of the $H \times (2L_r)$ region indicates that in the $L_r/H \to 0$ limit the horizontal velocity u vanishes. This scale analysis is not shown because it is well known as the defining statement of a fully developed flow (e.g., Ref. 6, p. 97). Equations (7.31)–(7.33) reduce themselves to

$$\frac{\partial v}{\partial x} = \frac{Kg\beta}{\nu}\frac{\partial T}{\partial x}, \tag{7.34}$$

$$v\frac{\partial T}{\partial y} = \alpha_m\frac{\partial^2 T}{\partial x^2}, \tag{7.35}$$

which can be solved exactly for v and T. The boundary conditions are $\partial T/\partial x = 0$ at $x = \pm L_r$ and the requirement that the extreme (corner) temperatures of the counterflow region are dictated by the top and bottom walls, $T(-L_r, H) = T_c$ and $T(L_r, 0) = T_b$. The solution is

$$v(x) = \frac{\alpha_m}{2H}\left[\text{Ra}_p - \left(\frac{\pi H}{2L_r}\right)^2\right]\sin\left(\frac{\pi x}{2L_r}\right), \tag{7.36}$$

$$T(x, y) = \frac{\nu}{Kg\beta}v(x) + \frac{\nu}{Kg\beta}\left(2\frac{y}{H} - 1\right)\frac{\alpha_m}{2H}\left[\text{Ra}_p - \left(\frac{\pi H}{2L_r}\right)^2\right] + (T_b - T_c)\left(1 - \frac{y}{H}\right), \tag{7.37}$$

where the porous-medium Rayleigh number $\text{Ra}_p = Kg\beta H(T_b - T_c)/(\alpha_m\nu)$ is a specified constant. The right-hand side of Fig. 7.11 shows the temperature distribution along the vertical boundaries of the flow region ($x = \pm L_r$): The vertical temperature gradient $\partial T/\partial y$ is independent of altitude. The transversal (horizontal) temperature difference (ΔT_t) is also a constant:

$$\Delta T_t = T(x = L_r) - T(x = -L_r) = \frac{\nu}{Kg\beta}\frac{\alpha_m}{H}\left[\mathrm{Ra}_p - \left(\frac{\pi H}{2L_r}\right)^2\right]. \quad (7.38)$$

The counterflow convects heat upward at the rate q', which can be calculated with Eqs. (7.36) and (7.37):

$$q' = \int_{-L}^{L} (\rho c_P)_f \nu T \, dx. \quad (7.39)$$

The average heat flux convected in the vertical direction is $q'' = q'/2L_r$; hence the thermal conductance expression

$$\frac{q''}{\Delta T} = \frac{k_m}{8H\mathrm{Ra}_p}\left[\mathrm{Ra}_p - \left(\frac{\pi H}{2L_r}\right)^2\right]^2. \quad (7.40)$$

This result is valid provided the vertical temperature gradient does not exceed the externally imposed gradient, $(-\partial T/\partial y) < \Delta T/H$. This condition translates into

$$\frac{L_r}{H} > \frac{\pi}{2}\mathrm{Ra}_p^{-1/2}, \quad (7.41)$$

which in combination with the assumed limit $L_r/H \to 0$ means that the domain of validity of Eq. (7.40) is wide when Ra_p is large. In this domain the thermal conductance $q''/\Delta T$ decreases monotonically as L_r decreases; see Fig. 7.17.

As L_r increases, the number of rolls decreases and the vertical counterflow is replaced with a horizontal counterflow in which the thermal resistance between T_h and T_c is dominated by two horizontal boundary layers, as in Fig. 7.13. Let δ be the scale of the thickness of the horizontal boundary layer. The thermal conductance $q''/\Delta T$ can be deduced from the heat transfer solution for natural convection boundary-layer flow over a hot isothermal horizontal surface facing upward or a cold surface facing downward. The similarity solution for the horizontal surface

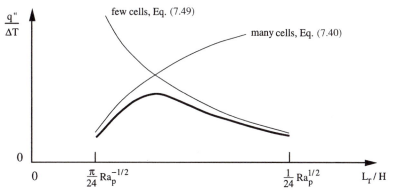

FIGURE 7.17. The geometric maximization of the thermal conductance of a fluid-saturated porous layer heated from below [13].

with power-law temperature variation [35] can be used to develop an analytical result, as is shown at the end of this section.

A simpler analytical solution can be developed in a few steps with the integral method. Consider the slender flow region $\delta \times (2L_r)$, where $\delta \ll 2L_r$, and integrate Eqs. (7.31)–(7.33) from $y = 0$ to $y \to \infty$, i.e., into the region just above the boundary layer. The surface temperature is T_h, and the temperature outside the boundary layer is T_∞ (constant). The origin $x = 0$ is set at the tip of the wall section of length $2L_r$. The integrals of Eqs. (7.31) and (7.33) yield

$$\frac{d}{dx} \int_0^\infty u(T - T_\infty)\, dy = -\alpha_m \left(\frac{\partial T}{\partial y} \right)_{y=0}. \tag{7.42}$$

The integral of Eq. (7.32), in which we neglect $\partial v / \partial x$ (see the boundary-layer theory), leads to

$$u_0(x) = \frac{Kg\beta}{\nu} \frac{d}{dx} \int_0^\infty T\, dy, \tag{7.43}$$

where u_0 is the velocity along the surface, $u_0 = u(x, 0)$. Reasonable shapes for the u and the T profiles are the exponentials

$$\frac{u(x, y)}{u_0(x)} = \exp \left[-\frac{y}{\delta(x)} \right] = \frac{T(x, y) - T_\infty}{T_h - T_\infty}, \tag{7.44}$$

which transform Eqs. (7.42) and (7.43) into

$$\frac{d}{dx}(u_0 \delta) = \frac{2\alpha_m}{\delta}, \tag{7.45}$$

$$u_0 = \frac{Kg\beta}{\nu}(T_h - T_\infty) \frac{d\delta}{dx}. \tag{7.46}$$

These equations can be solved for $u_0(x)$ and $\delta(x)$. Necessary in heat transfer calculations is the thickness

$$\delta(x) = \left[\frac{9\alpha_m \nu}{Kg\beta(T_h - T_\infty)} \right]^{1/3} x^{2/3}. \tag{7.47}$$

The corresponding solution for $u_0(x)$ is of the type $u_0 \sim x^{-1/3}$, which means that the horizontal velocities are large at the start of the boundary layer and decrease as x increases. This is consistent with the geometry of the $H \times 2L_r$ roll sketched in Fig. 7.13, in which the flow generated by one horizontal boundary layer turns the corner and flows vertically as a relatively narrow plume (narrow relative to $2L_r$), to start, with high velocity (u_0), a new boundary layer along the opposite horizontal wall.

We determine the thermal resistance of the geometry of Fig. 7.13 by estimating the local heat flux $k_m(T_h - T_\infty)/\delta(x)$ and averaging it over the total length $2L_r$:

$$q'' = \left(\frac{3}{4} \right)^{1/3} \frac{k_m \Delta T}{H} \left(\frac{T_h - T_\infty}{\Delta T} \right)^{4/3} \mathrm{Ra}_H^{1/3} \left(\frac{H}{L_r} \right)^{2/3}. \tag{7.48}$$

The symmetry of the sandwich of boundary layers requires that $T_h - T_\infty = \frac{1}{2}\Delta T$, such that

$$\frac{q''}{\Delta T} = \frac{3^{1/3} k_m}{4H} \mathrm{Ra}_p^{1/3} \left(\frac{H}{L_r}\right)^{2/3}. \qquad (7.49)$$

To check the goodness of this result we used the similarity solution for a hot horizontal surface that faces upward in a porous medium and has an excess temperature that increases as x^λ. The only difference is that the role that was played by $T_h - T_\infty$ in the preceding analysis is now played by the excess temperature averaged over the surface length $2L_r$. If we use $\lambda = 1/2$, which corresponds to uniform heat flux, then it can be shown that the solution of Cheng and Chang [35] leads to the same formula as Eq. (7.49) except that the factor $3^{1/3} = 1.442$ is replaced with $0.816(3/2)^{4/3} = 1.401$. The difference between the results of the two methods is only 3%.

Equation (7.49) is valid when the specified Ra_p is such that the horizontal boundary layers do not touch. We write this geometric condition as $\delta(x = 2L_r) < H/2$ and, using Eq. (7.47), we obtain

$$\frac{L_r}{H} < \frac{1}{24} \mathrm{Ra}_p^{1/2}. \qquad (7.50)$$

Because in this analysis L_r/H was assumed to be very large, we conclude that the L_r/H domain in which Eq. (7.49) is valid becomes wider as the specified Ra_H increases. The important feature of Eq. (7.49) is that in the few-rolls limit the thermal conductance decreases as the horizontal dimension L_r increases. This second asymptotic trend has been added to Fig. 7.17.

Figure 7.17 presents a bird's-eye view of the effect of flow shape on thermal conductance. Even though we did not draw $q''/\Delta T$ completely as a function of L_r, the two asymptotes tell us that the thermal conductance is maximum at an optimal L_r value that is close to their intersection. There is a family of such curves, one curve for each Ra_p. The $q''/\Delta T$ peak of the curve rises, and the L_r domain of validity around the peak becomes wider as Ra_p increases. Looking in the direction of small Ra_p values we see that the domain vanishes (and the cellular flow disappears) when the following requirement is violated:

$$\frac{1}{24} H\mathrm{Ra}_p^{1/2} - \frac{\pi}{2} H\mathrm{Ra}_p^{-1/2} > 0. \qquad (7.51)$$

This inequality means that the flow exists when $\mathrm{Ra}_p > 12\pi = 37.7$. This conclusion is extraordinary: It agrees very well with the stability criterion for the onset of two-dimensional convection [6, 34], $\mathrm{Ra}_p > 4\pi^2 = 39.5$, which is derived based on a lengthier analysis and the assumption of initial disturbances.

We obtain the optimal shape of the flow, $2L_{r,\mathrm{opt}}/H$, by intersecting asymptotes (7.40) and (7.49):

$$\pi^2 \left(\frac{H}{2L_{r,\mathrm{opt}}} \mathrm{Ra}_p^{-1/2}\right)^2 + 2^{5/6} 3^{1/6} \left(\frac{H}{2L_{r,\mathrm{opt}}} \mathrm{Ra}_p^{-1}\right)^{1/3} = 1. \qquad (7.52)$$

Over most of Ra_p domain (7.51), Eq. (7.52) is approximated well by its high Ra_p asymptote:

$$\frac{2L_{r,\mathrm{opt}}}{H} \cong \pi \mathrm{Ra}_p^{-1/2}. \tag{7.53}$$

We obtain the maximum thermal conductance by substituting the $L_{r,\mathrm{opt}}$ value of Eq. (7.52) into either Eq. (7.49) or Eq. (7.40). This estimate is an upper bound, as explained under relation (7.29). In high Ra_p limit (7.53) this upper bound assumes the analytical form

$$\left(\frac{q''}{\Delta T}\right)_{\max} \frac{H}{k_m} \lesssim \frac{3^{1/3}}{2^{4/3}\pi^{2/3}} \mathrm{Ra}_p^{2/3}. \tag{7.54}$$

Toward lower Ra_p values the slope of the $(q''/\Delta T)_{\max}$ curve increases such that the exponent of Ra_p approaches 1. This behavior is in excellent agreement with the large volume of experimental data collected for Bénard convection in saturated porous media [36]. The "less than 1" exponent of Ra_p in the empirical $\mathrm{Nu}(\mathrm{Ra}_p)$ curve and the fact that this exponent decreases as Ra_p increases have attracted considerable attention from theoreticians during the past two decades [34]. Relation (7.54) is the first instance in which this behavior is predicted in purely theoretical fashion.

7.9 Natural Structure in Multiphase Flow Systems

In this chapter we showed that if we start from the basic idea of constructal theory – the optimization of access to flow by optimizing the internal geometry of constrained systems – it is possible to anticipate the architectural features of flow patterns. We demonstrated these deterministic steps for two classes of fluid systems: shear layers and layers heated from below. We extended constructal theory to momentum transport and heat transport.

The occurrence of the first sign of organized motion (stream, roll, cell) has been connected theoretically to the shape changes (multiplication of rolls) that are observed in larger systems (Fig. 7.7) and at higher heat fluxes. The same theoretical viewpoint explains why organized motion (rolls) must always coexist with disorganized motion (diffusion), such that the latter is assigned to cover length scales smaller than the smallest roll.

In all the flows treated in this chapter the fluid was single phase. We have every reason to expect that the constructal principle governs the generation of structure in multiphase flow as well. The multiphase counterpart of Bénard convection is pool boiling [37]. To the boiling flow of cooking spaghetti (Fig. 7.16) we may add an equally common flow: the boiling of rice (Fig. 7.18). The flow is complicated, especially early when the rice is covered by water. In the end the flow structure is cast in solid by the surprisingly equidistant chimneys constructed by the steam flow through the rice layer.

This way of looking at boiling also leads to a somewhat different perspective on boiling stability, which now emerges as *the need* of a system to organize itself

FIGURE 7.18. The confusing image of the flow during the boiling of rice (left) and the surprisingly regular pattern of vertical ducts dug by the stem flow (right).

in order to optimize its geometry. At the same time, chaos can be restated as the inability of a flow system to find a stationary optimal configuration, i.e., the continual, dynamic search for a configuration that does not exist [37].

7.10 Dendritic Crystals

We end with a look at dendritic crystals, which are structures that form during rapid solidification. Although these structures visualize the flow of heat, not fluid, and might fit better in Chap. 4, we discuss them here because their stepwise growth is conceptually similar to that of turbulent shear flows (Sections 7.3 and 7.4).

In 1611 Kepler drew attention to the shapes, numbers, and geometric similarities exhibited by snowflakes [38]. In this century, the study of dendritic crystals has grown into a major field that deals mainly with two aspects: the shape and the growth of dendritic crystals. Not questioned was the *necessity* of the dendrite. Why are the needles necessary? Why don't we see a solid sphere growing around each nucleation site?

Consider the solidification of a single-component substance and assume that the liquid and the solid phases have the same density. This means that there is no liquid motion in the vicinity of the solidification front: What flows is heat, which starts from the solidification front and diffuses into the liquid. The solid phase is isothermal at T_s. Sufficiently far from the solidification front the liquid is metastable (subcooled) at the temperature T_∞ ($<T_s$). Solidification starts at $t=0$.

Why plane? Kepler asked one question about the geometry of the snowflake: Why six needles? This question was answered after the geometry of the molecular arrangement of ice became known. An equally important question is this: Why is the snowflake plane? This question cuts to the heart of the meaning of rapid solidification.

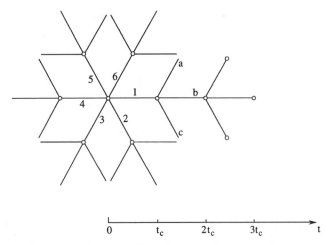

FIGURE 7.19. The formation of new needles after each time interval t_c, as the repeated manifestation of the mechanism shown in Fig. 7.20(a) [39].

The classical solutions for unidirectional solidification in plane, cylindrical, and spherical geometries show that in all cases the solid thickness (R) and the thickness of thermal diffusion into the liquid (r_1) increase as $(\alpha t)^{1/2}$, where α is the thermal diffusivity of the liquid. The three geometries are quite different with regard to their ability to fill the space with solid: specifically, (solid volume)/(total volume) $\sim (R/r_1) < 1$ (plane), $\sim (R/r_1)^2 \ll 1$ (cylinder), and $\sim (R/r_1)^3 \lll 1$ (sphere). Clearly, the most effective arrangement for solidifying a volume in the shortest time is the planar one. That snowflakes are plane is overwhelming evidence that the maximization of the speed of volumetric solidification is an integral part of the nature of the process.

Why needles? The preceding observation brings us to the most important problem, which is that of predicting the necessity of the dendrite. Let us follow, in time, the growth of the snowflake beginning with its birth ($t = 0$) at a point-size nucleation site (Fig. 7.19). Molecular structure and surface-tension-based stability arguments indicate that ice needles will begin to grow with the velocity U (constant) in the six directions shown; but this is only one heat transfer regime of the solidification process that just started. At the same time ($t = 0$), the temperature of the nucleation site jumped to T_s ($> T_\infty$) and triggered a spherical wave of thermal diffusion (warmth), the radius of which increases as $r_1 \sim 2(\alpha t)^{1/2}$. The initial speed of propagation of the liquid heating effect is infinite – that is, larger than any constant speed (U) that the tip of the needle might have. In time, the speed dr_1/dt decreases as $t^{-1/2}$ and is eventually overtaken by U.

Figure 7.20 shows how the warmed liquid sphere and the needle length grow in time. A critical time t_c is reached when the needle length L overtakes the radial length scale of the warmed liquid, $Ut_c \sim 2(\alpha t_c)^{1/2}$, which yields $t_c \sim 4\alpha/U^2$. At times slightly greater than t_c, the needle of length Ut_c just sticks its tip out of the warmed liquid sphere. The tip is once again surrounded by isothermal subcooled liquid from the half-space that lies in front of it. The situation at $t \sim t_c$ is the same as that

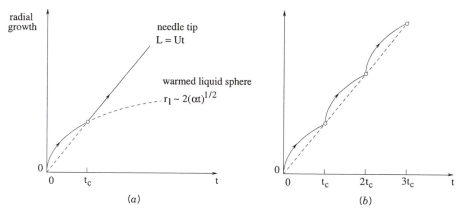

FIGURE 7.20. The simultaneous growth of the needle and the warm liquid sphere [39].

at $t = 0$, except that the new nucleation site (the needle tip) can send new needles only forward, because the trailing half-space is already warmed and/or solidified. In conclusion, at $t \sim t_c$, each tip serves as nucleation site for three forward-leaning fresh needles. There is no difference among the ages of these fresh needles. Because of the 60° angular symmetry, however, the middle needle (b) in Fig. 7.19 looks like a continuation of the original needle, and, consequently, the other two needles (a and c) look like branches.

From $t \sim t_c$ until $t \sim 2t_c$, the new generation of needles experiences the "growth inside the warm liquid sphere" process, which we saw between $t = 0$ and $t = t_c$. One difference is that the liquid spheres of the side needles (a, c) interfere eventually with the spheres of the side needles of the adjacent original directions. Consequently, at $t \sim 2t_c$ all the side needles become suffocated by the warm (saturated) liquid environment, and their needlelike growth ceases. Each middle needle (b), however, pierces its warmed liquid sphere and generates another group of three forward-leaning fresh needles. This third generation and its fully grown version at $t \sim 3t_c$ are illustrated in Fig. 7.19, which is based on repeating the argument of Fig. 7.20(a) three times, as shown in Fig. 7.20(b).

With this, the task of predicting the architecture (existence) of dendrites has been accomplished. Needles are necessary for the same reason that eddies are necessary: to provide internal paths such that entire volumes approach internal equilibrium in the shortest time possible. Geometric structure is necessary so that heat currents have maximum access as they spread through the volume. This is most evident in the solidification of alloys, where the dendrite layer (mushy zone [34]) smears the concentration difference over a much larger distance than the one-dimensional (planar) diffusion front [40]. This is analogous to how the eddies maximize the spreading of momentum in shear flow (Section 7.3).

Imagine now that you are a visitor from another planet, someone who knows biology but not heat transfer. You may describe Fig. 7.19 as follows. The organism is born at the time $t = 0$, and its innermost morphology (e.g., the number 6) is a reflection of information stored at the molecular level. The organism grows and expands into its surroundings. The growth is very fast when the organism is young, and it slows down with age. There comes a time – the time of death, t_c – when the organism becomes disposable. The fossil (solid dendrite) is a reminder of the flows that were present in the living organism (heat flows). Continuity is ensured in the form of several offspring, which repeat the life cycle of the original organism. The offspring may or may not be attached to their ancestor. In sum, the periodic regeneration and multiplication of the organism is an expression of the natural tendency toward the generation of geometry for optimal (or fastest) access for internal currents and external spreading.

PROBLEMS

7.1 Two masses (m_1, m_2) travel in the same direction at different velocities (V_1, V_2). They happen to touch and rub against each other, and after a sufficiently long time they acquire the same velocity (V_∞). Consider the system composed of m_1 and m_2, and also consider the process (a)–(b) illustrated in Fig. P7.1. There are no forces between the system and its environment. Determine the ratio $\eta = KE_b/KE_a$ as a function of m_2/m_1 and V_2/V_1, where KE is the kinetic-energy inventory of the system. Show that $\eta < 1$ when $V_2 \neq V_1$ and that η is of the order of 1 when m_2/m_1 is of the order of 1. Next, assume that the masses m_1 and m_2 are incompressible substances and that (a) and (b) are states of thermal equilibrium with the ambient temperature T_0. Determine an expression for the heat transfer between the system and the environment.

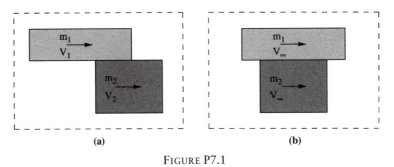

(a)   (b)

FIGURE P7.1

7.2 Show that the transition condition listed for boundary-layer flow in Table 7.1 corresponds to a critical Reynolds number of the order of 10^2 if the critical Reynolds number is based on either the displacement or the momentum thickness of the boundary layer. The boundary-layer scales can be found in Ref. 6.

7.3 Show that for the vertical natural convection boundary-layer flow, the local Reynolds number based on vertical velocity scale and velocity boundary-layer thickness (Table 7.1) is of the order of $Pr^{-1/2}Ra_y^{1/4}$ if $Pr > 1$. Prove that during transition this local Reynolds number is of the order of 10^2. The natural convection scales are listed in Ref. 6.

7.4 Experiments with cigarette smoke show that the buoyant plume becomes turbulent at the height y indicated by $Ra_q \sim 10^{10}$, where $Ra_q = g\beta y^2 q/(\alpha\nu k)$ and q (in watts) is the strength of the heat source that generates the plume. Use the velocity and thickness scales of the laminar plume [6] to show that this Ra_q transition criterion is equivalent to a local Reynolds number of the order of 10^2.

REFERENCES

1. A. Bejan, *Convection Heat Transfer*, Wiley, New York, 1984, Chap. 6.
2. A. Bejan, *Entropy Generation through Heat and Fluid Flow*, Wiley, New York, 1982, Chap. 4.
3. A. Bejan, On the buckling property of inviscid jets and the origin of turbulence, *Lett. Heat Mass Transfer*, Vol. 8, No. 3, 1981, pp. 187–194.
4. A. Bejan, *Advanced Engineering Thermodynamics*, 2nd ed., Wiley, New York, 1997.
5. A. Bejan, *Heat Transfer*, Wiley, New York, 1993, Appendix F.
6. A. Bejan, *Convection Heat Transfer*, 2nd ed., Wiley, New York, 1995.
7. H. Tennekes and J. L. Lumley, *A First Course in Turbulence*, MIT Press, Cambridge, MA, 1972, p. 20.
8. P. Bradshaw, ed., *Turbulence*, 2nd edition, Vol. 12 of Topics in Applied Physics, Series, Springer-Verlag, Berlin, 1978, p. 22.
9. A. Bejan, Predicting the pool fire vortex shedding frequency, *J. Heat Transfer*, Vol. 113, 1991, pp. 261–263.
10. P. J. Pagni, Pool vortex shedding frequencies, in L. M. Trefethen and R. L. Panton, Some unanswered questions in fluid mechanics, ASME Paper 89-WA/FE-5, American Society of Mechanical Engineers, New York, 1989.
11. B. B. Mikic, Lj. B. Vujisic, and J. Kapat, Turbulent transition and maintenance of turbulence; implication to heat transfer augmentation, *Int. J. Heat Mass Transfer*, Vol. 37, Suppl. 1, 1994, pp. 425–431.
12. G. L. Brown and A. Roshko, On density effects and large scale structures in turbulent mixing layers, *J. Fluid Mech.*, Vol. 64, 1974, pp. 775–816.
13. R. A. Nelson, Jr. and A. Bejan, Constructal optimization of internal flow geometry in convection, *J. Heat Transfer*, Vol. 120, 1998, pp. 357–364.
14. W. V. R. Malkus, The heat transport and spectrum of thermal turbulence, *Proc. R. Soc. Ser. A*, Vol. 225, 1954, pp. 196–212.
15. P. Glansdorff and I. Prigogine, *Thermodynamic Theory of Structure, Stability and Fluctuations*, Wiley, London, 1971.
16. A. Bejan, A synthesis of analytical results for natural convection heat transfer across rectangular enclosures, *Int. J. Heat Mass Transfer*, Vol. 23, 1980, pp. 723–726.
17. A. Bejan and C-. L. Tien, Laminar natural convection heat transfer in a horizontal cavity with different end temperatures, *J. Heat Transfer*, Vol. 100, 1978, pp. 641–647.
18. D. E. Cormack, L. G. Leal, and J. Imberger, Natural convection in a shallow cavity with differentially heated end walls, *J. Fluid Mech.*, Vol. 65, 1974, pp. 209–229.
19. A. Bejan, Note on Gill's solution for free convection in a vertical enclosure, *J. Fluid Mech.*, Vol. 90, 1979, pp. 561–568.
20. M. E. Newell and F. W. Schmidt, Heat transfer by natural convection within rectangular enclosures, *J. Heat Transfer*, Vol. 92, 1970, pp. 159–167.
21. J. T. Han, M. A. Sc. thesis, Department of Mechanical Engineering, University of Toronto, 1967.
22. J. W. Elder, Laminar free convection in a vertical slot, *J. Fluid Mech.*, Vol. 23, 1965, pp. 77–98.
23. A. E. Gill, The boundary layer regime for convection in a rectangular cavity, *J. Fluid Mech.*, Vol. 26, 1966, pp. 515–536.

24. R. L. Frederick, On the aspect ratio for which heat transfer in differentially heated cavities is maximum, *Int. Commun. Heat Mass Transfer*, Vol. 26, 1999, pp. 549–558.

25. M. Landon and A. Campo, Optimal shape for laminar natural convective cavities containing air and heated from the side, *Int. Commun. Heat Mass Transfer*, Vol. 26, 1999, pp. 389–398.

26. B. Lartigue, S. Lorente, and B. Bourret, Multicellular natural convection in a high aspect ratio cavity: experimental and numerical results, *Int. J. Heat Mass Transfer*, Vol. 43, 2000, pp. 3159–3170.

27. B. Lartigue, Contribution à l'étude thermique et dynamique de double vitrages courbés. Approche numerique et expérimentale, Ph.D. thesis, Institut National des Sciences Appliqués de Toulouse, France, November 5, 1999.

28. G. Ledezma, A. M. Morega, and A. Bejan, Optimal spacing between fins with impinging flow, *J. Heat Transfer*, Vol. 118, 1996, pp. 570–577.

29. A. Bejan, A general variational principle for thermal insulation system design, *Int. J. Heat Mass Transfer*, Vol. 22, 1979, pp. 219–228.

30. S. Weinbaum and L. M. Jiji, A new simplified bioheat equation for the effect of blood flow on local average tissue temperature, *J. Biomech. Eng.*, Vol. 107, 1985, pp. 131–139.

31. J. R. Lloyd and W. R. Moran, Natural convection adjacent to horizontal surfaces of various platforms, ASME Paper 74-WA/HT-66, American Society of Mechanical Engineers, New York, 1974.

32. S. Globe and D. Dropkin, Natural convection heat transfer in liquids confined by two horizontal plates and heated from below, *J. Heat Transfer*, Vol. 81, 1959, pp. 24–28.

33. K. R. Blake, D. Poulikakos, and A. Bejan, Natural convection near 4°C in a horizontal water layer heated from below, *Phys. Fluids*, Vol. 27, 1984, pp. 2608–2616.

34. D. A. Nield and A. Bejan, *Convection in Porous Media*, Springer–Verlag, New York, 1992, p. 224.

35. P. Cheng and J. D. Chang, On buoyancy induced flows in a saturated porous medium adjacent to impermeable horizontal surfaces, *Int. J. Heat Mass Transfer*, Vol. 19, 1976, pp. 1267–1272.

36. P. Cheng, Heat transfer in geothermal systems, *Adv. Heat Transfer*, Vol. 14, 1978, pp. 1–105.

37. R. A. Nelson, Jr. and A. Bejan, Self-organization of the internal flow geometry in convective heat transfer, in *Proceedings of the 7th AIAA/ASME Joint Thermophysics and Heat Transfer Conference*, ASME HTD-Vol. 357-3, 1998, pp. 149–161.

38. J. Kepler, *The Six-Cornered Snowflake*, Oxford Univ. Press, Oxford, UK, 1966 (original in Latin, 1611).

39. A. Bejan, How nature takes shape: extensions of constructal theory to ducts, rivers, turbulence, cracks, dendritic crystals, and spatial economics, *Int. J. Therm. Sci. (Rev. Gén. Therm.)*, Vol. 38, 1999, pp. 653–663.

40. S. Govender, Coriolis effect on the stability of convection in mushy layers during the solidification of binary alloys, Ph.D. thesis, University of Durban-Westville, South Africa, 1999.

CHAPTER EIGHT

CONVECTIVE TREES

8.1 Convection in the Interstices versus Convection in the Tree Branches

In this chapter we turn our attention to the more complex configurations in which the flow of heat between a finite volume and one point is aided by the flow of a fluid. The resulting structures are trees in which convection plays an important role, but not the complete role. In every elemental volume, convection is intimately coupled with pure conduction, in a phenomenon known as conjugate heat transfer [1]. The key to developing the optimal flow architecture, from the smallest elemental volumes to larger and larger constructs, is to find this optimal coupling, or optimal balance between convection and conduction. In this work of geometric optimization we retrace the steps made in earlier chapters: A convective tree can be viewed as two trees that cohabitate in the same volume, a heat tree and a fluid tree. Even though we speak of trees for heat convection, the same method and geometric results apply to trees, the purpose of which is to convect mass (chemical species); see the analogy between heat and mass transfer [1, 2].

Convective flow architectures are of two kinds, depending on which portions of the structure are reserved for convection. In the first part of this chapter we optimize structures with convection in the interstices, which is coupled with pure conduction in the solid branches of the tree. Every interstitial space serves as sink or source for the current that passes through the root of the tree. Numerous applications for this flow structure are found in the design of heat transfer-enhanced surfaces for heat exchangers and for cooling small-scale electronics. In the latter, the tree structures are better known as fin trees and fin bushes.

The second configuration type is considered in the closing sections of this chapter. The interstices are occupied by a solid that generates heat at every point and conducts the heat by diffusion in the manner of the k_0 material analyzed in Fig. 4.1. Convection is located in the branches of a tree formed by ducts filled with flowing fluid. The ramifications of two trees of this type visit each elemental volume. One tree delivers cold fluid to each element. The other tree collects the fluid heated by the element and reconstitutes it into a single stream that eventually leaves the volume.

The circulatory system performs its mass transfer function and secondary heat transfer function by using a convective double-tree structure of the second type (Section 10.6). In the respiratory system the two flow trees are superimposed, as they rely on the same network of bronchial tubes, one tree during inhaling and the other during exhaling. Engineering applications of trees of convective channels abound in the cooling of virtually every enclosed electrical heat generating system, e.g., windings of electrical machines, computers, and electrical and electronic packages of many types and sizes. All must be cooled at every point, by forced or natural convection.

As in the earlier chapters on simpler trees, the objective of our constructal optimizations is not to model, refine, and in this way reproduce a particular convective tree observed in a living or not-living system. The objective is to *deduce* the tree structure from theory. It is to reason the tree's existence as a necessary feature – a result – of invoking the purpose and constraints principle consistently, in every step. The observation that the deduced tree structure also occurs in nature and the use of this observation are for later, and for others. It can be used in physiology and geophysics to enhance substantially the theoretical character of specific models that continue to be perfected.

8.2 Two-Dimensional T-Shaped Plate Fins

Consider an assembly of rigidly connected fins that fills a specified space and has a specified amount of fin material. The heat current touches every point of this volume by convection, as the volume is being bathed steadily by a stream of fluid. This current is continued by conduction through the solid links of the fin assembly and passes through the root of the assembly into the supporting wall. For simplicity, we neglect the heat transfer through the remaining (bare, unfinned) surface of the base wall.

Individual fins and assemblies of fins have long been recognized as effective means to augment heat transfer. The literature on this subject is sizable, as shown by the most current reviews [3, 4]. The new aspect that is contributed by the constructal method is the complete geometric optimization of the assembly of fins when the total volume inhabited by the assembly is fixed. To illustrate this aspect in the most transparent terms, we begin with some of the simplest assembly types that have been recognized in practice [3–8].

The simplest is the T-shaped assembly of plate fins shown in Fig. 8.1. The flow is parallel to the width W. Temperature variations are neglected based on the assumption that the thermal conductivity of the fin material k_p is sufficiently large. Simplicity in analysis also recommends the assumption that the heat transfer coefficient between each swept surface and the fluid is a constant, h [2, 10]. The dimensionless heat transfer coefficients known as Biot numbers ($ht_1/k_p, ht_2/k_p$) are assumed to be smaller than 1, so that the conduction current through each plate is oriented almost longitudinally [2], along L_0 and L_1. The overall temperature range spanned by the system is fixed by the specified temperature of the approaching fluid (T_∞) and the temperature of the base wall (T_1).

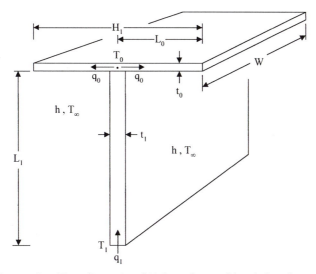

FIGURE 8.1. Two-dimensional T-shaped assembly of plate fins [9].

The geometry of the T-shaped assembly has 2 degrees of freedom, the external shape parameter L_1/L_0 and the internal ratio of plate thicknesses t_1/t_0. Constrained are the total volume, which is represented by the cross-sectional flow area $A = 2L_0L_1$, and the volume of fin material, represented by the frontal area $A_f = 2L_0t_0 + t_1L_1$. The latter is also expressed by the volume fraction $\phi_1 = A_f/A \ll 1$. The global performance of the assembly is described by the dimensionless thermal conductance,

$$\tilde{q}_1 = \frac{q_1}{k_p W(T_1 - T_\infty)}, \tag{8.1}$$

where q_1 is the total heat current through the root of the assembly. We can determine the relationship between $\tilde{q}_1$ and L_1/L_0 and t_1/t_0 by accounting for unidirectional conduction along the three plates and requiring the continuity of temperature and heat current through the T junction, where the temperature T_0 is allowed to vary with the design. This analysis is left as an exercise (Problem 8.1). It is found that in addition to the 2 geometric degrees of freedom of the assembly, the global conductance $\tilde{q}_1$ depends on the Biot number based on $A^{1/2}$ as length scale, $hA^{1/2}/k_p$. What is more useful is the square root of this group,

$$a = \left(\frac{hA^{1/2}}{k_p} \right)^{1/2}. \tag{8.2}$$

The global conductance can be maximized twice, with respect to L_1/L_0 and t_1/t_0. In the range $0.01 \leq \phi_1 \leq 1$ and $0.1 \leq a \leq 1$, the twice-maximized conductance is correlated within 12% by the power law [9]

$$\tilde{q}_{1,mm} = 0.89a^{1.08}\phi_1^{0.41}. \tag{8.3}$$

The range chosen for parameter a is consistent with practical values encountered for h and k_p. For example, in forced convection to gas flow the order of magnitude of h is 10^2 W/(m^2 K) (see Ref. 2), whereas the thermal conductivities of commercial

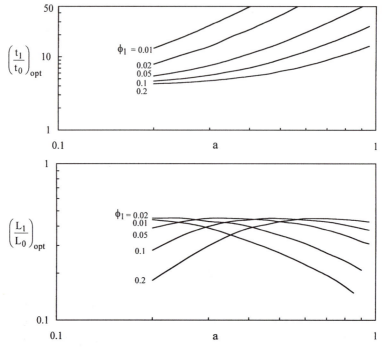

FIGURE 8.2. The optimized geometry of the T-shaped assembly of plate fins [9].

aluminum and copper are of the order of 10^2 W/(m K). Substituting these values and $A \sim 1$ cm^2 into Eq. (8.2) we obtain $a \sim 0.1$. This order of magnitude also validates the important assumption that the Biot numbers based on plate thicknesses are smaller than 1 because t_0 and t_1 are smaller than $A^{1/2}$.

Figure 8.2 shows the corresponding results for the optimized geometry of the construct. The internal aspect ratio $(t_1/t_0)_{opt}$ increases monotonically as a increases and as ϕ_1 decreases; however, these effects are weak when a becomes small and ϕ_1 becomes large. The external aspect ratio $(L_1/L_0)_{opt}$ has a more interesting behavior when ϕ_1 is fixed: This ratio exhibits a maximum with respect to parameter a. Note that when $(L_1/L_0)_{opt}$ is known, the individual lengths $L_{1,opt}$ and $L_{0,opt}$ follow immediately from the volume constraint A. Similarly, when the ratio $(t_1/t_0)_{opt}$ is known, the individual thicknesses $t_{1,opt}$ and $t_{0,opt}$ can be calculated easily from the material constraint, A_f or ϕ_1.

A numerical example of the optimized structure produced by this method is presented in Table 8.1 and the lower part of Fig. 8.3. This example corresponds to a case optimized in an earlier study by Kraus [3] (Fig. 8.3, top), who used

TABLE 8.1. Numerical Examples of Optimized T-Shaped Fin Assemblies ($\phi_1 = 0.086, a = 0.185$) [9]

	$L_0/A^{1/2}$	$L_1/A^{1/2}$	$t_0/A^{1/2}$	$t_1/A^{1/2}$	$\tilde{q}_1$
Constructal	1.33	0.376	0.0194	0.091	0.052
Kraus [3]	0.71	0.689	0.0191	0.086	0.040

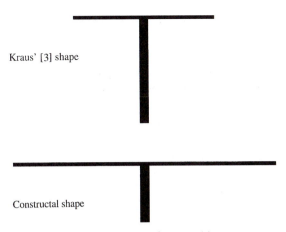

FIGURE 8.3. Examples of optimized T-shaped fin assemblies ($\phi_1 = 0.086$, $a = 0.185$) [9].

$k = 200$ W/(m K), $h = 60$ W/(m² K) and fin lengths and thicknesses that required the total frontal area $A = 32.4$ cm² and solid volume fraction $\phi_1 = 0.086$. In this case $a = 0.185$; see Eq. (8.2). Table 8.1 shows that the constructal thicknesses are nearly the same as those in Kraus' design and that the geometric differences result from the fin lengths. In the constructal case the elemental fins (L_0) are considerably longer. The contribution of the constructal design is the 30 percent increase in the global thermal conductance of the T-shaped assembly.

The slenderness of the elemental fins in the constructal design may turn into a disadvantage if the assembly experiences flow-induced vibrations. One way of extending the applicability of the constructal approach is to bend the ends of the L_0 fins, as shown in the tau-shaped assembly of Fig. 8.4. This technique was also described by Kraus [3]. It is particularly important in constructal design because the bending of the elemental ends allows the structure to "fill better" its allotted volume. Filling volumes in an optimal geometric way (with objective, or purpose) is the essence of the constructal method.

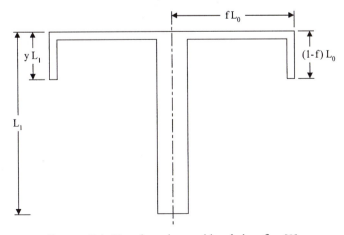

FIGURE 8.4. Tau-shaped assembly of plate fins [9].

The bending of the L_0 fins introduces a new dimensionless parameter: the fraction f, such that the turned end is of length $(1-f)L_0$ and the portion that is in thermal contact with the stem is of length fL_0. The frontal area constraint (1) is replaced with $A=2fL_0L_1$. The rest of the analysis is the same as for the T-shaped assembly (Problem 8.1). The new parameter f can take values in the range $f^* < f \leq 1$, where $f=1$ represents the T-shaped assembly, and $f^*=1-L_1/L_0$ represents the extreme where the bent end is as long as the stem, $(1-f)L_0 = L_1$.

The important question for the tau-shaped design is how the f parameter influences the optimal geometry and performance of the assembly. In other words, it is important to determine the thermal-design impact of increasing the stiffness of the assembly. This question is answered in Fig. 8.5. The three f values used in this figure mean that the lengths of the turned end of the L_0 fin are 25%, 50%, and, finally,

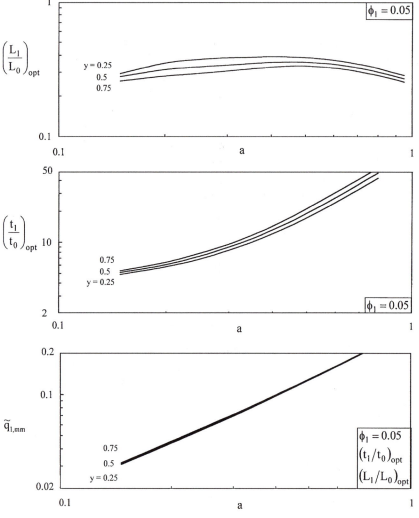

FIGURE 8.5. The optimized geometry and global conductance of the tau-shaped assembly of plate fins [9].

FIGURE 8.6. Narrower tau-shaped constructs mounted on the same wall [9].

75% of the stem length L_1. Note the graphic definition of the y fraction on Fig. 8.4, namely $f = 1 - yL_1/L_0$. The T-shaped designs of Fig. 8.2 correspond to $y = 0$.

The bottom frame of Fig. 8.5 shows that the overall conductance of the assembly decreases just slightly when the ends of the elemental fins are bent. The optimized aspect ratios $(L_1/L_0)_{opt}$ and $(t_1/t_0)_{opt}$ are also relatively insensitive to bending the ends. In conclusion, the optimized geometry and performance of T-shaped fins are robust and can be used as a good approximation for tau-shaped constructs that occupy the same frontal area. The bending of the elemental fins introduces a small thermal conductance penalty, which may be acceptable in view of the increased stiffness of the assembly.

The tau-shaped configuration needs an additional adjustment if several such fins are to be mounted one next to the other on the same wall. The uniform-h assumption makes it necessary to leave a space between the bent ends of consecutive elemental fins. This means that the rectangle circumscribed to each tau must be narrower than the volume (dashed line, Fig. 8.6) allocated and kept constant for that assembly. The numerical optimization of this modified geometry shows that the space left between adjacent assemblies has almost no effect on the optimized aspect ratios of the construct [9]. Furthermore, the global conductance decreases only slightly relative to that of the fins optimized in Fig. 8.2 and Table 8.1.

In summary, the optimized assembly is robust not only with respect to the bending of the tips of the elemental (L_0) fins, but also with respect to the spacing left between the bent tip and the margin of the frontal flow area allocated to the assembly.

8.3 Umbrellas of Cylindrical Fins

Section 8.2 outlined several classes of results that are generated by the same principle. In this section we consider an additional example of how the method may be applied to fin assemblies: the umbrella arrangement shown in Fig. 8.7. A cylindrical fin of length L_1 and diameter D_1 serves as the central stem for the spokes of the wheel formed by n_1 elemental fins of length L_0 and diameter D_0. The cases illustrated in Fig. 8.7 are for $n_1 = 2, 4$, and 6. The total space allocated to this construct is the cylinder of radius L_0 and height L_1 indicated by the dashed line, $V = \pi L_0^2 L_1$, constant. The total volume occupied by the fin material is also constrained, $V_f = (\pi/4)D_1^2 L_1 + n_1(\pi/4)D_0^2 L_0$. The dimensionless alternative to the V_f constraint is the solid volume fraction $\phi_1 = V_f/V$, which is fixed.

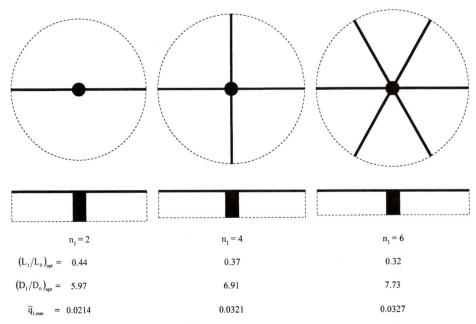

FIGURE 8.7. Examples of optimized umbrella assemblies of cylindrical fins ($\phi_1 = 0.01$, $b = 0.8$) [9].

The objective is to determine the optimal umbrella geometry such that the global thermal conductance $q_1/(T_1 - T_\infty)$ is maximum. The analysis follows the same steps as in the analysis of Fig. 8.1. The role of plate thicknesses t_0 and t_1 is now played by the cylinder diameters D_0 and D_1, respectively. The key is the continuity of temperature and heat current through the umbrella hub of temperature T_0. We report only the emerging dimensionless groups and the optimization results. In place of parameter a, Eq. (8.2), we now have

$$b = \left(\frac{4h V^{1/3}}{k_p} \right)^{1/2}. \tag{8.4}$$

Dimensionless global conductance definition (8.3) is replaced with

$$\tilde{q}_1 = \frac{q_1}{k_p V^{1/3}(T_1 - T_\infty)}. \tag{8.5}$$

The optimized geometry for the case with only two spokes is condensed in Fig. 8.8. The geometry is represented by the aspect ratios $(D_1/D_0)_{\text{opt}}$ and $(L_1/L_0)_{\text{opt}}$, which are functions of b and ϕ_1. The ratio $(D_1/D_0)_{\text{opt}}$ approaches the constant 1.82 as b drops below 0.1. The ratio $(L_1/L_0)_{\text{opt}}$ is nearly constant (~ 0.4) when b is of the order of 1. The twice-maximized conductance behaves as a power law in both b and ϕ_1: The results are correlated within 12% by an expression similar to Eq. (8.3) [9]:

$$\tilde{q}_{1,mm} = 0.77 b^{1.47} \phi_1^{0.61} \quad (n_1 = 2). \tag{8.6}$$

The double-optimization procedure was repeated for larger n_1 values. The results covering the range $2 \le n_1 \le 15$ are reported in Ref. 9 for constant ϕ_1 and varying

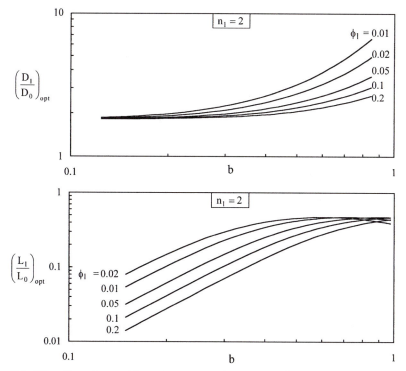

FIGURE 8.8. The effect of ϕ_1 and b on the optimized geometry of the case $n_1 = 2$ of Fig. 8.7 [9].

b and for constant b and varying ϕ_1. The effect of n_1 on the twice-maximized global conductance is very weak. The ratio $(D_1/D_0)_{opt}$ increases as n_1 increases, i.e., the spokes become relatively thinner when they are more numerous. The ratio $(L_1/L_0)_{opt}$ is more sensitive to changes in b than to changes in ϕ_1 and n_1.

The three scale drawings of Fig. 8.7 show how the number of spokes influences the optimized geometry when $\phi_1 = 0.01$ and $b = 0.8$. The dashed contour indicates the allocated space (V), which is the same in all three cases. In sum, the optimization of the umbrella construct leads to concrete geometric aspect ratios and the observation that the optimized geometry is robust with respect to changes in some design parameters. The same conclusion is reached in many other cases detailed in this book.

8.4 Fin Trees with Optimal Plate-to-Plate Spacings

More complex trees fill larger volumes more efficiently. When we contemplate adding more elemental plate fins to the T-shaped design of Fig. 8.1, that is, more L_0 blades on the same L_1 stem, we run into an important limitation of the simplified analysis employed until now. The assumption that the heat transfer coefficient h is uniform and constant breaks down when the flow sweeping one surface is affected by the shapes and proximity of the neighboring surfaces. That assumption is abandoned in this section. One less constraint means one more opportunity for optimization: the spacing between adjacent fins (see also Section 12.2).

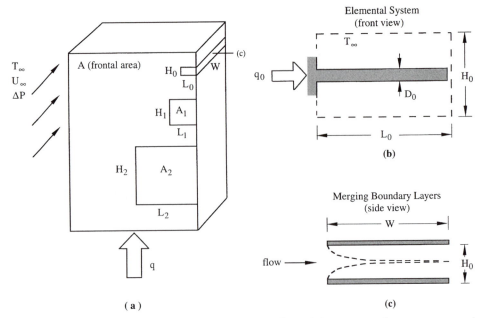

FIGURE 8.9. (a) Volume (AW) serving as convective heat sink or source for a concentrated heat current, (b) the smallest volume element associated with a single plate fin, and (c) how the boundary layers touch when the fin-to-fin spacing is optimal [11].

Let us review the constructal fin-tree problem statement, this time with reference to the general geometry sketched in Fig. 8.9. We consider the two-dimensional volume of frontal area A and fixed length W, where W is aligned with the free stream (U_∞, T_∞). The problem consists of distributing optimally through this volume a fixed amount of high-conductivity (k_p) material, which takes heat from one spot on the boundary and discharges it throughout the volume. We may think of the boundary spot (root) as the external surface of an electronic module that must be cooled. In this case the volume AW is the space that is allocated for the purpose of cooling the module by forced convection. This volume constitutes an overall constraint.

As in the pure-conduction applications of the constructal method (Chap. 4), we start the space-filling optimization sequence from the smallest, finite-size scale. The smallest system (the elemental system in constructal terminology) consists of a two-dimensional volume $H_0 L_0 W$, in which there is only one blade of k_p material [Fig. 8.9(b)]. The thickness of this blade is D_0. Heat is transferred from one boundary spot (T_0, at the root of the fin) to the entire elemental volume. If we neglect the heat transfer through the fin tip and if we use the unidirectional fin conduction model [2, 10] the elemental heat current is

$$\frac{q_0}{(T_0 - T_\infty)W} = (2k_p D_0 h_0)^{1/2} \tanh\left[\left(\frac{2h_0}{k_p D_0}\right)^{1/2} L_0\right]. \qquad (8.7)$$

Unlike in Chap. 4, where it was possible to optimize the shape of the elemental volume, in the present problem the thickness of the elemental volume (H_0) is fixed because it is the same as the optimal spacing between two successive D_0-thick plate fins (review Section 3.3). The spacing is optimal when the laminar boundary layers that develop over the swept length W became thick enough to touch at the trailing edge of each plate fin [12, 13]. Conversely, the spacing is optimal when the time of fluid travel along W matches the thermal diffusion time across the D_0 channel (Section 3.5). This feature is illustrated in Fig. 8.9(c). Specifically, the optimal spacing $(H_0 - D_0 \sim H_0)$ is determined uniquely by the length W and the pressure difference Δp maintained across the swept volume [2, 13] [see Eq. (3.22)]:

$$\frac{H_0}{W} \cong 2.7 \, \mathrm{Be}^{-1/4} = 2.7 \left(\frac{\mu \alpha}{W^2 \Delta p} \right)^{1/4}. \tag{8.8}$$

Using the results of Section 3.3, we can show that the minimized thermal resistance that corresponds to this spacing is also characterized by an average heat transfer coefficient that is given approximately by [11]

$$h_0 \cong 0.55 \frac{k}{W} \left(\frac{W^2 \Delta p}{\mu \alpha} \right)^{1/4}. \tag{8.9}$$

Here k is the thermal conductivity of the fluid. The factor Δp that appears in Eq. (8.9) refers to the pressure difference that is maintained in the W direction (e.g., by a fan), where $H_0 \times L_0$ is the cross section of one duct. If the H_0-wide channel is open to one side (the side that would connect the tips of two successive fins) and if the entire assembly is immersed in a stream of velocity U_∞, then Eqs. (8.8) and (8.9) are adequate if Δp is replaced with the dynamic pressure associated with the free stream, $\Delta p \cong \rho U_\infty^2 / 2$. The correctness of this approach in correlating the optimal spacing between parallel plates immersed in a free stream has been demonstrated based on complete numerical simulations of the flow field [14].

We now proceed toward larger scales by recognizing that at the elemental level the h_0 and H_0 values provided by Eqs. (8.8) and (8.9) are two known constants. The next volume subsystem is the first assembly of constant frontal area $H_1 L_1 = A_1$, which is shown in Fig. 8.10. The shape of this volume (H_1 / L_1) is free to vary. The assembly is defined by a central blade of thickness D_1, which is connected to all the elemental volumes that are needed to fill the $A_1 W$ volume. The D_1 blade connects the roots of all the D_0 fins. When the number of elemental volumes in this assembly is large, the cooling effect provided by the D_0 fins is distributed almost uniformly along the D_1 stem. In this limit the D_1 stem functions as a fin immersed in a convective medium with a constant heat transfer coefficient. We can deduce the effective heat transfer coefficient of this medium (h_1) from Eq. (8.7) by noting that each q_0 current flows out of the D_1 blade through an area of size $H_0 W$. In other words, we combine $h_1 = q_0 / [H_0 W (T_0 - T_\infty)]$ with Eq. (8.7) and $L_0 = H_1 / 2$ and obtain

$$h_1 = \frac{1}{H_0} (2 k_p D_0 h_0)^{1/2} \tanh \left[\left(\frac{2 h_0}{k_p D_0} \right)^{1/2} \frac{H_1}{2} \right]. \tag{8.10}$$

The D_1 blade functions as a fin with insulated tip; therefore we write [see Eq. (8.7)]

$$\frac{q_1}{(T_1 - T_\infty) W} = (2 k_p D_1 h_1)^{1/2} \tanh \left[\left(\frac{2 h_1}{k_p D_1} \right)^{1/2} L_1 \right]. \tag{8.11}$$

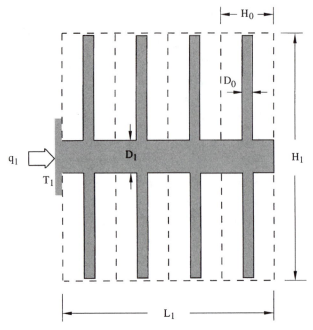

FIGURE 8.10. First construct consisting of a large number of elemental plate-fin volumes [11].

Next, we find that the geometry of the first assembly has 2 degrees of freedom. The search for the optimal geometry is aided by the introduction of the dimensionless quantities,

$$\tilde{q}_1 = \frac{q_1}{(T_1 - T_\infty)Wb_0 H_0}, \qquad \tilde{b}_1 = \frac{b_1}{b_0}, \tag{8.12}$$

$$\tilde{D}_0 = \frac{D_0 k_p}{b_0 H_0^2}, \qquad \tilde{D}_1 = \frac{D_1 k_p}{b_0 H_0^2}, \tag{8.13}$$

$$\tilde{H}_1 = \frac{H_1}{H_0}, \qquad \tilde{L}_1 = \frac{L_1}{H_0}, \tag{8.14}$$

such that Eqs. (8.10) and (8.11) become

$$\tilde{b}_1 = (2\tilde{D}_0)^{1/2}\, \tanh\left[(2\tilde{D}_0)^{-1/2}\,\tilde{H}_1\right], \tag{8.15}$$

$$\tilde{q}_1 = (2\tilde{D}_1\tilde{b}_1)^{1/2}\, \tanh\left[\left(\frac{2\tilde{b}_1}{\tilde{D}_1}\right)^{1/2}\tilde{L}_1\right]. \tag{8.16}$$

There are two constraints that must be satisfied. First, the volume constraint $H_1 L_1 = A_1$ now reads

$$\tilde{H}_1\tilde{L}_1 = \tilde{A}_1, \tag{8.17}$$

where $\tilde{A}_1 = A_1/H_0^2$. The second constraint is the total frontal area of all the k_p blades,

$$A_{p,1} = D_1 L_1 + n_1 D_0 L_0, \tag{8.18}$$

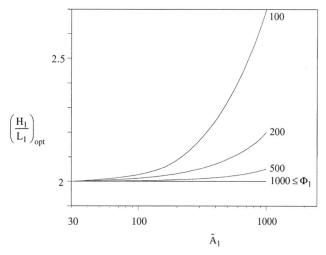

FIGURE 8.11. The optimal external shape of the first construct of Fig. 8.10 [11].

where n_1 is the number of elemental systems contained by the first assembly, $n_1 = 2L_1/H_0$, which is assumed large. The k_p material constraint assumes the dimensionless form

$$\Phi_1 = \frac{\tilde{D}_1}{\tilde{H}_1} + \tilde{D}_0, \tag{8.19}$$

in which Φ_1 accounts for the volume fraction $\phi_1 = A_{p,1}/A_1$ occupied by the high-conductivity material:

$$\Phi_1 = \phi_1 \frac{k_p}{h_0 H_0}. \tag{8.20}$$

The objective is to maximize the $\tilde{q}_1$ expression of Eq. (8.16) subject to Eqs. (8.15), (8.17), and (8.19), and to determine the optimal geometry represented by the aspect ratios H_1/L_1 and D_1/D_0. Figure 8.11 shows that the optimal aspect ratio of the assembly is approximately 2 in the entire $\tilde{A}_1$, Φ_1 range considered. This is a robust optimization result, which also agrees numerically with the optimal external shape of first constructs in tree networks for heat conduction (Section 4.3).

The large $\tilde{A}_1$ values used on the abscissa of Fig. 8.11 are required by the geometric relation $A_1 \gg H_0^2$. The large Φ_1 values are due to the factor $k_p/(h_0 H_0)$ on the right-hand side of Eq. (8.20), in spite of the fact that the volume fraction ϕ_1 is generally small. To see this, assume that the fluid and its velocity are such that Eq. (8.8) yields $H_0/W = 0.1$. The corresponding result deduced from Eq. (8.9), using for k the thermal conductivity of room-temperature air, is $h_0 W = 0.38$ W/m K. Next, in Eq. (8.20) we substitute for k_p the thermal conductivity of commercial copper and obtain $\Phi_1 \cong \phi_1 \times 10^4$.

The second geometric feature that can be optimized in the first construct is the ratio of the two plate thicknesses, which is shown in the upper half of Fig. 8.12. This ratio varies as $(D_1/D_0)_{\text{opt}} \cong 1.4 \, \tilde{A}_1^{1/2}$ when $\Phi_1 \geq 10^3$. The effect of the volume fraction Φ_1 is weak.

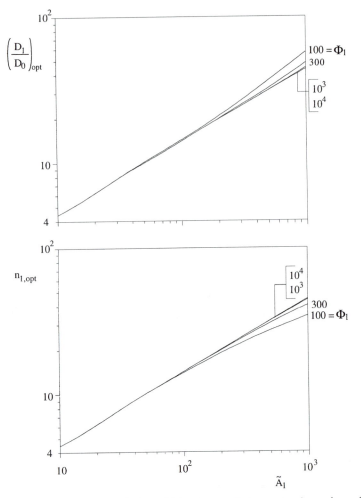

FIGURE 8.12. The optimal ratio of plate thicknesses and the optimal number of elemental volumes for the first construct of Fig. 8.10 [11].

The lower half of Fig. 8.12 shows the corresponding optimal number of elemental plate fins. This graph is an alternate presentation of the external-shape results of Fig. 8.11, because $n_1 = 2L_1/H_0$. Most interesting is that the two frames of Fig. 8.12 are very similar: The optimal ratio of fin thicknesses is almost equal to the optimal number of elemental systems assembled. This means that in the first construct the cross section for fin conduction is conserved in going from the elemental constituents to the root of the central stem, i.e., $n_1 D_0 \cong D_1$.

Figure 8.13 shows the results obtained for the twice-maximized overall heat transfer rate of the first assembly, $\tilde{q}_{1,mm}$. This quantity is almost proportional to $\tilde{A}_1$, i.e., proportional to the size of the volume allocated for the removal of the heat current q_1 by convection. The Φ_1 effect is weaker but not negligible: The heat transfer rate decreases as the amount of high-conductivity material decreases. The effects of the volume size $\tilde{A}_1$ and fin material Φ_1 on the heat transfer rate are expected: More space and more fin material mean larger heat currents.

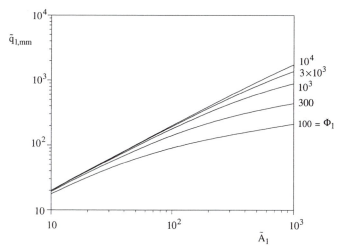

FIGURE 8.13. The twice-maximized global thermal conductance of the first construct [11].

In Fig. 8.14 we see the evolution of the twice-optimized geometry of the first assembly, while the size parameter $\tilde{A}_1$ increases at constant volume fraction of fin material ($\Phi_1 = 1000$). In other words, the amount of fin material increases in proportion to the volume occupied by the fin tree. The drawing was made with the assumption that $\phi_1 = 0.1$ or $\Phi_1 = 10^4 \phi_1$, as in the preceding numerical example. Each frame was drawn to scale to show the growth of the volume occupied by the assembly. The role of unit length is played by the thickness of the elemental

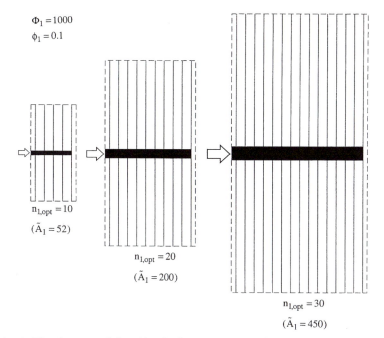

FIGURE 8.14. The changes exhibited by the first-construct architecture as the occupied volume increases ($\phi_1 = 0.1$, $\Phi_1 = 1000$) [11].

volume (H_0), which is constant according to the discussion that followed Eqs. (8.8) and (8.9). The exhibited structures were selected to correspond to even values of n_1; the corresponding values of the assembly size $\tilde{A}_1$ are also listed.

Two invariant features of the fin trees of Fig. 8.14 are worth noting. The overall shape of the volume occupied by the assembly does not change much as the volume size increases. This feature is an illustration of the $(H_1 L_1)_{opt}$ results reported in Fig. 8.11. Another feature is the thickness of the elemental fin $D_{0,opt}$, which in Fig. 8.14 does not vary from frame to frame. It can be verified that when the elemental spacing H_0 and the volume fraction Φ_1 are fixed, the optimized thickness of the elemental fins is also fixed. If in the preceding numerical example $H_0 = 0.5$ cm, then $D_{0,opt} = 0.025$ cm in each of the frames of Fig. 8.14. The corresponding thickness of the trunk, $D_{1,opt}$, takes the values 0.25, 0.5, and 0.75 cm as $n_{1,opt}$ increases.

Future work may address various improvements and refinements of the classical fin conduction model that was used in setting up the present analysis and optimization. For example, the effects of temperature-dependent thermal conductivity, radiative heat transfer, and spatially varying heat transfer coefficient can be incorporated at the elemental level, although a penalty is paid through the increased complexity of the analysis and the need for numerical work even at the elemental level. Equation (8.8) will most definitely be replaced with a numerical procedure in the refined model.

In summary, the maximized heat transfer rate accommodated by the first assembly increases monotonically as the volume increases (Fig. 8.13). Can this increase continue forever? No, because the invariant thickness of the elemental fins leads eventually to mechanical difficulties (bending, vibrations, warpage). The alternative is to stiffen the construct by increasing its complexity. The next step toward greater complexity is to form a second construct by mounting a number (n_2) of first constructs on a new central plate fin of thickness D_2 and length L_2. The details of this construction and optimization are presented in Ref. 11.

It was found that the global conductance of the second construct is the largest when only two first constructs are combined into a second construct, $n_2 = 2$, as in the top of Fig. 8.15. In this case the length of the D_2 stem is $L_2/2$, because the stem has to reach only the center of the frontal area ($A_2 = H_2 L_2$) of the second construct.

The second optimized feature is the thickness of the central stem. The numerical results are correlated to within 4% by the power law

$$\tilde{D}_{2,opt} \cong 0.9 \tilde{A}_2^{1/2} \Phi_2, \qquad (8.21)$$

where

$$\tilde{D}_2 = \frac{D_2 k_p}{h_0 H_0^2}, \qquad \tilde{A}_2 = \frac{A_2}{H_0^2}, \qquad (8.22)$$

$$\phi_2 = \frac{A_{p,2}}{A_2}, \qquad \Phi_2 = \frac{\phi_2 k_p}{h_0 H_0}. \qquad (8.23)$$

In the ϕ_2 definition, $A_{p,2}$ is the area of the k_p material cross section, $A_{p,2} = (L_2 - H_1/2)D_2 + n_2 A_{p,1}$. What is interesting is the ratio $(D_2/D_1)_{opt}$ reported in Fig. 8.16,

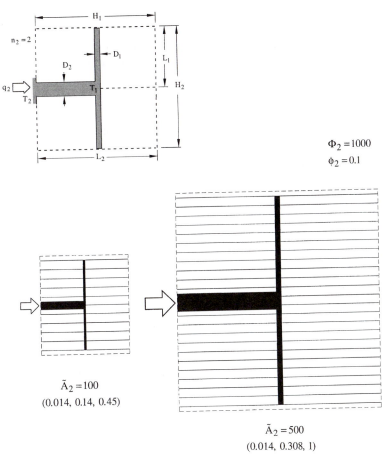

FIGURE 8.15. The changes exhibited by the optimized architecture of the second construct as the total volume increases ($\phi_2 = 0.1$, $\Phi_2 = 1000$) [11].

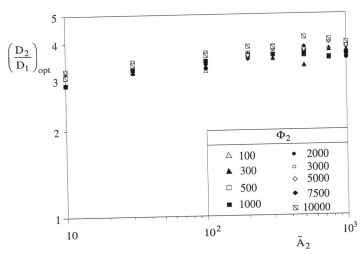

FIGURE 8.16. The optimal ratio of stem thicknesses for the second construct [11].

which is a replotting of the thickness optimization results that led to Eq. (8.21). We arrive in this way at a new invariant in the geometry of the second assembly. The stem thickness increases by a nearly constant factor in going from the first assembly to the second assembly. A similar characteristic was found in the optimization of the second construct of the conduction tree (Section 4.4). In the present case, the ratio $(D_2/D_1)_{opt}$ is equal to 3.54 with a standard deviation of only 0.048.

The lower half of Fig. 8.15 provides a pictorial summary of the quantitative results developed for the second construct [11]. The figure is a scale drawing of the optimized architecture when the volume fraction and properties of the fin material are characterized by $\Phi_2 = 10^3$ and $\phi_2 = 0.1$. As in Fig. 8.14, the construction is based on the assumption that the external flow is such that the elemental spacing is $H_0 = 0.5$ cm. This leads to $D_{0,opt} = 0.014$ cm. The two lower frames of Fig. 8.15 show how the architecture changes as the tree of fins spreads over a larger volume. The shape of the occupied volume remains almost the same (note the nearly square frontal area), and the thickness of the central stem increases relative to the other fin thicknesses. The calculated fin thicknesses $(D_0, D_1, D_2)_{opt}$ are indicated in centimeters under each frame.

The performance of the second construct is described further in Ref. 11, which also reports a fully numerical simulation and optimization of the first-construct configuration.

Practical constraints must be taken into account before continuing the optimization with even more complex constructs. Manufacturing constraints will certainly rule the decision with regard to the complexity of the fin design (e.g., first construct vs. second construct). In this regard, the practical value of the work summarized in this section has two facets. First, the constructal optimization represents a *strategy* (a road map) for pursuing optimal structure in a design subjected to volume constraint. Second, many of the geometrical features that are determined by this method are *robust* (relatively insensitive to other design parameters): Examples are the optimized shapes of the frontal areas and the ratios between successive fin thicknesses.

8.5 Trees of Circular Fins

Constructal trees of fins can also be pursued in cylindrical geometries [15]. For example, consider the first construct defined in the upper part of Fig. 8.17. The solid is the k_p material that fills the D_0 disks and the D_1 stem. Heat transfer occurs between the root disk D_1 and the fluid (T_∞) that flows along the faces of the D_0 disks. The total volume [$V_1 = (\pi/4)H_1^2 L_1$] and the volume fraction occupied by the solid ($\phi_1 = V_{p1}/V_1$) are fixed. The number of D_0 fins installed on the stem is n_1. We seek to maximize the overall conductance by varying the external and the internal geometric features of the assembly.

The overall conductance is calculated in two steps. First, we evaluate the heat transfer rate collected by the stem from each disk,

$$q_0 = \eta \frac{\pi}{2}(H_1^2 - D_1^2)h_0(T_s - T_\infty), \qquad (8.24)$$

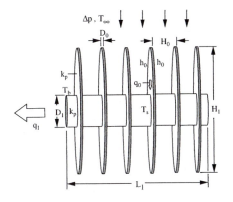

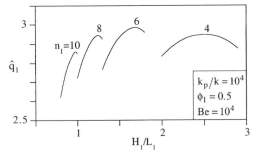

FIGURE 8.17. First construct of circular fins with convection in the interstices and the double maximization of the global resistance [15].

where the constant h_0 is the heat transfer coefficient on both sides of the D_0-thin plate. The fin efficiency η is available in terms of Bessel functions [16] and is a function of two dimensionless parameters,

$$\eta = \text{function} \left[\left(\frac{H_1}{2} - \frac{D_1}{2} \right) \left(\frac{2h_0}{k_p D_0} \right)^{1/2}, \frac{H_1}{D_1} \right]. \tag{8.25}$$

In the second step we average the q_0 effect of one circular fin over the stem surface of length H_0, which is allocated to one fin. The heat flux $q_0/(\pi D_1 H_0)$ can be used to define an equivalent heat transfer coefficient h_1,

$$\frac{q_0}{\pi D_1 H_0} = h_1(T_s - T_\infty), \tag{8.26}$$

as if the stem, alone, were transferring heat to the flow. If we assume that the circular plate fins are sufficiently numerous and close to each other, then the D_1 stem performs as a one-dimensional fin with insulated end and uniform heat transfer coefficient h_1. The total heat transfer rate through the base (T_b) is given by the classical formula

$$q_1 = (T_b - T_\infty)(k_p A_c h_1 p)^{1/2} \tanh(mL_1), \tag{8.27}$$

where $A_c = (\pi/4)D_1^2$, $p = \pi D_1$, and $m = (h_1 p / k_p A_c)^{1/2}$. Equation (8.27) can be nondimensionalized with $V_1^{1/3}$ as length scale and $\hat{h} = h_1/h_0$. The resulting overall thermal conductance,

$$\hat{q}_1 = \frac{q_1}{(T_b - T_\infty)(k_p V_1 h_0)^{1/2}} = \frac{\pi}{2} \hat{D}_1^{3/2} \hat{h}^{1/2} \tanh \left[2\hat{L}_1 \left(\frac{\hat{h}}{\hat{D}_1} \right)^{1/2} \frac{h_0^{1/2} V_1^{1/6}}{k_p^{1/2}} \right], \quad (8.28)$$

reveals the dimensionless group $h_0^{1/2} V_1^{1/6}/k_p^{1/2}$. We seek an estimate for the heat transfer coefficient h_0 and note that it will vary with the spacing between two consecutive fins. We make the same assumption as in the preceding section: The space between consecutive fins has already been optimized for maximum conductance. In laminar forced convection the optimal spacing (in our case, $H_0 - D_0$) is proportional to the swept flow length (H_1) [see Eq. (8.8)],

$$\frac{H_0 - D_0}{H_1} \sim \left(\frac{\mu \alpha}{H_1^2 \Delta p} \right)^{1/4}. \quad (8.29)$$

The pressure difference Δp is maintained by a fan in order to drive the flow through the assembly of Fig. 8.17. Nondimensionalized, approximation (8.29) becomes

$$\frac{\hat{H}_0 - \hat{D}_1}{\hat{H}_1^{1/2}} \sim \mathrm{Be}^{-1/4}, \quad (8.30)$$

where $\mathrm{Be} = \Delta p V_1^{2/3}/(\mu \alpha)$ is the pressure-drop number [2, 17]. As in the preceding section, this design with optimal interstices is characterized by a maximum conductance [1, 13] that can be used to estimate the order of magnitude of the interstitial heat transfer coefficient:

$$h_0 \sim \left(\frac{\rho \Delta p}{\mathrm{Pr}} \right)^{1/2} c_p. \quad (8.31)$$

In this way the heat transfer coefficient h_0 emerges as a constant set by the fan (Δp), which is why h_0 was used as a constant in the definitions of $\hat{q}_1$ and $\hat{h}$. The dimensionless group revealed by Eq. (8.28),

$$b = \frac{h_0^{1/2} V_1^{1/6}}{k_p^{1/2}} = \left[(c_p V_1^{1/3}/k_p)(\rho \Delta p/\mathrm{Pr})^{1/2} \right]^{1/2}, \quad (8.32)$$

is a parameter proportional to $\mathrm{Be}^{1/4}$:

$$b = \left(\frac{k}{k_p} \right)^{1/2} \mathrm{Be}^{1/4}. \quad (8.33)$$

The rest of the problem statement can be put in the same nondimensional notation as that of Eq. (8.28). Equations (8.24) and (8.26) provide the needed expression for $\hat{h}$, which is

$$\hat{h} = \frac{2\eta}{\hat{D}_1 \hat{H}_0} (\hat{H}_1^2 - \hat{D}_1^2). \quad (8.34)$$

The fin efficiency given by Eq. (8.25) assumes the dimensionless form:

$$\eta = \text{function} \, [(\hat{H}_1 - \hat{D}_1) b / (2\hat{D}_0)^{1/2}, \hat{H}_1 / \hat{D}_1]. \qquad (8.35)$$

The geometric relations are the total-volume constraint, the solid-fraction volume constraint, and the number of circular plate fins:

$$\hat{H}_1^2 \hat{L}_1 = 4/\pi, \qquad (8.36)$$

$$\phi_1 = n_1 \frac{\pi}{4} (\hat{H}_1^2 - \hat{D}_1^2) \hat{D}_0 + \frac{\pi}{4} \hat{D}_1^2 \hat{L}_1, \qquad (8.37)$$

$$n_1 = \hat{L}_1 / \hat{H}_0. \qquad (8.38)$$

In summary, there are seven equations [namely, (8.28), (8.30), (8.34)–(8.38)] that contain nine dimensionless variables: $\hat{H}_1$, $\hat{D}_1$, $\hat{L}_1$, $\hat{H}_0$, $\hat{D}_0$, n_1, $\hat{q}_1$, $\hat{h}$, and η. The design has 2 degrees of freedom, one external and the other internal: H_1/L_1 and n_1, or $H_1 L_1$ and D_1/D_0. The physical parameters ϕ_1, Be, and k_p/k are fixed and play the role of constraints. The objective is to maximize $\hat{q}_1$ in two dimensions subject to the constraints. This procedure is illustrated for one case in the lower part of Fig. 8.17, where it is found that the best geometry is represented by $n_{1,\text{opt}} = 6$ and $(H_1 L_1)_{\text{opt}} = 1.7$.

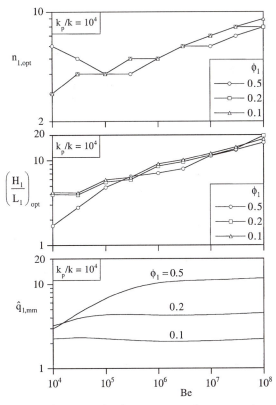

FIGURE 8.18. The number of elemental volumes, external aspect ratio, and maximized conductance of the first construct of circular fins [15].

This double optimization was repeated for other combinations of solid volume fraction ϕ_1 and pressure-drop number Be. All this work was conducted by setting $k_p/k = 10^4$, which is the correct order of magnitude for the combination of copper in the solid part and room air in the fluid part. The results are presented in Fig. 8.18. The twice-maximized conductance $\hat{q}_{1,mm}$ increases monotonically with both ϕ_1 and Be, which is the expected trend. The optimal number of elements increases slowly as Be increases: This trend is almost insensitive to ϕ_1 when ϕ_1 is of the order of 0.1 or smaller. The external aspect ratio $(H_1 L_1)_{opt}$ is more sensitive to changes in Be because the interstitial spaces H_0 become narrower as Be increases [see approximation (8.29)].

The body of work on the geometric optimization of electronic packages [14, 18] shows that the effect of the pressure-drop number Be can be expressed more readily in terms of the free-stream velocity of the stream. The order of magnitude of Δp is ρU_∞^2, and this transforms the Be definition into

$$Be \cong Re^2 Pr, \tag{8.39}$$

where $Re = U_\infty V_1^{1/3}/\nu$ and $Pr = \nu/\alpha$. The pressure-drop number Be accounts for the free-stream velocity squared.

8.6 Conduction in Interstitial Spaces and Convection in Channels

In this and the following sections, the locations of conduction and convection are reversed: conduction covers all the interstitial volumes, while convection proceeds along channels or ducts that are distributed through the volume. The heat current is generated volumetrically, at every point in the solid material. It is collected by the channels and led out of the system through one port. This new configuration may be viewed as the inside–out counterpart of the convective fin constructs treated until now in this chapter.

The new fundamental problem that we propose can be stated with reference to Fig. 8.19. The two-dimensional space HL generates heat volumetrically at the rate q''' (W/m^3), which is uniform. The temperature in this space cannot exceed a prescribed level, T_{peak}. The coolant is a stream of single-phase fluid $\dot{m}'$ (kg/s m) and initial temperature T_0. The objective is to cool every point of the volume, to

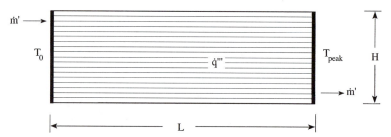

FIGURE 8.19. Space with uniform volumetric heat generation and unidirectional permeating flow [19].

maximize the overall thermal conductance $q'''HL/(T_{peak} - T_0)$, and to accomplish this task with *minimal pumping power*. This new problem and its geometric solution have fundamental implications not only in heat transfer (e.g., electric windings, electronic packages) but also in biology and the thermodynamics of nonequilibrium systems: Every lung and vascularized tissue serves a similar function and is subjected to similar constraints.

To see the direction of the geometric approach that we use, imagine the ultimate design in which the $\dot{m}'$ stream is distributed uniformly over H and flows in the L direction while bathing every point of the HL space. This arrangement requires the use of two headers (the larger fluid volume indicated with thick black lines in Fig. 8.19), one upstream, to spread the $\dot{m}'$ stream, and the other downstream, to reconstitute the $\dot{m}'$ stream. It also requires a sufficiently refined porous structure (e.g., a set of small parallel channels) through which the stream sweeps the entire HL space, from left to right. Such a flow pushes the hot spots (T_{peak}) of the heat generating material to the right extremity of the HL space. When the porous structure is sufficiently fine, the peak temperature of the material is approximately the same as the peak bulk temperature of the permeating fluid. The latter occurs in the exit plane, such that the first law for the HL control volume requires that

$$\dot{m}' c_p(T_{peak} - T_0) = q'''HL. \qquad (8.40)$$

This form is correct if the coolant exhibits ideal-gas behavior with nearly constant c_p or incompressible fluid behavior with moderate pressure changes. Equation (8.40) shows that the overall conductance $q'''HL/(T_{peak} - T_0)$ is synonymous with the capacity flow rate $\dot{m}'c_p$. In an electronic package (HL), the largest amount of circuitry installed (q''') and the highest permissible temperature are dictated by materials and manufacturing and electrical engineering considerations. Once fixed, the overall thermal conductance fixes the coolant flow rate or the distributed bathing flow rate $\dot{m}'' = \dot{m}'/H$.

In sum, the ultimate design described by Eq. (8.40) is a constant-$\dot{m}''$ design that would require a fine structure and, necessarily, a high pressure drop. In the following geometric construction we investigate the possibility of meeting the objective of Eq. (8.40) by using flow paths with smaller pressure drops and with hot spots that are distributed *more uniformly* over the volume. We base this search on the constructal method, which showed that paths with lower overall resistance are achieved when their geometry (layout) is optimized, and when their internal complexity is increased. Such paths form tree networks.

8.7 Parallel-Plate Channels

Regardless of which fluid-flow network is chosen, the first leg of the path followed by the flow of heat will always be one of conduction (thermal diffusion) through the solid heat generating material. An elemental volume scale exists where the heat conducted out of the material is swept away by the first stream of coolant.

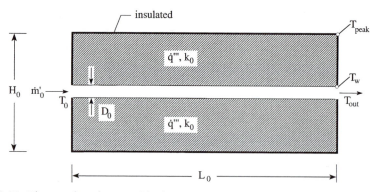

FIGURE 8.20. Elemental volume with heat generation, conduction, and single-stream cooling [19].

This smallest scale is the rectangular elemental volume, shown in Fig. 8.20. Only one parallel-plate channel of thickness D_0 transverses this volume. The smallness of the elemental size (the area $A_0 = H_0 L_0$) is fixed by manufacturing considerations: Smaller $H_0 L_0$ sizes lead to better designs. The external shape (H_0/L_0) and the internal opening (D_0/H_0) are the 2 degrees of freedom of the elemental geometry.

All the heat generated inside the elemental volume ($q''' H_0 L_0$) is convected away by the elemental stream that flows through the D_0 channel. The flow rate of this stream is $\dot{m}_0' = \dot{m}'' H_0$, where $\dot{m}''$ is a constant. In Fig. 8.20 the rectangular boundary $H_0 \times L_0$ is assumed to be adiabatic with the exception of the coldest spot (T_0) that occurs in the immediate vicinity of the inlet to the D_0 channel. The hot spots (T_{peak}) occur in the two corners that are situated the farthest from the inlet.

An analytical expression for the peak excess temperature ($T_{\text{peak}} - T_0$) can be developed when k_0 is small and the aspect ratio H_0/L_0 is sufficiently smaller than 1 such that the conduction through the heat generating material (k_0) is oriented perpendicularly to the fluid channel. If we also assume that $D_0 \ll H_0$, the temperature drop between the hot-spot corner (T_{peak}) and the wall spot near the channel outlet is $T_{\text{peak}} - T_w = q''' H_0^2/(8k_0)$, in accordance with the solution for steady conduction represented by the first term on the right-hand side of Eq. (4.1). The increase experienced by the bulk temperature of the stream from the inlet to the outlet is $T_{\text{out}} - T_0 = q''' H_0 L_0/(\dot{m}_0' c_p)$. There is also a temperature difference between the bulk temperature T_{out} and the duct wall temperature (T_w) in the plane of the outlet: temperature differences of this kind are neglected in this study based on the assumption that the flow is fully developed and the channel spacing is sufficiently small. The necessary validity condition is discussed immediately under Eqs. (8.45).

In conclusion, the peak excess temperature is given by a two-term expression $[T_{\text{peak}} - T_0 = (T_{\text{peak}} - T_w) + (T_w - T_0)]$ that can be nondimensionalized in the form of the overall resistance of the elemental volume:

$$\Delta \tilde{T}_0 = \frac{\tilde{H}_0}{8\tilde{L}_0} + \frac{1}{M\tilde{H}_0}, \tag{8.41}$$

where

$$(\tilde{H}_0, \tilde{L}_0) = \frac{(H_0, L_0)}{A_0^{1/2}}, \qquad \Delta \tilde{T}_0 = \frac{T_{\text{peak}} - T_0}{q''' A_0 / k_0}, \tag{8.42}$$

$$M = \dot{m}'' c_p A_0^{1/2} / k_0, \quad \text{constant.} \tag{8.43}$$

In this notation the elemental size constraint $H_0 L_0 = A_0$ becomes $\tilde{H}_0 \tilde{L}_0 = 1$. The right-hand side of Eq. (8.41) is equal to $(\tilde{H}_0^2/8) + 1/(M\tilde{H}_0)$ and shows that the overall resistance $\Delta \tilde{T}_0$ can be minimized with respect to the external-shape parameter $\tilde{H}_0$. The results are

$$\tilde{H}_{0,\text{opt}} = \left(\frac{4}{M}\right)^{1/3}, \qquad \tilde{L}_{0,\text{opt}} = \left(\frac{M}{4}\right)^{1/3}, \tag{8.44}$$

$$\left(\frac{H_0}{L_0}\right)_{\text{opt}} = \left(\frac{4}{M}\right)^{2/3}, \qquad \Delta \tilde{T}_{0,\text{min}} = \frac{3}{2^{5/3} M^{2/3}}. \tag{8.45}$$

What is noteworthy is the optimal external shape $(H_0/L_0)_{\text{opt}}$, which is independent of the channel size D_0. The elemental volume is more elongated when the flow parameter M is large, i.e., when $\dot{m}''$ and A_0 are large and k_0 is small. The starting assumption that the volume element is slender ($H_0/L_0 < 1$) means that the above solution is valid when $M > 4$.

The assumption that $T_w - T_{\text{out}}$ is negligible (relative to $T_{\text{peak}} - T_w$) means that $D_0/H_0 \ll (\text{Nu}/8)(k/k_0)$, where k is the thermal conductivity of the fluid and $\text{Nu}/8$ is of the order of 1 (Nu is the Nusselt number for fully developed laminar flow in a parallel-plate channel [1, 2]). This inequality states that $D_0/H_0 \ll 1$, and comes from writing $T_w - T_{\text{out}} \ll T_{\text{peak}} - T_w$ and the definitions $\text{Nu} = 2D_0 h/k$, $h = q''/(T_w - T_{\text{out}})$, and $q'' = q''' H_0/2$.

The original space HL of Fig. 8.19 can be filled with the necessary number (HL/A_0) of optimized elemental volumes. The remaining question is how to connect these building blocks so that each is bathed by a portion of the original stream $\dot{m}'$. The challenge is to connect the elements in a way that minimizes the overall pressure drop experienced by the $\dot{m}'$ stream.

The first step in this direction is shown in Fig. 8.21. We take a number (n_1) of elemental volumes and stack them into a first construct of dimensions $H_1 = L_{0,\text{opt}}$ and $L_1 = n_1 H_{0,\text{opt}}$. The elements are fed with cold fluid from a channel of length $L_1 - H_0/2$ and thickness $D_1/2$. A similar channel collects the elemental streams and reconstitutes the total stream of the first construct, $\dot{m}'_1 = n_1 \dot{m}'_0$. The total pressure difference between the inlet and the outlet of the first construct is

$$\Delta P_1 = \Delta P_0 + \Delta P_{D_1/2}. \tag{8.46}$$

If the flow through each elemental channel is in the Hagen–Poiseuille regime, the pressure drop across one elemental volume is $\Delta P_0 = 12\nu L_{0,\text{opt}} \dot{m}'_0 / D_0^3$. The second term in Eq. (8.46) refers to the pressure drop along one of the $D_1/2$-wide channels. For reasons that will be made clear shortly after Eqs. (8.52), the outer side of each $D_1/2$ channel is modeled as a zero-shear surface. The flow through the $D_1/2$

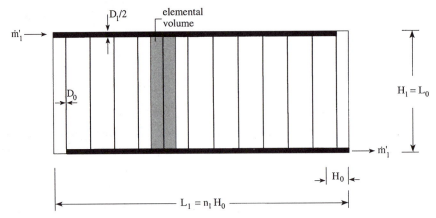

FIGURE 8.21. First construct containing n_1 elemental volumes [19].

channel is also in the Hagen–Poiseuille regime. When n_1 is sufficiently large, the flow rate varies linearly along the channel, from the total value $\dot{m}_1'$ at one end to zero at the opposite end. Under these circumstances the pressure drop integrated along the $D_1/2$ channel is equal to $12\nu L_1\dot{m}_1'/D_1^3$. The total pressure drop of Eq. (8.46) can be rewritten as a dimensionless overall flow resistance:

$$\Delta \tilde{P}_1 = \frac{\Delta P_1 A_0}{12\nu \dot{m}_1'} = \frac{\tilde{L}_{0,\text{opt}}}{n_1 \tilde{D}_0^3} + \frac{n_1 \tilde{H}_{0,\text{opt}}}{\tilde{D}_1^3}. \tag{8.47}$$

Consider now the effect of the number of constituents (n_1) on the overall resistance of the first construct. Both ΔP_1 and $\dot{m}_1'$ increase as n_1 increases. There is an optimal n_1 – an optimal size – for minimal resistance at the first-construct level:

$$n_{1,\text{opt}} = \left(\frac{L_0}{H_0}\right)_{\text{opt}}^{1/2} \left(\frac{D_1}{D_0}\right)^{3/2}, \qquad \Delta \tilde{P}_{1,\text{min}} = 2\left(\frac{A_0}{D_1 D_0}\right)^{3/2}. \tag{8.48}$$

In this first optimization step the minimization of resistance was achieved through optimal growth, i.e., through the aggregation of an optimal number of elements. The expression for $\Delta \tilde{P}_{1,\text{min}}$ can be minimized further by an increase in D_1 and D_0: These changes are not without limits, because the space occupied by fluid channels is taken away from space that could have been occupied by heat generating material. We account for this limitation through a total fluid volume constraint,

$$A_{1,\text{fluid}} = n_1 D_0 L_{0,\text{opt}} + 2\frac{D_1}{2} L_1, \tag{8.49}$$

where $A_{1,\text{fluid}}$ is a specified (small) fraction of the total volume of the first construct, $\phi_1 = A_{1,\text{fluid}}/(H_1 L_1)$,

$$\phi_1 = \frac{D_0}{H_{0,\text{opt}}} + \frac{D_1}{L_{0,\text{opt}}}, \quad \text{constant.} \tag{8.50}$$

The minimization of $\tilde{P}_{1,\text{min}}$ expression of Eqs. (8.48) with respect to either D_1 or

D_0 and subject to constraint (8.50) yields

$$D_{0,\mathrm{opt}} = \frac{\phi_1}{2} H_{0,\mathrm{opt}}, \qquad D_{1,\mathrm{opt}} = \frac{\phi_1}{2} L_{0,\mathrm{opt}}, \tag{8.51}$$

$$\left(\frac{D_1}{D_0}\right)_{\mathrm{opt}} = \left(\frac{M}{4}\right)^{2/3}, \qquad \left(\frac{H_1}{L_1}\right)_{\mathrm{opt}} = \left(\frac{M}{4}\right)^{-2/3}, \qquad \Delta \tilde{P}_{1,mm} = \frac{16}{\phi_1^3}. \tag{8.52}$$

The subscript $()_{mm}$ indicates that the overall resistance $\Delta \tilde{P}_{1,mm}$ has been minimized twice. Combining $(D_1/D_0)_{\mathrm{opt}}$ with Eqs. (8.48) and (8.45), we find the optimal number of constituents in the first construct, $n_{1,\mathrm{opt}} = (M/4)^{4/3}$. Because the analysis is valid for $M \gg 4$, we conclude that $n_{1,\mathrm{opt}} \gg 1$, and, in this way we justify the assumption on which Fig. 8.21 is based.

Even larger portions of the original volume (Fig. 8.19) can be filled by connecting together a number (n_2) of optimized first constructs. The resulting geometry is the second construct shown in Fig. 8.22, where $H_2 L_2 = n_2 H_1 L_1$, $H_2 = L_1$, and $L_2 = n_2 H_1$. Note the alternating pattern in which we combined first constructs: we flipped the design of Fig. 8.21 so that in the second construct two $D_1/2$-wide channels come together to form a single D_1-wide channel. The midplane of each D_1 channel is the outer (zero-shear) surface assumed in Fig. 8.21.

The total flow rate of the second construct is equal to the sum of the flow rates handled by the first constructs, $\dot{m}_2' = n_2 \dot{m}_1'$. The $\dot{m}_2'$ stream is first distributed and later is collected by channels of length L_2 and width $D_2/2$. The outer planes of

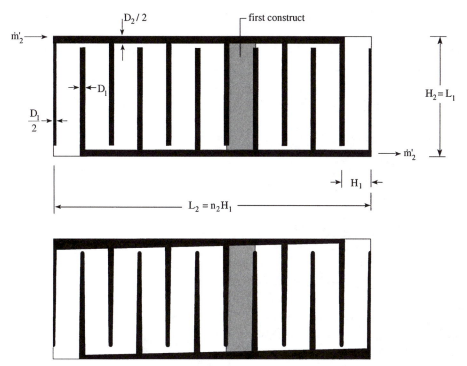

FIGURE 8.22. Second construct containing n_2 first constructs: constant-thickness channels (top), and tapered channels (bottom) [19].

the two $D_2/2$ channels are modeled as zero-shear surfaces. The total pressure drop experienced by the $\dot{m}_2'$ stream is

$$\Delta P_2 = \Delta P_{1,mm} + \Delta P_{D_2/2}, \tag{8.53}$$

where $\Delta P_{1,mm}$ can be derived from Eq. (8.52). The second term on the right-hand side of Eq. (8.53) can be derived as in the first-construct analysis, and the result is $12\nu L_2 \dot{m}_2'/D_2^3$. The nondimensional resistance that follows from Eq. (8.53),

$$\Delta \tilde{P}_2 = \frac{\Delta P_2 A_0}{12\nu \dot{m}_2'} = \frac{16}{n_2 \phi_1^3} + \frac{n_2 H_1 A_0}{D_2^3}, \tag{8.54}$$

can be minimized with respect to n_2:

$$n_{2,\text{opt}} = \left(\frac{16 D_2^3}{\phi_1^3 A_0 H_1}\right)^{1/2}, \qquad \Delta \tilde{P}_{2,\text{min}} = \frac{8(M/4)^{1/6}}{\tilde{D}_2^{3/2} \phi_1^{3/2}}, \tag{8.55}$$

where $\tilde{D}_2 = D_2/A_0^{1/2}$. The second minimization of the overall flow resistance is conducted subject to the total fluid space constraint,

$$A_{2,\text{fluid}} = n_2 A_{1,\text{fluid}} + 2\frac{D_2}{2} L_2, \tag{8.56}$$

or, in terms of the specified fluid-volume fraction in the second construct $\phi_2 = A_{2,\text{fluid}}/(H_2 L_2)$,

$$\phi_2 = \phi_1 + \frac{4}{M}\tilde{D}_2, \quad \text{constant.} \tag{8.57}$$

The minimization of $\Delta \tilde{P}_{2,\text{min}}$ with respect to $\tilde{D}_2$ and ϕ_1 and subject to constraint (8.57) yields

$$\phi_{1,\text{opt}} = \frac{\phi_2}{2}, \qquad D_{2,\text{opt}} = \frac{M}{8}\phi_2, \qquad \Delta \tilde{P}_{2,mm} = \frac{512}{2^{1/3} M^{4/3} \phi_2^3}. \tag{8.58}$$

In view of Eq. (8.56) and the double optimization performed at the first-construct level, the optimized second construct is also characterized by

$$n_{2,\text{opt}} = 4^{-1/3} M^{4/3} = 4n_1, \tag{8.59}$$

$$\left(\frac{H_2}{L_2}\right)_{\text{opt}} = \frac{1}{2}\left(\frac{D_1}{D_2}\right)_{\text{opt}} = \frac{1}{4}\left(\frac{4}{M}\right)^{2/3} = \frac{1}{4}\left(\frac{H_1}{L_1}\right)_{\text{opt}}. \tag{8.60}$$

8.8 Optimally Tapered Parallel-Plate Channels

In Section 8.7 we assumed that the central channel of each new construct has a uniform thickness D_1 and, subsequently, D_2. We made this choice for the sake of simplicity, in order to highlight each step of assembly (growth, aggregation), which is followed by geometric optimization. The constant-thickness assumption is not compatible with the requirement that the collecting channel must receive a uniform flow rate per unit of channel length. To see why, recall that in the first construct of Fig. 8.21 the flow rate must be the same through each D_0 channel, and, consequently, the flow rate through each $D_1/2$ channel must vary linearly along L_1, from $\dot{m}'_1$ to zero:

$$\dot{m}'_{1,x} = \dot{m}'_1\left(1 - \frac{x}{L_1}\right). \tag{8.61}$$

This expression refers to the top $D_1/2$ channel of Fig. 8.21, where $x = 0$ marks the entrance. The local pressure gradient along this channel is

$$-\frac{dP_H}{dx} = 24\nu\frac{\dot{m}'_{1,x}}{D_1^3}. \tag{8.62}$$

The distribution of high-pressure $P_H(x)$ depends on the function $D_1(x)$, which must be determined. If D_1 is a constant, then $P_H(x)$ has a parabolic shape. If D_1 is tapered as $(1 - x/L_1)^{1/3}$, then $P_H(x)$ varies linearly, as shown by the solid lines in Fig. 8.23.

Relations similar to Eqs. (8.61) and (8.62) hold for the distribution of low pressure $P_L(x)$ along the receiving $D_1/2$ channel. The pressure difference that drives the flow through each elemental D_0 channel is $P_H(x) - P_L(x)$. The elemental flow rate is independent of x only when $P_H(x) - P_L(x)$ is a constant, i.e., when $P_H(x)$ and $P_L(x)$ vary linearly. In conclusion, the ratio $\dot{m}_{1,x}/D_1^3$ must be a constant in Eq. (8.62) and, in view of Eq. (8.61),

$$D_1(x) = C\left(1 - \frac{x}{L_1}\right)^{1/3}. \tag{8.63}$$

The constant C accounts for the volume of the D_1 channel or for the result of averaging $D_1(x)$ from $x = 0$ to $x = L_1$:

$$\bar{D}_1 = \frac{1}{L_1}\int_0^{L_1} D_1(x)dx = \frac{3}{4}C. \tag{8.64}$$

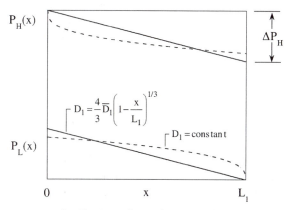

FIGURE 8.23. The pressure distribution along the $D_1/2$-thick channels of the first construct [19].

The pressure drop along the channel length L_1, which results from Eq. (8.62), is

$$\frac{\Delta P_H}{L_1} = \frac{81}{8} \nu \frac{\dot{m}_1'}{\bar{D}_1^3}. \qquad (8.65)$$

The optimal power-law tapering of the channel applies to every subsequent-level channel that must receive uniform flow rate per unit length. The lower part of Fig. 8.22 shows the tapered versions of $D_1(x)$ and $D_2(x)$, which can be compared directly with the uniform (D_1, D_2) drawing shown in the upper part of the figure. The optimal dimensions and aspect ratios of the constructs with tapered collecting channels agree closely with the results obtained in Section 8.7. The analysis starts with the line above Eq. (8.47), in which the pressure drop along the $D_1/2$ channel of the first construct $[12\nu L_1 \dot{m}'/D_1^3$, or Eq. (8.62)] is replaced now with $2(3/2)^4 \nu \dot{m}_1' / \bar{D}_1^3$; see Eq. (8.65). In other words, the place of D_1 is taken by $\bar{D}_1$, and the factor 12 is replaced with $2(3/2)^4 = 10.13$. The main features of the optimized first construct are now described by

$$\left(\frac{\bar{D}_1}{D_0}\right)_{opt} = \left(\frac{M}{4}\right)^{2/3}, \qquad \left(\frac{H_1}{L_1}\right)_{opt} = 2^{1/2}\left(\frac{3}{4}\right)^{3/2}\left(\frac{4}{M}\right)^{2/3}, \qquad (8.66)$$

$$n_{1,opt} = 2^{-1/2}\left(\frac{4}{3}\right)^{3/2}\left(\frac{M}{4}\right)^{4/3}, \qquad \Delta \tilde{P}_{1,mm} = 2^{1/2}\left(\frac{3}{4}\right)^{3/2}\frac{16}{\phi_1^3}. \qquad (8.67)$$

These results can be compared with Eqs. (8.52) to see the close agreement: The uniform D_1 analysis is an approximate and more direct method of determining the optimized geometry.

The optimization of the second construct starts with the line above Eq. (8.54), in which $12\nu L_2 \dot{m}_2' / D_2^3$ is replaced with $(81/8)\nu L_2 \dot{m}_2' / \bar{D}_2^3$. Equations (8.58)–(8.60) are replaced with

$$\left(\frac{\bar{D}_2}{\bar{D}_1}\right)_{opt} = \frac{2^{7/2}}{3^{3/2}}\left(\frac{M}{4}\right)^{2/3}, \qquad \left(\frac{H_2}{L_2}\right)_{opt} = \frac{3^{3/2}}{2^{9/2}}\left(\frac{4}{M}\right)^{2/3}, \qquad (8.68)$$

$$n_{2,opt} = \frac{2^7}{3^3}\left(\frac{M}{4}\right)^{4/3}, \qquad \Delta \tilde{P}_{2,mm} = \frac{2^4 3^3}{2^{1/3} M^{4/3} \phi_2^3}. \qquad (8.69)$$

Combining the revised analyses of the first construct and the second construct, we find

$$n_{2,opt} = \frac{2^{5/2}}{3^{3/2}} 4 n_{1,opt}, \qquad \left(\frac{H_2}{L_2}\right)_{opt} = \left(\frac{2}{3}\right)^{3/2}\left(\frac{\bar{D}_1}{\bar{D}_2}\right)_{opt} = \frac{1}{4}\left(\frac{H_1}{L_1}\right)_{opt}. \qquad (8.70)$$

These results agree to within 20% with the uniform (D_1, D_2) results of Eqs. (8.59) and (8.60).

Figure 8.24 shows qualitatively the layout of the first, second, and third constructs. Blue indicates colder channels that bring in the coolant and distribute it through the given volume. Red indicates the heated fluid that is collected and led out of the volume. The elemental channels change color in the middle, from blue to red, to suggest that at the elemental level each stream warms up while absorbing the generated heat that diffuses through the solid material, which is shown in yellow.

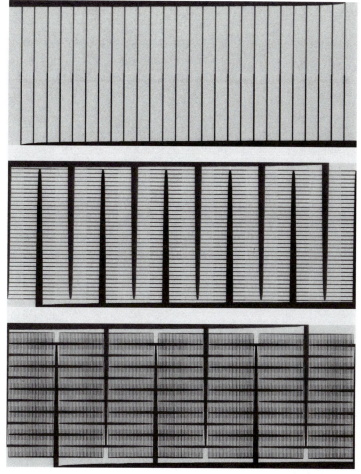

FIGURE 8.24. Color displays of inflowing (blue) and outflowing (red) streams in the first construct (top), second construct (middle), and third construct (bottom) [19] (see Plate IX).

As the constructs compound themselves, the cooled volume acquires a structure similar to that of a vascularized tissue, in which one tree network (arteries) meets another tree (veins) in every volume element of the tissue. Each element is served by two trees, which are intertwined.

The optimization sequence can be continued toward constructs of higher order, which cover increasingly larger volumes. The geometric results would continue the trends that became visible as early as the second construct. Specifically, because the simplified analytical course exhibited here is valid when $M/4 > 1$, each new channel will be larger than the largest preceding channel size, with a width magnification factor of the order of $\bar{D}_{i+1}/\bar{D}_i \sim (M/4)^{2/3}$ for $i \geq 2$. The stepwise increase in channel width agrees with the behavior of the corresponding solution for tree networks in pure heat conduction (Chap. 4).

A trend that goes against the trees of heat conduction is the stepwise increase in the external slenderness ratio. In heat conduction the higher-order constructs were square or close to square. In view of Eqs. (8.45), (8.66), and (8.70), we anticipate that the ratios H_{i+1}/L_{i+1} will continue to be of the order of $(4/M)^{2/3} < 1$ and that they will decrease in small steps as the constructs become larger. The point–volume–point flow schemes will have to fit in more and more slender spaces, and this will interfere with the outer constraints (boundary, shape) of the total volume that must be cooled. The only course of action left is to optimize the geometry of the shape-constrained system numerically, as shown for pure conduction in Ref. 20.

Another trend that differs from pure heat conduction is shown by the first of Eqs. (8.70): The number of constituents increases as the order of the construct increases. In the heat conduction tree the number of constituents decreased until it reached two (dichotomy).

The decision to connect the two fluid trees end to end (canopy to canopy, Fig. 8.24) was made here only for illustration. An alternative with more immediate relevance to the thermal insulation function served by vascularized tissues is to superimpose the cold and hot trees, in counterflow. This architecture is discussed and optimized in Sections 9.4 and 10.6.

8.9 Round Tubes

We illustrated the geometric optimization of the point–volume–point flow path in two dimensions by assuming that each channel is a gap with parallel walls. The same method can be used to optimize designs with channels of round cross section. The start of this optimization sequence is based on Fig. 8.25. The elemental level is defined by the smallest channel, which has diameter D_0 and length L_0. As elemental volume we consider the cylinder of diameter H_0. Heat is generated in the solid annulus of thermal conductivity k_0.

To construct the analytical expression for the thermal resistance of the elemental volume we assume again that $H_0 < L_0$ and note that $T_{peak} - T_0 = (T_{peak} - T_w) + (T_w - T_0)$. The temperature difference in the radial direction is derived from the solution to the problem of steady conduction in an annular space with a uniform

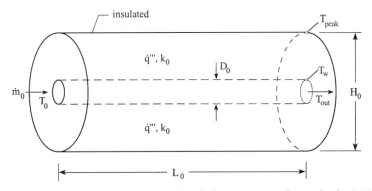

FIGURE 8.25. Elemental volume with radial symmetry and round tube [19].

heat generation rate:

$$T_{\text{peak}} - T_w = \frac{q'''H_0}{16k_0}\left[2\ln\frac{H_0}{D_0} + \left(\frac{D_0}{H_0}\right)^2 - 1\right].\qquad(8.71)$$

The flow rate of the elemental stream is $\dot{m}_0 = \dot{m}''(\pi/4)H_0^2$. The longitudinal temperature increase experienced by the stream is

$$T_{\text{out}} - T_0 = \frac{\pi}{4}\left(H_0^2 - D_0^2\right)L_0 q'''/(\dot{m}_0 c_p).\qquad(8.72)$$

Next, we assume that $T_{\text{out}} - T_0 \cong T_w - T_0$ and add Eqs. (8.71) and (8.72). The resulting expression,

$$\Delta\hat{T}_0 = \frac{\hat{L}_0}{\hat{M}}\left(\frac{H_0^2 - D_0^2}{H_0^2}\right) + \frac{\hat{H}_0^2}{16}\left[2\ln\frac{\hat{H}_0}{\hat{D}_0} + \left(\frac{\hat{D}_0}{\hat{H}_0}\right)^2 - 1\right],\qquad(8.73)$$

has been nondimensionalized by use of $V_0^{1/3}$ as length scale, where V_0 is the constant volume of the elemental system, $V_0 = (\pi/4)H_0^2 L_0$:

$$(\hat{H}_0, \hat{L}_0, \hat{D}_0) = \frac{(H_0, L_0, D_0)}{V_0^{1/3}}, \qquad \Delta\hat{T}_0 = \frac{T_{\text{peak}} - T_0}{q'''V_0^{2/3}/k_0},\qquad(8.74)$$

$$\hat{M} = \dot{m}''c_p V_0^{1/3}/k_0, \quad \text{constant.}\qquad(8.75)$$

The volume constraint $(\pi/4)\hat{H}_0^2\hat{L}_0 = 1$ and the known size of the smallest tube $(\hat{D}_0)$ leave only 1 degree of freedom in the minimization of $\Delta\hat{T}_0$. The free parameter is $\hat{L}_0$ or $\hat{H}_0$, or the slenderness ratio H_0/L_0. The results of minimizing $\Delta\hat{T}_0$ numerically are shown in Fig. 8.26. The ratio $(H_0/L_0)_{\text{opt}}$ varies almost as $\hat{M}^{-1/2}$, which agrees qualitatively with the result for parallel-plate channels, Eqs. (8.45). The data of Fig. 8.26 and the corresponding minimum resistance are correlated to

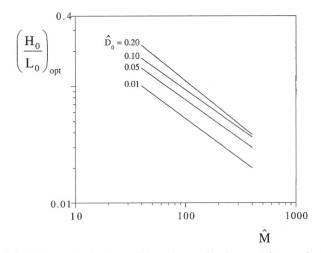

FIGURE 8.26. The optimal shape of the elemental volume with round tube [19].

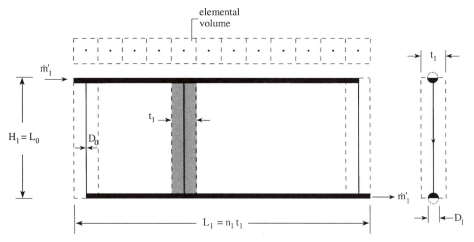

FIGURE 8.27. First construct with round tubes [19].

within 2% and, respectively, 1% by the expressions [19]

$$\left(\frac{H_0}{L_0}\right)_{\text{opt}} \cong 2.37\,\hat{M}^{-0.556}\,\hat{D}_0^{0.248}, \tag{8.76}$$

$$\Delta\hat{T}_{0,\text{min}} = 0.575\,\hat{M}^{-0.536}\,\hat{D}_0^{-0.264}. \tag{8.77}$$

The first construct is shown in Fig. 8.27. The collected stream $\dot{m}_1$ flows through half of a channel of diameter D_1. The elements form a slab of length L_1, height H_1 ($= L_0$), and thickness $t_1 = H_0$. The latter is an approximation, because n_1 parallel cylinders do not fill a parallelepiped completely. A better estimate for the slab thickness t_1 is obtained from the volume conservation argument $L_1 H_1 t_1 = n_1 (\pi/4) H_0^2 L_0$ and $L_1 = n_1 t_1$, which yields $t_1 = (\pi/4)^{1/2} H_0$.

The calculation of the pressure drop ΔP_1 across the first construct follows the steps outlined in Section 8.7. We assume that D_1 and D_2 are fixed during the first phase of the optimization procedure. The pressure drop between the inlet and the outlet of the first construct (Fig. 8.27) is $\Delta P_1 = \Delta P_0 + \Delta P_{D_1/2}$. The Hagen–Poiseuille flow overcomes the overall resistance:

$$\Delta\hat{P}_1 = \frac{\Delta P_1 \pi V_0}{128 \dot{m}_1 \nu} = \frac{\hat{L}_{0,\text{opt}}}{n_1 \hat{D}_0^4} + n_1 \frac{\hat{t}_{1,\text{opt}}}{\hat{D}_1^4}. \tag{8.78}$$

We can minimize the resistance by selecting the number of constituents,

$$n_{1,\text{opt}} = \left(\frac{\hat{D}_1}{\hat{D}_0}\right)^2 \left(\frac{\hat{L}_{0,\text{opt}}}{\hat{t}_{1,\text{opt}}}\right)^{1/2}, \qquad \Delta\hat{P}_{1,\text{min}} = \frac{2}{\hat{t}_{1,\text{opt}} \hat{D}_0^2 \hat{D}_1^2}. \tag{8.79}$$

We may replace $\hat{t}_{1,\text{opt}}$ with $(\pi/4)^{1/2} \hat{H}_{0,\text{opt}}$ to express $\Delta\hat{P}_{1,\text{min}}$ as a function of the shape, since $\hat{H}_{0,\text{opt}} = [(4/\pi)(\hat{H}_0/\hat{L}_0)_{\text{opt}}]^{1/3}$: we find that $\Delta\hat{P}_{1,\text{min}}$ varies nearly as $\hat{M}^{1/5}$ for a given $\hat{D}_0$.

Equations (8.79) also show that we may reduce the resistance further by increasing $\hat{D}_0$ and $\hat{D}_1$. These diameters are related through the fluid-volume constraint,

$$\phi_1 = \frac{V_{1,\text{fluid}}}{V_1} = \frac{\pi}{4}\hat{D}_0^2 \hat{L}_{0,\text{opt}} + \frac{\pi}{4}\hat{D}_1^2 \hat{t}_{1,\text{opt}}, \quad \text{constant.} \tag{8.80}$$

The results of minimizing $\Delta\hat{P}_{1,\text{min}}$ with respect to $\hat{D}_0$ and $\hat{D}_1$ subject to the ϕ_1 constraint are

$$\hat{D}_{0,\text{opt}} \cong 0.714\,\phi_1^{1/2}\hat{H}_{0,\text{opt}}, \tag{8.81}$$

$$\hat{D}_{1,\text{opt}} \cong 0.7\left(\frac{4}{\pi}\right)^{3/4}\phi_1^{1/2}\hat{H}_{0,\text{opt}}^{-1/2}. \tag{8.82}$$

These results can be combined with $\hat{H}_{0,\text{opt}}$ relation (8.76), such that the geometry and flow resistance assume the forms

$$\hat{D}_{0,\text{opt}} \cong \hat{M}^{-0.2}\phi_1^{0.55}, \qquad \hat{D}_{1,\text{opt}} \cong 0.7\hat{M}^{0.1}\phi_1^{0.48}, \tag{8.83}$$

$$\hat{H}_{0,\text{opt}} \cong 1.1\hat{M}^{-0.2}\phi_1^{0.045}, \qquad \Delta\hat{P}_{1,mm} \cong 3.7\hat{M}^{0.4}\phi_1^{-2.1}. \tag{8.84}$$

Other features of the optimized construct are

$$\left(\frac{D_1}{D_0}\right)_{\text{opt}} \cong \left(\frac{\hat{M}}{4}\right)^{0.3}\phi_1^{-0.07}, \qquad \left(\frac{H_1}{L_1}\right)_{\text{opt}} \cong 1.2\left(\frac{\hat{M}}{4}\right)^{-0.3}\phi_1^{0.07}, \tag{8.85}$$

$$n_{1,\text{opt}} \cong 2\left(\frac{\hat{M}}{4}\right)^{0.9}\phi_1^{-0.2}. \tag{8.86}$$

These features agree qualitatively with the results determined in Eqs. (8.52) for the first construct with parallel-plate channels.

8.10 Two Fluid Trees in Counterflow are One Tree for Convection

The most important convective tree configuration is the one in which the two fluid trees are superimposed perfectly and in which, in each pair of parallel tubes, the streams are in counterflow. We can visualize this configuration by imagining that the blue and the red channels of Fig. 8.24 are placed on top of each other. The elemental channels (blue and red) and elemental volumes (yellow) will fill the interstitial spaces. Each elemental channel would be shaped like a hair pin.

This configuration is most important in living systems, tissues, and animals. The structure of vascularization is dominated by pairs of trees of large vessels (arteries, veins) superimposed in counterflow. The lung too works as two fluid trees in counterflow, one tree during inhaling and the other during exhaling. The often overlooked property of two streams in counterflow is that they constitute one longitudinal path for convection: The convective heat current flows in the direction of the warmer stream and is proportional to the longitudinal temperature gradient of the assembly [21]. This property will be analyzed in detail in Section 9.4.

The counterflow leaks heat longitudinally toward lower temperatures – less heat when the thermal contact between the two streams is better. The counterflow serves as a thermal insulation function, and the tree of counterflow pairs works as a *thermal insulation structure.*

The astonishing discovery that results from the study of convective trees, which is due to fluid counterflows, is that the metabolic rate of an animal must be proportional to the body size raised to the power 3/4 [22]. This purely theoretical development is reviewed in Section 10.6. The fluid flow is laminar through straight tubes. The ratios of successive tube diameters and tube lengths are deduced from the minimization of flow resistance subject to two constraints, total volume and total tube volume (Problems 5.1 and 6.1) [23]. From the optimization of geometric arrangement follows the deduction that the total heat current convected by the double tree must be proportional to the total volume or body mass raised to the power 3/4 (see Section 10.6).

The general conclusion is this: The resistance to the loss of body heat governs the relation between metabolic rate and body size in all animals, warm blooded and cold blooded. In mammals and birds, the resistance is dominated by superimposed convective trees of fluid flow, and the predicted power-law exponent is the observed 3/4. In the limit of small body sizes of warm-blooded animals, heat flows mainly through the conductive tissue and the exponent decreases to 1/3. In lizards and amphibians, the resistance is dominated by convection on the outside of the body surface, and the power-law exponent is the observed 2/3. We return to this unifying theory of body heat loss and body size effects in Section 10.6.

PROBLEMS

8.1 Consider the T-shaped assembly of fins sketched in Fig. 8.1. Two elemental fins of thickness t_0 and length L_0 serve as tributaries to a stem of thickness t_1 and length L_1. The configuration is two dimensional, with the third dimension (W) sufficiently long in comparison with L_0 and L_1. The heat transfer coefficient h is uniform over all the exposed surfaces. The temperatures of the root (T_1) and the fluid (T_∞) are specified. The temperature at the junction (T_0) is one of the unknowns and varies with the geometry of the assembly.

The objective is to determine the optimal geometry ($L_1/L_0, t_1/t_0$) that is characterized by the maximum global thermal conductance $q_1/(T_1 - T_\infty)$, where q_1 is the heat current through the root section. The optimization is subjected to two constraints, the total-volume (i.e., front area) constraint, $A = 2L_0L_1$, and the fin-material volume constraint, $A_f = 2L_0t_0 + t_1L_1$. The latter can be expressed in dimensionless form as the solid volume fraction $\phi_1 = A_f/A$, which is a specified constant considerably smaller than 1. The analysis that delivers the global conductance as a function of the assembly geometry consists of accounting for conduction along the L_0 and L_1 fins and invoking the continuity of temperature and heat current at the T junction. The unidirectional conduction model [2] is recommended for the analysis of each fin.

Develop the global conductance as the dimensionless function $\tilde{q}_1(\tilde{L}_0, \tilde{t}_0, \tilde{L}_1, \tilde{t}_1, a)$, where $\tilde{q}_1$ and a are defined in Eqs. (8.1) and (8.2), and $(\tilde{L}_0, \tilde{t}_0, \tilde{L}_1, \tilde{t}_1) = (L_0, t_0, L_1, t_1)/A_1^{1/2}$. Set $a = 1$ and $\phi_1 = 0.1$. Use the dimensionless versions of the A and A_f constraints and recognize that the geometry of the assembly has only 2 degrees of freedom. Choose L_1/L_0 and t_1/t_0 as degrees of freedom. First, maximize numerically $\tilde{q}_1$ with respect to L_1/L_0 while holding t_1/t_0 constant, and your result will be the curve $\tilde{q}_{1,m}(t_1/t_0)$. Finally, maximize $\tilde{q}_{1,m}$ with respect to t_1/t_0 and report the twice-maximized conductance $\tilde{q}_{1,mm}$ and the corresponding architecture $(L_1/L_0)_{opt}$ and $(t_1/t_0)_{opt}$. Make a scale drawing of the T-shaped geometry that represents this design.

8.2 The temperature distribution along the two-dimensional fin with sharp tip shown in Fig. P8.2 is linear, $\theta(x) = (x/L)\theta_b$, where the local temperature difference is $\theta(x) = T(x) - T_\infty$ and $\theta_b = T_b - T_\infty$. The fin temperature is $T(x)$, and the fluid temperature is T_∞. The fin width is W. The tip is at the temperature of the surrounding fluid: $\theta(0) = 0$.

Show that the thickness δ of this fin must be parabolic in x: $\delta = (h/k)x^2$. Derive an expression for the total heat transfer rate through the base of this fin, $\dot{Q}_b$.

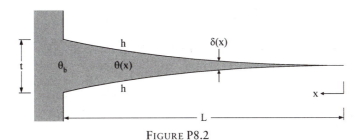

FIGURE P8.2

8.3 The sizing procedure for a fin generally requires the selection of more than one physical dimension. In a plate fin, for example, there are two such dimension, the length of the fin, L, and the thickness, t. One of these dimensions, or the shape of the fin profile t/L can be selected optimally when the total volume of fin material is fixed. Such a constraint is often justified by the high cost of the high-thermal-conductivity metals that are employed in the manufacture of finned surfaces (e.g. copper, aluminum), and by the cost associated with the weight of the fin.

Consider a plate fin of constant thickness t, conducivity k, and length L extending from a wall of temperature T_b into a fluid flow of temperature T_∞. The plate-fin dimension perpendicular to $t \times L$ is W. The heat transfer coefficient h is uniform on all the exposed surfaces. The profile is slender $(t \ll L)$, the Biot number is small $(ht/k \ll 1)$, and the unidirectional fin condution model applies. Assume that the heat transfer through the fin tip is negligible so that the total heat transfer rate through the fin base is $q_b = \theta_b(kA_chp)^{1/2}\tanh(mL)$, where the notation [2] is $\theta_b = (T_b - T_\infty)$, $m = (hp/kA_c)^{1/2}$, $A_c = tW$, and $p = 2W$.

The global conductance of the fin is q_b/θ_b. Maximize this function subject to the constraint that the fin volume is fixed, $V = tLW$. Report analytically the optimal shape of the fin profile and the maximized global conductance.

REFERENCES

1. A. Bejan, *Convection Heat Transfer*, 2nd ed., Wiley, New York, 1995.

2. A. Bejan, *Heat Transfer*, Wiley, New York, 1993.

3. A. D. Kraus, Developments in the analysis of finned arrays, Donald Q. Kern Award Lecture, National Heat Transfer Conference, Baltimore, MD, Aug. 11, 1997; *Int. J. Transport Phenomena*, Vol. 1, 1999, pp. 141–164.

4. A. Aziz, Optimum dimensions of extended surfaces operating in a convective environment, *Appl. Mech. Rev.*, Vol. 45(5), 1992, pp. 155–173.

5. W. R. Hamburgen, Optimal, finned heat sinks, WRL Research Report 86/4, Digital, Western Research Laboratory, Palo Alto, CA, 1986.

6. D. J. Lee and W. W. Lin, Second-law analysis on a fractal-like fin under crossflow, *AIChE. J.*, Vol. 41, 1995, pp. 2314–2317.

7. W. W. Lin and D. J. Lee, Diffusion-convection process in a branching fin, *Chem. Eng. Commun.*, Vol. 158, 1997, pp. 59–70.

8. W. W. Lin and D. J. Lee, Second-law analysis on a pin-fin array under cross-flow, *Int. J. Heat Mass Transfer*, Vol. 40, 1997, pp. 1937–1945.

9. A. Bejan and M. Almogbel, Constructal T-shaped fins, *Int. J. Heat Mass Transfer*, Vol. 43, 2000, pp. 141–164.

10. K. A. Gardner, Efficiency of extended surfaces, *Trans. ASME*, Vol. 67, 1945, pp. 621–631.

11. A. Bejan and N. Dan, Constructal trees of convective fins, *J. Heat Transfer*, Vol. 121, 1999, pp. 675–682.

12. J. C. Watson, N. K. Anand, and L. S. Fletcher, Mixed convective heat transfer between a series of vertical parallel plates with planar heat sources, *J. Heat Transfer*, Vol. 118, 1996, pp. 984–990.

13. A. Bejan and E. Sciubba, The optimal spacing of parallel plates cooled by forced convection, *Int. J. Heat Mass Transfer*, Vol. 35, 1992, pp. 3259–3264.

14. Al. M. Morega and A. Bejan, The optimal spacing of parallel boards with discrete heat sources cooled by laminar forced convection, *Num. Heat Transfer, Part A*, Vol. 25, 1994, pp. 373–392.

15. A. Alebrahim and A. Bejan, Constructal trees of circular fins for conductive and convective heat transfer, *Int. J. Heat Mass Transfer*, Vol. 42, 1999, pp. 3585–3597.

16. H. S. Carslaw and J. C. Jaeger, *Conduction of Heat in Solids*, Oxford Univ. Press, Oxford, UK, 1959.

17. S. Petrescu, Comments on the optimal spacing of parallel plates cooled by forced convection, *Int. J. Heat Mass Transfer*, Vol. 37, 1994, p. 1283.

18. S. J. Kim and S. W. Lee, *Air Cooling Technology for Electronic Equipment*, CRC, Boca Raton, FL, 1995, Chap. 1.

19. A. Bejan and M. R. Errera, Convective trees of fluid channels for volumetric cooling, *Int. J. Heat Mass Transfer*, Vol. 43, 2000, pp. 3105–3118.

20. M. Almogbel and A. Bejan, Conduction trees with spacings at the tips, *Int. J. Heat Mass Transfer*, Vol. 42, 1999, pp. 3739–3756.

21. A. Bejan, *Entropy Generation through Heat and Fluid Flow*, Wiley, New York, 1982, pp. 180–182.

22. A. Bejan, The tree of convective streams: its thermal insulation function and the predicted 3/4-power relation between body heat loss and body size, *Int. J. Heat Mass Transfer*, Vol. 43, 2000, to appear.

23. A. Bejan, L. A. O. Rocha, and S. Lorente, Thermodynamic optimization of geometry: T- and Y-shaped constructs of fluid streams, *International Journal of Thermal Sciences*, Vol. 40, 2001, to appear.

CHAPTER NINE

STRUCTURE IN POWER SYSTEMS

The constructal principle that generated so many shapes and structures in the preceding chapters manifests itself in even larger and more complicated systems, engineered and natural. In this chapter we examine a few examples of systems, the purposes of which are power generation and the judicious use of power, e.g., refrigeration and powered flight. The established view on the development and time evolution of such systems is that improvements occur when thermodynamic irreversibility is reduced. Indeed, the energy-conversion efficiencies of power-generating plants have been creeping upward, toward the Carnot ceiling (Fig. 9.1). We have every reason to expect that this trend will continue, in time.

Power plants are systems with internal flows of heat, fluid, species, and electricity. Irreversibility is reduced, and power production per unit of fuel used is maximized, when the resistances encountered by these flows are reduced. Resistance minimization can be achieved by increasing size, for example, by an increase in the cross sections of the "ducts" through which the various currents must pass: heat transfer surfaces for heat flow, duct diameters for fluid flow, and cable cross sections for electric current. Size is always constrained, if not by total cost, which is the most common constraint, then by weight and volume, as in naval and airborne power plants.

In summary, the established view on the improvement of power plant performance is entirely consistent with the constructal optimization philosophy of this book. Performance improvements follow from the application of the purpose and constraints principle. Resistances are minimized together, when the balance between them is improved – more balances in more complex systems with more internal currents. Optimal balance is another name for the optimal allocation of the constrained size (hardware) among the various paths for internal flow [2]. Optimal allocation of hardware means optimized architecture – the physical configuration of the power system. Geometric structure is the visible and palpable result of the purpose and constraints principle.

The topics assembled in this chapter illustrate how the geometric structure of power systems is derived from principle. Most of the discussion is about systems for power generation, although the thinking holds for refrigeration systems as well.

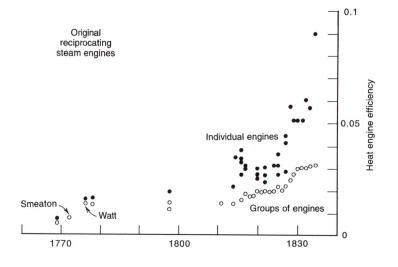

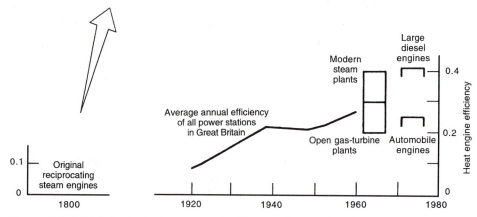

FIGURE 9.1. Highlights in the improvement of heat engines and modern power plants (after Ref. 1).

In refrigeration the objective is the minimization of the power input required by the system, which is directly proportional to the system irreversibility and intimately related to how the various internal resistances are arranged and their sizes balanced.

9.1 Allocation of Heat Exchange Inventory

We start with the system that triggered the industrial revolution and the science of thermodynamics: the heat engine. The system is closed and operates steadily, or in an integral number of cycles, between two fixed temperatures, high (T_H) and low (T_L). Its purpose is to convert the specified heat input (Q_H) into mechanical power (W), Fig. 9.2. The lateral boundaries of the system are insulated perfectly such that heat is forced to flow vertically. Better performance means more W for the same unit of Q_H.

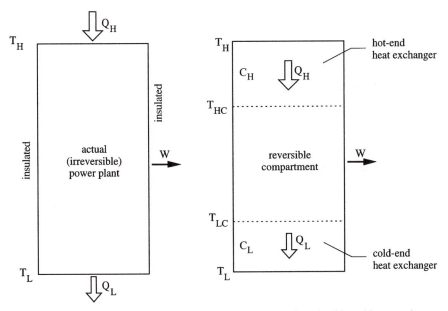

FIGURE 9.2. Model of irreversible power plant with hot-end and cold-end heat exchangers.

Irreversible operation is a characteristic of real power plants, and a reflection of the physical constraints (e.g., finite sizes) that designs must face. Consequently, the energy-conversion ratio W/Q_H is expectedly lower than the Carnot efficiency,

$$\frac{W}{Q_H} < 1 - \frac{T_L}{T_H}, \qquad (9.1)$$

although in time the strength of the inequality sign decreases, Fig. 9.1. A simple way to illustrate the origin of the inequality sign is to recognize that the heat input Q_H and the rejected heat Q_L are driven in the indicated directions by *finite* temperature differences [3–9]. These differences are located inside the power plant, as shown in Fig. 9.2. One way to account for the relationship between heat transfer rate and temperature difference is to assume the proportionalities

$$Q_H = C_H(T_H - T_{HC}), \qquad Q_L = C_L(T_{LC} - T_L). \qquad (9.2)$$

In these relations C_H and C_L are the thermal conductances of the hot-end and the cold-end heat exchangers, $C_H = (UA)_H$ and $C_L = (UA)_L$, where A is the respective heat transfer surface and U is the overall heat transfer coefficient based as A.

Since each thermal conductance is proportional to the size of the heat exchanger surface, i.e., proportional to the size of the piece of hardware, it makes sense to regard C_H and C_L as partners in an overall size constraint [10–13]

$$C_H + C_L = C, \qquad \text{constant.} \qquad (9.3)$$

The rest of the power plant – the compartment shown between T_{HC} and T_{LC} in Fig. 9.2 – operates reversibly. This means that the rate of entropy generation in this compartment is zero:

$$\frac{Q_L}{T_{LC}} - \frac{Q_H}{T_{HC}} = 0. \tag{9.4}$$

Combining the second law (9.4) with the first law for the same compartment ($W = Q_H - Q_L$), we arrive at the energy-conversion ratio of the power plant:

$$\frac{W}{Q_H} = 1 - \frac{T_{LC}}{T_{HC}}. \tag{9.5}$$

This ratio is smaller than its Carnot limit, because $T_{HC} < T_H$ and $T_{LC} > T_L$. Inequality (9.1) is satisfied.

The W/Q_H ratio of Eq. (9.5) may be rewritten with the help of Eqs. (9.2)–(9.3) to show explicitly the effect of distributing the C inventory to the two heat exchangers:

$$\frac{W}{Q_H} = 1 - \frac{T_L/T_H}{1 - \frac{Q_H}{T_H C}\left(\frac{1}{x} + \frac{1}{1-x}\right)}. \tag{9.6}$$

In this expression $x = C_H/C$ is the conductance allocation fraction, such that $C_L/C = 1 - x$. It is evident that when C_H and C_L vary subject to size constraint (9.3) the power output and conversion ratio are maximized when $x = 1/2$, i.e., when [10–13]

$$C_H = C_L. \tag{9.7}$$

This optimum is illustrated in Fig. 9.3, which shows three possible configurations, $C_H < C_L$, $C_H = C_L$, and $C_H > C_L$, while the heat input is held fixed. The size of each conductance (heat transfer surface) is indicated by the length of the

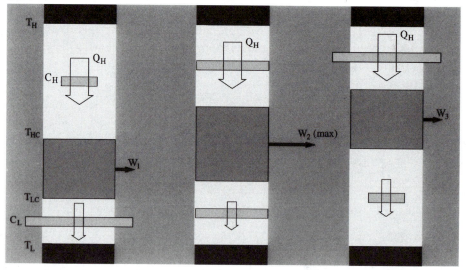

FIGURE 9.3. The allocation of heat transfer surface and its effect on temperature differences and power output (see Plate X).

respective yellow bar. Small surfaces require large temperature differences. The reversible compartment is shown in green: When this compartment appears "tall" on the temperature scale, the power output is large; see Eq. (9.5). The figure shows that the height of the reversible compartment reaches a peak somewhere between the extremes $C_H \ll C_L$ and $C_H \gg C_L$, i.e., where C_H is comparable with C_L. The analytical maximization of objective function (9.6) pinpoints the optimal configuration, $C_H = C_L$. Global optimization subject to size constraint generates structure.

The same structure is deduced if, instead of maximizing W, we minimize the total entropy generated by the power plant. The result, $C_H = C_L$, is astonishing because of its simplicity and robustness. The engineering literature continues to show that $C_H = C_L$ holds at least approximately for increasingly complex models of irreversible power plants, starting for example with the model of Fig. 9.2 in which the heat input Q_H may vary, in addition to the variable size ratio C_H/C_L [10]. The same principle also holds for the corresponding class of models of irreversible refrigeration plants [14, 15]; see Problem 9.1.

9.2 Distribution of Insulation

Hand in hand with the balancing and minimization of resistances to internal flow comes the requirement to *guide* each current along a certain path. In a power plant the heat input must be forced to flow through the machine in order to do the most work possible, because otherwise it would leak directly into the ambient, bypassing the machine. This is why in the simple model of Fig. 9.2 the sides of the system were insulated perfectly. The guiding of the heat current is achieved through the distribution of *insulation*, in configurations in which the insulated surfaces are nonisothermal (e.g., the surface $T_H - T_{HC} - T_{LC} - T_L$ in Fig. 9.2).

In this section we examine the fundamental question [16] of how a finite amount of insulation material can best be distributed over a wall with nonuniform temperature in order to minimize the total heat loss from the wall to the ambient. In addition to the heat current guiding function outlined above, this question is important because energy *conservation* is a major concern in many industrial applications in which nonisothermal walls must be covered with insulation. An example is the outer wall of a long reheating oven in which steel laminates are being heated while riding slowly on a conveyor belt. Other examples are the outer walls of virtually all heat exchangers, storage tanks with thermally stratified liquids (e.g., solar thermal applications), and the lateral surfaces of mechanical supports connecting regions with different temperatures.

Next to the task of conserving energy, the idea that the supply of insulation material is finite is always on the mind of the designer. The purchase, installation, and maintenance of an insulation can be expensive. In some cases even the size (weight, volume) of the used insulation material cannot exceed a certain limit. Examples of this kind are airborne applications and installations in which the integrity of the mechanical supports is threatened by the weight of the insulated system (e.g., the suspended insulated duct in Problem 9.2).

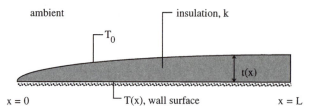

FIGURE 9.4. Wall covered by a thin layer of insulation with varying thicknesses [16].

This fundamental question of optimal spreading of a finite amount of material is interesting, even if we discount its many applications. The process of one-dimensional heat transfer through a layer of insulation is too simple and too well known [17] to hide any more subtleties at the start of the twenty first century. Indeed, insulations are visualized routinely as layers and shells of uniform thickness. The question of optimal spreading is interesting because, once again, it acts as mechanism (principle) for the generation of geometric form.

The simplest configuration in which this question can be examined is shown in Fig. 9.4. A plane wall of length L and width W (perpendicular to the figure) has the known temperature distribution $T(x)$. A layer of insulating material of low thermal conductivity k and unknown thickness $t(x)$ covers the wall. The temperature of the ambient is T_0. For simplicity we assume that the thermal resistance (e.g., convection) between the outer surface of the insulation and the ambient is negligible relative to the resistance of the insulating layer. This is equivalent to assuming that the temperature of the outer surface of the insulating layer is essentially uniform and equal to T_0.

When the layer is relatively thin ($t \ll L$), the flow of heat through the insulating material is unidirectional, perpendicular to the wall. In this limit, the following analysis holds not only for plane walls but also for thin insulations covering curved walls [17]. From the law of Fourier we know the local heat flux $k[T(x) - T_0]/t$ and, after integration, the total heat transfer rate

$$q = \int_0^L kW \frac{T(x) - T_0}{t(x)} \, dx. \tag{9.8}$$

The amount of insulating material, which is constrained, is given by the volume integral:

$$V = \int_0^L t(x) W \, dx, \qquad \text{constant.} \tag{9.9}$$

An alternative constraint is the insulation thickness averaged over the wall surface:

$$t_{\text{avg}} = \frac{1}{L} \int_0^L t(x) \, dx = \frac{V}{WL}, \qquad \text{constant.} \tag{9.10}$$

The objective is to find the optimal distribution of insulation $t_{\text{opt}}(x)$ that minimizes heat leak (9.8) subject to volume constraint (9.10). This problem has a simple solution based on variational calculus. The problem is equivalent to seeking the extremum of the aggregate integral,

$$\Phi = \int_0^L \left[kW\frac{T(x) - T_0}{t(x)} + \lambda\frac{t(x)}{L} \right] dx, \tag{9.11}$$

in which λ is a Lagrange multiplier. Let F be the integrand of Φ, and note that F is a linear combination of the integrands of integrals (9.8) and (9.10). The optimal $t(x)$ function for which Φ reaches an extremum is the solution to the Euler equation, which in this case is $\partial F/\partial t = 0$. The solution has the form

$$t_{\mathrm{opt}}(x) = K[T(x) - T_0]^{1/2}, \tag{9.12}$$

in which K is shorthand for the constant $(kWL/\lambda)^{1/2}$. We determine this constant by substituting Eq. (9.12) back into volume constraint (9.10), so that in the end the optimal thickness function is

$$t_{\mathrm{opt}}(x) = \frac{t_{\mathrm{avg}}L[T(x) - T_0]^{1/2}}{\int_0^L [T(x) - T_0]^{1/2}\, dx}. \tag{9.13}$$

The corresponding minimum heat transfer rate through the insulated surface $L \times W$ is

$$q_{\mathrm{min}} = \frac{kW}{t_{\mathrm{avg}}L} \left\{ \int_0^L [T(x) - T_0]^{1/2}\, dx \right\}^2. \tag{9.14}$$

The same $t_{\mathrm{opt}}(x)$ distribution is obtained when the amount of insulation material is minimized subject to a fixed rate of heat loss to the ambient. In other words, the same geometric form is obtained when integral (9.10) is minimized while integral (9.8) is held fixed.

In conclusion, the optimal geometric form is represented by an insulation thickness that is proportional to the square root of the local temperature difference across the insulating layer. For example, if the wall temperature increases linearly from T_0 at $x = 0$ to T_L at $x = L$,

$$T(x) = T_0 + \frac{x}{L}(T_L - T_0), \tag{9.15}$$

the optimal insulation thickness must increase in proportion with $x^{1/2}$, as sketched in Fig. 9.2. In this case Eqs. (9.13) and (9.14) yield

$$t_{\mathrm{opt}}(x) = \frac{3}{2}t_{\mathrm{avg}}\left(\frac{x}{L}\right)^{1/2}, \qquad q_{\mathrm{min}} = \frac{4}{9}kWL\frac{T_L - T_0}{t_{\mathrm{avg}}}. \tag{9.16}$$

We can evaluate the goodness of this insulation architecture by considering simpler shapes, all for the same linear distribution of wall temperature, Eq. (9.15). If the thickness $t(x)$ is uniform, Eqs. (9.10) and (9.14) become

$$t = t_{\mathrm{avg}}, \qquad q = \frac{1}{2}kWL\frac{T_L - T_0}{t_{\mathrm{avg}}}. \tag{9.17}$$

Another alternative is to increase the insulation thickness linearly from $x = 0$ to $x = L$, for which Eqs. (9.10) and (9.14) yield

$$t(x) = 2\frac{x}{L}t_{\text{avg}}, \qquad q = \frac{1}{2}kWL\frac{T_L - T_0}{t_{\text{avg}}}. \tag{9.18}$$

Compared with optimal geometry (9.16), designs (9.17) and (9.18) are inferior because their rates of heat loss exceed the minimum by 12.5%.

The difference between constant-t and optimal-t designs can be smaller or larger than this 12.5% difference, depending on the wall shape (plane versus cylindrical) and the wall temperature distribution (linear versus nonlinear). These generalizations of the geometry-generating principle are explored in Refs. 16 and 18. They reinforce the conclusion that a finite amount of material can be distributed optimally (unevenly) and that the maximization of the global insulation effect lends geometric form to the design.

9.3 Structure in Low-Temperature Machines

The field of refrigeration engineering is full of examples of structures generated by the purpose and constraints principle. The purpose of a refrigerating machine is to keep a cold space cold. The space cannot remain cold by itself because of the leakage of heat from the ambient. No insulation is perfect. Furthermore, in most applications the heat leak is only one of the heat currents that threaten the cold space: Additional examples are the heat currents that are due to electrical resistances, friction between moving parts, and the movement of warm materials into the cold space and cold materials out of the cold space (e.g., domestic refrigerators, helium liquefiers).

The heat currents that reach the cold space must be collected and removed. They must be "forced" to flow in the unnatural direction, toward room temperature. This takes work, or power. The purpose of the machine is not only to maintain the low temperature of the cold space, but to do so with the least power expenditure. We are reminded of an important observation made in Section 1.3 about the concept of objective, or purpose. To do something is not enough. To do it better and better, in spite of the constraints, is the real objective, the drive.

The origin of geometric structure in the objective and constraints principle can be illustrated on the basis of the simplest possible model of a refrigerating machine (Ref. 19, p. 46). The cold-space temperature is T_L and the ambient temperature is T_H. The only heat current that reaches the cold space is due to the leakage through the insulation of cross-sectional area A, thickness L, and effective thermal conductivity k. These parameters (T_L, T_H, A, L, k) are fixed and play the role of constraints in the discussion that follows.

For simplicity we also assume that the effective thermal conductivity is independent of temperature, so that for the configuration shown in the upper part of Fig. 9.5 we write

$$Q_0 = k\frac{A}{L}(T_H - T_L). \tag{9.19}$$

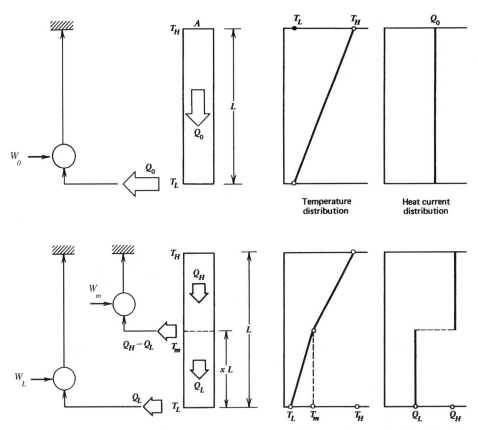

FIGURE 9.5. Insulation with cold-end cooling (top) and with intermediate and cold-end cooling (bottom) [19].

In this configuration the current Q_0 is conserved as it flows from T_H to T_L; this current must be removed as Q_0 from T_L. This task is accomplished by a reversible refrigerating machine, which in the steady state requires the power input

$$W_0 = Q_0 \left(\frac{T_H}{T_L} - 1 \right). \tag{9.20}$$

This result comes from writing the second law of thermodynamics for the machine $(W_0 + Q_0)/T_H = Q_0/T_L$, which states that the machine generates no entropy. It represents a first cut, a reference design, remarkable only for its simplicity.

The direction for improvements is indicated by Eq. (9.20). The power requirement W_0 is large because the temperature ratio T_H/T_L is large (T_L is small). The machine will use less power if its cold end is attached to the insulation at an intermediate temperature T_m, which is larger than T_L. This idea is tried in the lower part of Fig. 9.5. Now the insulation has *structure*: two layers, each with its own constant heat current, Q_H and Q_L. The position of the cooled plane (T_m) is indicated by the dimensionless parameter x, which is defined by the thickness (xL) of the layer of insulation between T_m and T_L. Both T_m and x are variable.

The T_m refrigerator removes the heat current difference $Q_H - Q_L$, requiring the power

$$W_m = (Q_H - Q_0)\left(\frac{T_H}{T_m} - 1\right), \tag{9.21}$$

where

$$Q_H = \frac{kA(T_H - T_m)}{(1-x)L}, \qquad Q_L = \frac{kA(T_m - T_L)}{xL}. \tag{9.22}$$

There is still a need for refrigeration at T_L, because the heat current Q_L reaches the cold space. The power required by the T_L refrigerator is

$$W_L = Q_L\left(\frac{T_H}{T_L} - 1\right). \tag{9.23}$$

What is important is the sum of the two power requirements $(W_m + W_L)$, which can be expressed in a form that shows the effect of varying x and T_m:

$$\frac{W_m + W_L}{kAT_H/L} = \frac{1}{1-x}\left(\frac{T_H}{T_m} - 2 + \frac{T_m}{T_H}\right) + \frac{1}{x}\left(\frac{T_m}{T_L} - 2 + \frac{T_L}{T_m}\right). \tag{9.24}$$

Minimizing this expression, we arrive at geometric structure (x_{opt}) and thermal structure $(T_{m,\mathrm{opt}})$:

$$x_{\mathrm{opt}} = \frac{1}{2}, \qquad T_{m,\mathrm{opt}} = (T_L T_H)^{1/2}, \tag{9.25}$$

$$\frac{(W_m + W_L)_{\min}}{kAT_H/L} = 4\left[\left(\frac{T_H}{T_L}\right)^{1/4} - \left(\frac{T_L}{T_H}\right)^{1/4}\right]^2. \tag{9.26}$$

We measure the goodness of this structure relative to the reference design shown in the upper part of Fig. 9.5 by dividing Eq. (9.26) by Eq. (9.20):

$$\frac{(W_m + W_L)_{\min}}{W_0} = \frac{4(T_H/T_L)^{1/2}}{[(T_H/T_L)^{1/2} + 1]^2} \leq 1. \tag{9.27}$$

This ratio decreases when T_L decreases. In conclusion, the structure revealed in the lower part of Fig. 9.5 is especially recommended for low-temperature systems.

The intermediate cooling example that we just completed is the start of an extensive body of work that has generated complex structures for low-temperature refrigerating machines. The idea of cooling the insulation at an intermediate temperature in a given interval ($T_H - T_L$ in Fig. 9.5) is valid for any other interval. Immediate candidates are the sections $T_H - T_m$ and $T_m - T_L$. After this optimization, the L-thick insulation acquires a more complex structure (four layers), which is sustained by intermediate cooling applied at three locations, in addition to the T_L end.

This optimization can be repeated several times until the intermediate cooling is distributed almost continuously along the insulation, from T_H to T_L. The optimal design for this limit is located much more rapidly based on variational calculus subject to fixed A/L [19, 20]. The conclusion is that the optimal heat current

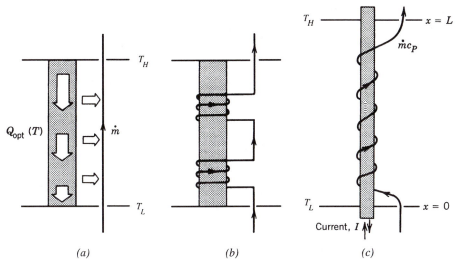

FIGURE 9.6. Examples of structure in low-temperature machine design: (a) continuous cooling provided by a single stream of cold gas, (b) cooling concentrated in two heat exchangers, and (c) continuous cooling of an electrical cable.

varies continuously through the insulation [Fig. 9.6(a)]:

$$Q_{\text{opt}}(T) = \left(\frac{A}{L} \int_{T_L}^{T_H} \frac{k^{1/2}}{T} \, dT \right) k^{1/2} T. \qquad (9.28)$$

This expression is valid for the more general case in which the effective thermal conductivity is a known function of temperature, $k(T)$, as for the structural materials commonly used at low temperatures. If the thermal conductivity is constant, or approximated by a constant, Eq. (9.28) requires a proportionality between Q_{opt} and T or a uniformly distributed intermediate cooling effect, dQ_{opt}/dT.

Structure is the engineer's way of implementing this result. Uniform cooling can be provided by a stream of cold gas flowing from T_L to T_H, as shown in Fig. 9.6(a). The capacity rate of this stream has a special value, which is pinpointed by $dQ_{\text{opt}}/dT = (\dot{m}c_p)_{\text{opt}}$:

$$(\dot{m}c_p)_{\text{opt}} = k \frac{A}{L} \ln \frac{T_H}{T_L} \qquad (k = \text{constant}). \qquad (9.29)$$

The design and the fabrication of the heat transfer surface between the stream and the insulation are difficult. When the insulation is a solid structural member, bringing the stream in close thermal contact with the solid means to weaken the solid by machining channels in it. An alternative is to approximate the uniform cooling scheme by building heat exchangers at a few locations, as shown in Fig. 9.6(b).

There are many other types of insulation that lend themselves to the intermediate cooling technique. A stack of concentric radiation shields built around the cold space works in the same way as the one-dimensional insulation (k, A, L) discussed until now. The role of A is played by the shield surface (the cross section of the radiation heat leak), the role of L is played by the number of shields, and the effective

k depends on temperature because it accounts for the temperatures and radiative properties of mutually facing surfaces. Electrical current leads and power cables, which are necessary in superconducting systems, must also be cooled continuously [Fig. 9.6(c)].

It is convincing that all these structures [Figs. 9.6(a)–9.6(c)] are commonplace in cryogenic engineering practice and that they have their origin in a single *principle*. Now a unifying theory accounts for the seemingly diverse and unrelated structures developed *independently* by engineers over many decades, in separate schools, design rooms, countries, and continents.

If we look at all these forms strictly from the point of view of today, assuming no knowledge of their historical development and the principle that generated them, we are struck by their combination of diversity, similarities, and continuity (persistence, survival). We are struck in the same way as when we admire the many living species and the many river basins. Engineering opens our eyes to the principle that generates diversity, similarities, and continuity throughout nature – the inanimate, the animate, and the engineered.

9.4 Streams in Counterflow

Even more prevalent in machines is the counterflow heat exchanger cooled along its length. The cold structure of every low-temperature refrigerator and liquefier is dominated by a vertical counterflow heat exchanger, which brings warm and compressed gas down to the coldest space and returns low-pressure gas up to room temperature.

The two streams are in intimate thermal contact: The descending stream is cooled by the rising stream. The stream-to-stream heat transfer is indicated by the horizontal arrows in Fig. 9.7. The large heat transfer area (A) between the two streams is responsible for the small but finite local temperature difference ΔT. Ideal-gas behavior is assumed on both sides of the heat exchanger. Invoking the first law of thermodynamics for one of the streams, for example, the cold stream rising from x to $x + dx$ on the left-hand side of Fig. 9.7, we obtain

$$\dot{m}c_p\,dT = U\Delta T\,dA. \tag{9.30}$$

In this expression, U is the overall stream-to-stream heat transfer coefficient based on A. The constraint is the contact area, which can be expressed as an integral derived from Eq. (9.30):

$$A = \int_{T_L}^{T_H} \frac{\dot{m}c_p}{U\Delta T}\,dT. \tag{9.31}$$

The constraint confirms the known rule that small stream-to-stream temperature differences require a large contact area.

What is less known about the counterflow is a feature first pointed out in 1979 in heat transfer engineering [19, 21]: Excellent stream-to-stream thermal contact means excellent insulation in the end-to-end direction. This alternative view holds that the counterflow is a path for net energy transfer in the longitudinal direction,

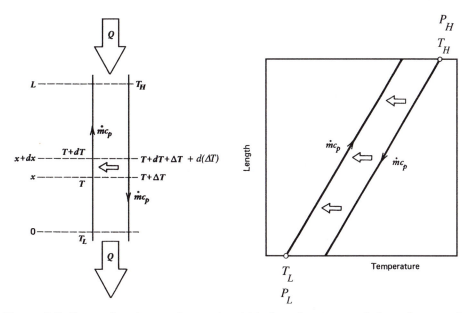

FIGURE 9.7. Counterflow heat exchanger in which the enhancement of thermal contact in the stream-to-stream direction (right) is equivalent to increasing the insulation effect in the end-to-end direction (left) [19].

i.e., an insulation system. The energy that flows is the convective heat current Q indicated by the vertical arrows in Fig. 9.7. The magnitude of Q is the difference between the enthalpy of the descending stream $[\dot{m}c_p(T + \Delta T)]$ that crosses the plane x and the enthalpy of the ascending stream $(\dot{m}c_p T)$ that crosses the same plane:

$$Q = \dot{m}c_p \Delta T. \qquad (9.32)$$

It was shown in a 1979 paper [21] and in 1982 reviews [19, 22] that when Eqs. (9.30) and (9.32) are combined the longitudinal heat current emerges as a quantity proportional to the longitudinal temperature gradient:

$$Q = \frac{(\dot{m}c_p)^2}{Up} \frac{dT}{dx}. \qquad (9.33)$$

In this expression, p is the contact area per unit length of stream flow, $p = dA/dx$.

The proportionality between the convective heat current and the longitudinal temperature gradient is analogous to the Fourier law (9.19), or $Q = kA \, dT/dx$, which served as the starting point in the development of structure from the objective and constraints principle. Consequently, every structural feature that was derived for the unidirectional insulations exhibited in Fig. 9.6 also holds for the column (control volume) that houses the counterflow. The intermediate cooling effect must be distributed uniformly along L, that is, along the temperature span $T_L - T_H$. This can be accomplished as in Fig. 9.6(a): This time we place an additional stream of cold gas in counterflow with the convective heat current Q. This cold stream is labeled $\dot{m}_e$ on the left-hand side of Fig. 9.8. It can be made available by bleeding an optimal fraction of the warm high-pressure stream and expanding it through a work-producing device (cylinder and piston, or turbine). The analysis is presented

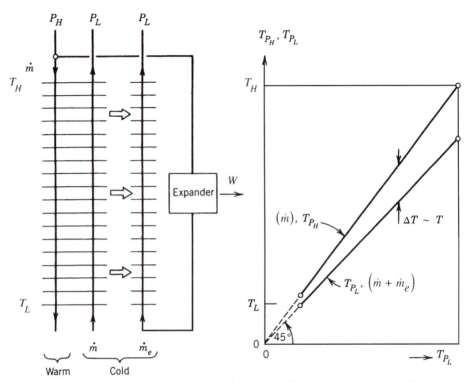

FIGURE 9.8. Optimal intermediate cooling of a counterflow heat exchanger and the corresponding temperature distribution along the high-pressure side (P_H) and the low-pressure side (P_L).

in detail in Refs. 1 and 13, although it can be reproduced based on Eqs. (9.30)–(9.33) and the analogy with the more complete analysis of Section 9.3. The optimal flow imbalance is

$$\frac{\dot{m}_{e,\text{opt}}}{\dot{m}} = \frac{\dot{m}c_p}{UA} \ln \frac{T_H}{T_L}. \tag{9.34}$$

Together, the three streams constitute a counterflow heat exchanger with slightly unbalanced capacity rates in which the cold side $(\dot{m} + \dot{m}_e)$ is larger than the warm side. Optimal imbalance (9.34) is reflected in a tapering of the temperature gap between the two streams, such that the smaller temperature differences are near the cold end (Fig. 9.8, right-hand side):

$$\left(\frac{\Delta T}{T}\right)_{\text{opt}} = \frac{\dot{m}c_p}{UA} \ln \frac{T_H}{T_L}. \tag{9.35}$$

This smooth taper can be produced when the cold stream is delivered by a single expander, i.e., when the pressure ratio P_H/P_L is large enough so that the expander embraces the entire temperature scale of the heat exchanger, $T_H - T_L$.

When the temperature span of one expander is too narrow relative to $T_H - T_L$, the tapering of the ΔT versus T distribution can be achieved by the installation of two or more expanders along the main heat exchanger. This technique is illustrated

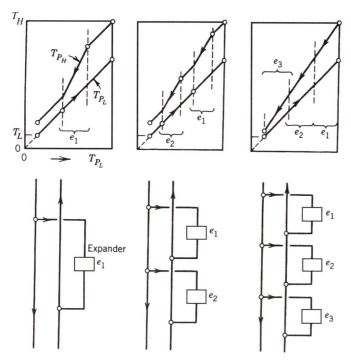

FIGURE 9.9. The approximate tapering of the temperature distribution along the counterflow when one, two, and three expanders are used.

in Fig. 9.9 for a sequence of one, two, and, finally, three expanders. Given the fixed-temperature span of one expander, there exists an optimal spatial position for inserting each expander along the $T_H - T_L$ scale: See Ref. 1. One effect of the string of expanders is that the flow rate handled by the counterflow heat exchanger decreases in the direction of lower temperatures. In other words, the cold-end flow rate ($\dot{m}_L$) that removes the refrigeration load from the T_L end is smaller than the flow rate processed by the room-temperature compressor.

There is an analogy between the use of a few expanders (Fig. 9.9) and the use of a few intermediate cooling stations along a conducting support [Fig. 9.6(b)]. Furthermore, the physical structures exhibited in the lower part of Fig. 9.9 can be observed in practically every machine built for operation at liquid-nitrogen temperature (77 K) and liquid-helium temperature (4.2 K). They were not built based on the theory presented in this section. The theory came much later [19–22]. Existing engineering structures, with their great diversity and striking similarities, call for a unifying form-generating principle as much as all the natural structures.

Nature offers many examples of structures with streams in counterflow. For these, natural scientists have offered the same purpose-based explanation as the design principle for engineered counterflows: Natural streams are organized in counterflow in order to reduce the convective current (heat, or mass species) that flows longitudinally.

Best known is the example of the blood counterflow in the long legs of wading birds and in whale flippers [23]. The body of the bird is warm and the foot is cold in

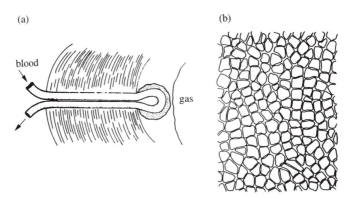

FIGURE 9.10. Counterflow heat exchanger in the wall of the fish's swim bladder. Blood vessels going inward through the wall are in close contact with those coming out. The right-hand side shows a cross section through these blood vessels ([25]; reprinted with permission from Princeton University Press).

the water. Warm oxygenated blood flows downward and is cooled in counterflow by the returning venous blood. In this way the loss of body heat through the foot is minimized. Otherwise, the body extension (foot) would act as any other extended surface (fin, Chap. 8) and maximize the loss of body heat.

Another beautiful example is the blood counterflow in the wall of the fish's swim bladder [24, 25]. In this case the objective of the counterflow is to minimize the convective leakage of oxygen. The blood flowing away from the bladder loses excess dissolved gas to the blood flowing toward the bladder (Fig. 9.10). Numerous other counterflows prevent the excessive loss of water from animals living in the desert [26].

In biomedical engineering, the longitudinal convective heat transfer expression (9.33) was rediscovered in 1985 by Weinbaum and Jiji [27]. This effect was incorporated in a heat transfer model of the vascularized tissue to account for the occurrence of countercurrent pairs of thermally significant blood vessels (Fig. 9.11). In this model, Eq. (9.33) makes an additional, convective contribution to the effect of conduction through the living tissue.

All these natural counterflows are structures with purpose. No matter how hard we try to avoid pronouncing the word purpose, substitute language such as "optimized," "minimized," and "in order to do this" hide one and only one meaning – purpose, objective, or better and better performance. Once again, the great geometric and functional similarities between the naturally occurring structures and the engineered ones strengthen the view that the constructal law of purpose and constraints makes sense in rationalizing the occurrence of structure in every compartment of nature, inanimate systems, animate systems, and engineered systems.

9.5 Flying Machines and Animals

More evidence that what generates form in engineering also generates form in nature comes from the study of powered flight. As in refrigeration and other

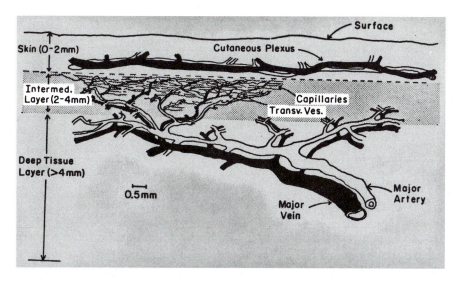

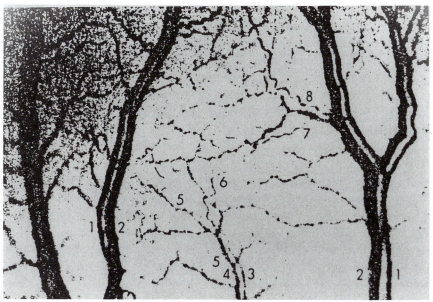

FIGURE 9.11. The top drawing shows the three layers of peripheral circulation modeled by Weinbaum, et al. [28]. The skin is a 2-mm-thick layer with a plexus of large vessels at the lowest part of the dermis. This plexus supplies the thermally insignificant vessels of the arterial and venous plexus of the upper dermal layers. The intermediate layer is an approximately 2-mm-thick region of muscle directly under the skin. In this layer most vessels are thermally insignificant and perfusion is lateral from transverse arteriole to transverse venule. In the deep tissue layer are the major countercurrent arteries and veins and their branches as well as the vasculature found in the intermedite layer. The bottom photograph covers 1 mm^2 of tissue [27] and shows (1, 2) countercurrent artery–vein pairs, (3, 4) terminal ends of countercurrent vessels, (5, 6) isotropically perfused tissue space, and (7, 8) directed perfusion from/to countercurrent artery of vein (reprinted with permission from the American Society of Mechanical Engineers International).

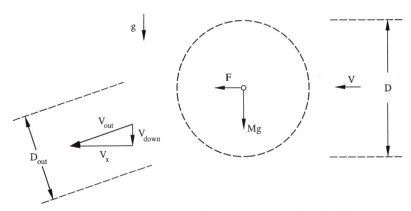

FIGURE 9.12. Simple model and interactions of a flying body [32].

power-consuming systems, our theoretical starting point is the objective – to do the most based on the power or fuel consumed. The flying structure that is generated by this principle has many features. More obvious are the shapes: bodies, wings, feathers, and flaps. More hidden are the internal parts such as the hollow bones of birds and the aluminum parts of most airplanes. Even more subtle is the speed of the flying body, on which we focus in this section.

The physics principles of powered flight are known in biology [29–31]: In this section they are reformulated by use of the engineering thermodynamics of open systems in steady flow [32]. The most basic features and needs of powered flight are retained in the simple model of Fig. 9.12. The flying body of mass M has the linear dimension D, density ρ_b, and horizontal speed V relative to the surrounding air. The air density ρ_a is much smaller than ρ_b. This leads to the global requirement that the net vertical body force $Mg \sim \rho_b D^3 g$ must be supported by other forces. The generation of the latter is achieved through the relative motion called flight.

Consider the conservation of mass and momentum in the control volume occupied by the flying body and the immediately close fluid regions affected by relative motion. In the steady state, an air stream of mass flow rate $\dot{m} \sim \rho_a D^2 V$ enters the control volume and the same stream exits ($\dot{m} \sim \rho_a D_{\text{out}}^2 V_{\text{out}}$). The exit velocity V_{out} must have a vertical component V_{down} in order to develop a vertical flow of momentum to support the body force:

$$V_{\text{out}} = \left(V_x^2 + V_{\text{down}}^2\right)^{1/2}. \tag{9.36}$$

The vertical momentum balance is $\rho_b D^3 g \sim \dot{m} V_{\text{down}}$, or

$$V_{\text{down}} \sim \frac{\rho_b g D}{\rho_a V}. \tag{9.37}$$

The conservation of horizontal momentum is the statement that the momentum generated by the outflow ($\dot{m} V_x$) must balance the retarding forces associated with momentum inflow ($\dot{m} V$) and drag (F):

$$\dot{m} V_x \sim \dot{m} V + F. \tag{9.38}$$

The drag force F is of order of $C_D D^2 \rho_a V^2$, where D^2 is the scale of the body cross section. The drag coefficient C_D is a relatively constant number of the order of 1 when the Reynolds number $(\rho_a V D/\mu_a)$ is greater than the order of 10^2 (e.g., Ref. 33, p. 325). This means that the ratio $F/\dot{m}$ scales as V:

$$\frac{F}{\dot{m}} \sim V. \tag{9.39}$$

Finally, the flying system must spend exergy or mechanical power $(\dot{W})$ in order to increase the kinetic energy of the air stream from the inlet $(\dot{m}V^2/2)$ to the outlet $(\dot{m}V_{\text{out}}^2/2)$:

$$\dot{W} \sim \frac{1}{2}\dot{m}\left(V_{\text{out}}^2 - V^2\right). \tag{9.40}$$

In an animal this power is delivered by the muscles and has its origin in metabolism. In a flying machine $\dot{W}$ is produced by the power plant installed on board. In both cases, $\dot{W}$ is drawn from the chemical exergy of the consumed food or fuel.

By using approximations (9.36)–(9.38), approximating $V_x^2 \sim (V + F/\dot{m})^2 \sim V^2 + 2VF/\dot{m}$, and neglecting all the factors of the order of 1, we can eliminate V_{down} and rewrite approximation (9.40) as

$$\dot{W} \sim \frac{\rho_b^2 g^2 D^4}{\rho_a V} + \rho_a D^2 V^3. \tag{9.41}$$

This two-term expression shows the power that is required for maintaining the body in the air (the first term) and overcoming the drag (the second term). Changes in the flying speed induce changes of opposing signs in the two terms. Power function (9.41) has a clear minimum with respect to V:

$$V_{\text{opt}} \sim 3^{-1/4}\left(\frac{\rho_b}{\rho_a}gD\right)^{1/2}, \tag{9.42}$$

$$\dot{W}_{\text{min}} \sim \frac{4\rho_b^{3/2}g^{3/2}D^{7/2}}{3^{3/4}\rho_a^{1/2}}. \tag{9.43}$$

At this optimum the power spent on lifting the body is three times larger than the power needed to overcome the drag. Here we have an example of optimal allocation or optimal partition, which is a common occurrence in thermodynamic optimization [1, 13]. When the flying speed is significantly less than the optimal, the power requirement is dominated by the need to hold the body in the air. In the opposite extreme the power is spent mainly on overcoming drag.

Astonishingly good agreement exists between the principle-generated flying speed formula and the flying speeds found in nature and engineering [29, 34]. Since the body length scale D is the same as $(M/\rho_b)^{1/3}$, the optimal speed of approximation (9.42) is proportional to the body mass raised to the power 1/6:

$$V_{\text{opt}} \sim \frac{\rho_b^{1/3}g^{1/2}}{\rho_a^{1/2}}M^{1/6}. \tag{9.44}$$

Figure 9.13 shows that the flying speeds of all the things we know align themselves on a power-law curve of the predicted type. The data plotted in Fig. 9.13 are from a

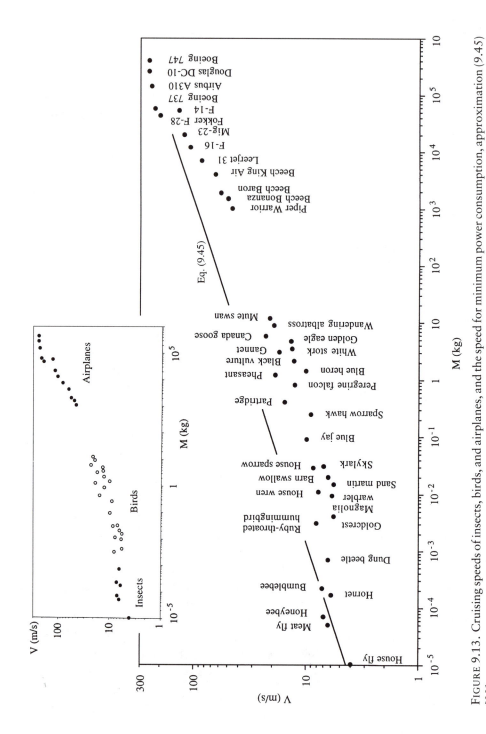

FIGURE 9.13. Cruising speeds of insects, birds, and airplanes, and the speed for minimum power consumption, approximation (9.45) [32].

more extensive compilation, from many independent sources, reported graphically by Tennekes [34]. The agreement between the predicted and the observed is not only qualitative but also quantitative. All flying animals and machines are represented by the orders of magnitude $\rho_b \sim 10^3$ kg/m^3 and $\rho_a \sim 1$ kg/m^3 (air at atmospheric conditions). The V_{opt} formula (9.42) becomes, roughly,

$$V_{\text{opt}} \sim 30(M/\text{kg})^{1/6} \text{m/s}, \tag{9.45}$$

which has been drawn on Fig. 9.13. We see how the minimization of power consumption or exergy* destruction unites all the flying systems, the animate with the engineered.

The minimization of the required mechanical power is not the only way of deriving from the constructal law the proportionality between V and $M^{1/6}$. Zoologists who contemplate the purpose of migratory birds also argue that the design is driven by the minimization of the work spent (W) to cover a distance (L) between two fixed points on the surface of the globe [29]. The same argument is being made by aircraft designers, who know much better the purpose of what they develop! The surprise of this argument is that it leads to the same predictions as those of the power-minimization analysis. For if t is the time needed to cover the distance L with the speed V, the work/distance ratio is the same as the power/speed ratio,

$$\frac{W}{L} \sim \frac{W/t}{L/t} \sim \frac{\dot{W}}{V}, \tag{9.46}$$

or, according to approximation (9.41),

$$\frac{W}{L} \sim \frac{\rho_b^2 g^2 D^4}{\rho_a V^2} + \rho_a D^2 V^2. \tag{9.47}$$

The minimization of this W/L expression with respect to V yields

$$V_{\text{opt}} \sim \left(\frac{\rho_b}{\rho_a} g D\right)^{1/2}, \tag{9.48}$$

$$\left(\frac{W}{L}\right)_{\text{min}} \sim 2g\rho_b D^3. \tag{9.49}$$

At this optimum the speed is of the same order as that of the optimal speed for minimum power, approximation (9.42). The difference between the two estimates is only 30%, the estimate of approximation (9.48) being the larger. Equipartition, the special case of optimal allocation, does occur when the exergy spent per distance traveled is minimized: The two terms of approximation (9.47) are equal when $V = V_{\text{opt}}$.

Another important result is the $(W/L)_{\text{min}}$ expression (9.49), which shows that the fuel or the food that must be used is proportional to the body mass $\rho_b D^3$.

---

* Exergy is the useful part, the work-producing content of a stream (e.g., fluid, heat) [1]. Exergy streams are the life blood of all power and refrigeration systems. Exergy is destroyed partially or completely when streams interact with each other and with components. The rate of exergy destruction is proportional to the rate of entropy generation. Thermodynamic performance is improved by minimizing entropy generation or exergy destruction subject to constraints (see Refs. 1, 13, and 19).

Throughout the aircraft industry this principle is known as the take-off gross weight criterion: The fuel penalty associated with adding a new component on board is proportional to the mass of that component [35]. What rules architectural design in the aircraft industry also rules the minimization of the food required by flying animals. Approximation (9.49) also shows that the minimum work required for flying the distance L scales as the weight of the body times the distance. It is as if the power plant must lift the body to a height comparable with L.

What we accomplished based on simple thermodynamic optimization in this section is only a sketch, a hint. The *complete* thermomechanical analysis of the flying body and the geometric optimization of the internal and external architecture will establish exactly what fraction of the used food or fuel is needed to produce the exergy ($\dot{W}$) needed to sustain the flight – the exergy fraction that is destroyed in the air motion caused by flight. The rest of the fuel exergy is destroyed in all the other flow systems and processes on board, e.g., combustion, energy conversion, avionics and environmental control.

9.6 Flying Carpets and Processions

Even though the theoretical line plotted in Fig. 9.13 is correct in only an order-of-magnitude sense, it comes surprisingly close to the observed cruising speeds of animals and machines. The scatter of the data in each group (insects, birds, airplanes) can be discussed further based on differences in body shape, wing slenderness, and life style (e.g., terrestrial versus migratory birds) [29]. More interesting is that within each group the shift toward larger bodies and larger speeds is accompanied by a form of *external* organization. Migratory birds fly in large groups organized in precise patterns. Note the synchronization of the pelicans in Fig. 9.14. From the single-seat airplanes of the World War I era to the commercial airliners of today, people have also proven that larger masses travel better.

The same external organization of movement is visible on the ground, in team bicycle racing, and in the slip stream shed by the lead race car. The principle is the same: Aerodynamic drag decreases when individual masses coalesce [36, 37], and in this way mass travels faster and farther, all the way to the formation of cellestrial bodies (see Ref. 1, pp. 804–805 and Ref. 38). Strings of racers on the ground and flying carpets (birds, people) in the air – they are all driven by the same principle.

Processions are visible everywhere. Schools of fish are beautiful because they display precise patterns of organization. The synchronized swimming and jumping of dolphins are also due to spatial organization. Ducklings paddle behind their mother arranged in a pattern, like pieces on a chess board.

Biologists and fluid mechanicists have long recognized the geometric relation between a patterned procession and the regular occurrence of vortices in the wake shed by the body that leads the procession. They are right, of course, but not in the sense that is meant in this book. It is the inanimate fluid in the wake of the leading body that organizes itself. It does so using no brain power whatsoever, so that it may travel and spread itself the fastest through the stationary fluid.

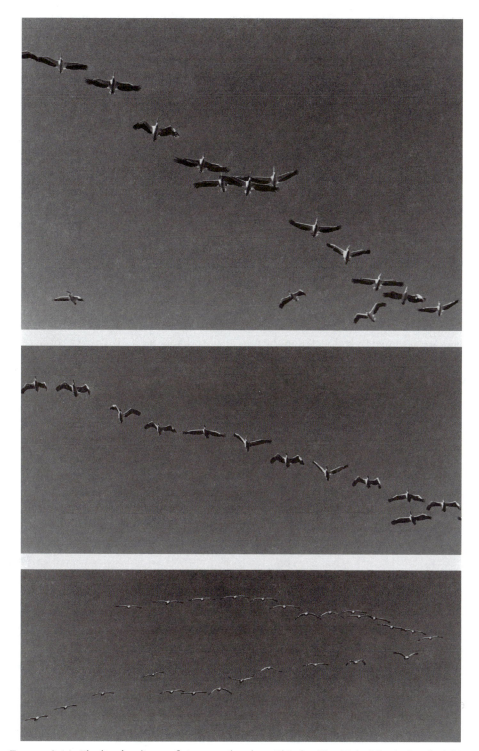

FIGURE 9.14. Flock of pelicans flying overhead at Phinda, KwaZulu-Natal, South Africa (photograph courtesy of William A. Bejan) (see Plate XI).

The fish and the ducklings, for their own power-minimization (constructal) reasons, act as markers in the flow. They feel the flow, and they follow it. In this way they visualize the principle that accounts not only for the eddies and turbulence in the fluid (Sections 7.3–7.5), but also for the animals' guiding objective to advance in regular processions.

As engineers, in this section we focused on the thermodynamic optimization of aircraft. The same principles apply to designs in which all the functions are driven by the exergy drawn from the limited fuel installed on board: ships, automobiles, military vehicles, environmental-control suits, portable power tools, etc. These extensions are waiting to be developed. The swimming of fish [39, 40] and humans [41] is already developed theoretically and can be presented in the manner of Section 9.5. The same can be done with the optimization (i.e., prediction) of the flapping frequency of wings and flippers. For these and other engineering applications the present book defines and illustrates the method.

PROBLEMS

9.1 In the simple model shown in Fig. P9.1 the irreversibilities of a refrigeration plant are assumed to be located in the two heat exchangers, $(UA)_H$ above room temperature (T_H, fixed) and $(UA)_L$ below the refrigeration load temperature (T_L, fixed). A reversible compartment between T_{HC} and T_{LC} completes the model. The heat transfer rates Q_H and Q_L are given by $\dot{Q}_H = (UA)_H(T_{HC} - T_H)$ and $\dot{Q}_L = (UA)_L(T_L - T_{LC})$. With reference to the right-hand figure of Fig. P9.1, minimize $UA = (UA)_H + (UA)_L$ while regarding the refrigeration load $\dot{Q}_L$ as specified. Show that in this case $(UA)_{min}$ is divided equally between the two heat exchangers. Derive an expression for $\dot{W}/\dot{Q}_L$ and comment on the effect of $(UA)_{min}$.

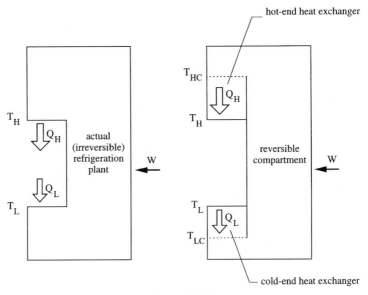

FIGURE P9.1

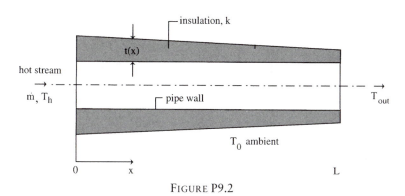

insulation, k

t(x)

hot stream

$\dot{m}, T_h$

pipe wall

T_{out}

T_0 ambient

0   x   L

FIGURE P9.2

9.2 As shown in Fig. P9.2, a hot stream originally at the temperature T_h flows through an insulated pipe suspended in a space at temperature T_0. The stream temperature $T(x)$ decreases in the longitudinal direction because of the heat transfer from $T(x)$ to T_0 that takes place everywhere along the pipe. The function of the pipe is to deliver the stream at a temperature (T_{out}) as close as possible to the original temperature T_h. The amount of insulation is fixed. Determine the best way of distributing the insulation along the pipe [16].

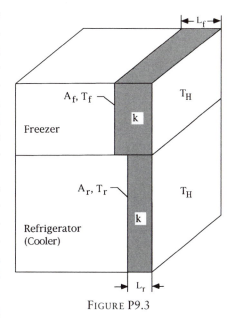

FIGURE P9.3

9.3 A domestic refrigerator and freezer have two inner compartments, the refrigerator or cooler of temperature T_r and the freezer of temperature T_f. The ambient temperature is T_H. This assembly is insulated with a fixed amount of insulation ($A_f L_f + A_r L_r =$ constant), where A_f and A_r are the surfaces of the freezer and the refrigerator enclosures. The thermal conductivity of the insulating material is k. The two heat currents that penetrate the insulation and arrive at T_f and T_r are removed by a reversible refrigerating machine that consumes the power W. Determine the optimal insulation architecture, e.g., the ratio L_f/L_r, such that the total power requirement W is minimum [42].

L_f

A_f, T_f T_H k

Freezer

A_r, T_r T_H k

Refrigerator (Cooler)

L_r

9.4 A fixed amount of insulation material of volume V must be distributed optimally over the outer surface of a cylinder of radius r. The cylinder wall temperature varies as $T(x) = T_0 + (x/L)(T_L - T_0)$. The thermal conductivity of the insulation material k and the length of the wall L are known. The outer surface of the insulation is at the ambient temperature T_0. Determine the optimal distribution of insulation, $t_{opt}(x)$, and the corresponding minimum heat transfer rate to the ambient. Show that the heat loss reduction that is due to using an insulation with optimal thickness on a cylindrical wall is smaller than on the corresponding plane wall [16].

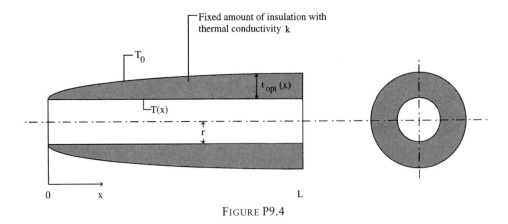

FIGURE P9.4

REFERENCES

1. A. Bejan, *Advanced Engineering Thermodynamics*, 2nd ed., Wiley, New York, 1997.

2. A. Bejan and D. Tondeur, Equipartition, optimal allocation, and the constructal approach to predicting organization in nature, *Rev. Gen. Therm.*, Vol. 37, 1998, pp. 165–180.

3. I. I. Novikov, The efficiency of atomic power stations, *J. Nuclear Energy II*, Vol. 7, 1958, pp. 125–128.

4. P. Chambadal, *Les Centrales Nucleaires*, Armand Colin, Paris, 1957.

5. P. Chambadal, *Évolution et Applications du Concept d'Éntropie*, Dunod, Paris, 1963.

6. M. M. El-Wakil, *Nuclear Power Engineering*, McGraw-Hill, New York, 1962.

7. M. M. El-Wakil, *Nuclear Energy Conversion*, International Textbook, Scranton, PA, 1971.

8. I. I. Novikov and K. D. Voskresenskii, *Thermodynamics and Heat Transfer*, Atomizdat, Moscow, 1977.

9. F. L. Curzon and B. Ahlborn, Efficiency of a Carnot engine at maximum power output, *Am. J. Phys.*, Vol. 43, 1975, pp. 22–24.

10. A. Bejan, *Advanced Engineering Thermodynamics*, Wiley, New York, 1988.

11. A. Bejan, Theory of heat transfer-irreversible power plants, *Int. J. Heat Mass Transfer*, Vol. 31, 1988, pp. 1211–1219.

12. A. Bejan, Theory of heat transfer-irreversible power plants – II. The optimal allocation of heat exchange equipment, *Int. J. Heat Mass Transfer*, Vol. 38, 1995, pp. 433–444.

13. A. Bejan, *Entropy Generation Minimization*, CRC, Boca Raton, FL, 1996.

14. A. Bejan, Theory of heat transfer-irreversible refrigeration plants, *Int. J. Heat Mass Transfer*, Vol. 32, 1989, pp. 1631–1639.

15. V. Radcenco, J. V. C. Vargas, A. Bejan, and J. S. Lim, Two design aspects of defrosting refrigerators, *Int. J. Refrig.*, Vol. 18, 1995, pp. 76–86.

16. A. Bejan, How to distribute a finite amount of insulation on a wall with nonuniform temperature, *Int. J. Heat Mass Transfer*, Vol. 36, 1993, pp. 49–56.

17. A. Bejan, *Heat Transfer*, Wiley, New York, 1993.

18. A. Bejan, G. Tsatsaronis, and M. Moran, *Thermal Design and Optimization*, Wiley, New York, 1996.

19. A. Bejan, *Entropy Generation through Heat and Fluid Flow*, Wiley, New York, 1982.

20. A. Bejan and J. L. Smith, Jr., Thermodynamic optimization of mechanical supports for cryogenic apparatus, *Cryogenics*, Vol. 14, 1974, pp. 158–163.

21. A. Bejan, A general variational principle for thermal insulation system design, *Int. J. Heat Mass Transfer*, Vol. 22, 1979, pp. 219–228.

22. A. Bejan, Second law analysis in heat transfer and thermal design, *Adv. Heat Transfer*, Vol. 15, 1982, pp. 1–58.

23. K. Schmidt-Nielsen, *How Animals Work*, Cambridge Univ. Press, Cambridge, UK, 1972.

24. G. N. Lapennas and K. Schmidt-Nielsen, Swimbladder permeability to oxygen, *J. Exp. Biol.*, Vol. 67, 1977, pp. 175–196.

25. S. Vogel, *Life's Devices*, Princeton Univ. Press, Princeton, NJ, 1988.

26. K. Schmidt-Nielsen, *Desert Animals: Physiological Problems of Heat and Water*, Oxford Univ. Press, Oxford, UK, 1964.

27. S. Weinbaum and L. M. Jiji, A new simplified bioheat equation for the effect of blood flow on the local average tissue temperature, *J. Biomech. Eng.*, Vol. 107, 1985, pp. 131–139.

28. S. Weinbaum, L. M. Jiji, and D. E. Lemons, Theory and experiment for the effect of vascular microstructure on surface tissue heat transfer – part I: anatomical foundation and model conceptualization, *J. Biomech. Eng.*, Vol. 106, 1984, pp. 321–330.

29. J. C. Pennyquick, *Animal Flight*, Edward Arnold, London, UK, 1972.

30. K. Schmidt-Nielsen, *Scaling: Why Is Animal Size so Important?*, Cambridge Univ. Press, Cambridge, UK, 1984.

31. Sir James Lighthill, *Mathematical Biofluiddynamics*, Society for Industrial and Applied Mathematics, Philadelphia, PA, 1975.

32. A. Bejan, A role for exergy analysis and optimization in aircraft energy-system design, in *Proceedings of the ASME Advanced Energy Systems Division – 1999*, S. M. Aceves, S. Garimella, and R. Peterson, eds., American Society of Mechanical Engineers, New York, 1999, pp. 209–218.

33. A. Bejan, *Convection Heat Transfer*, 2nd ed., Wiley, New York, 1995.

34. H. Tennekes, *The Simple Science of Flight*, MIT Press, Cambridge, MA, 1996.

35. D. L. Siems, Private communications, The Boeing Corporation, St. Louis, MO, 1997–1999.

36. A. M. Al Taweel, J. Militzer, J. M. Kan, and F. Hamdullahpur, Motion of hydrodynamic aggregates, *Powder Technology*, Vol. 59, 1989, pp. 173–181.

37. J. Militzer, J. M. Kan, F. Hamdullahpur, P. R. Amyotte, and A. M. Al Taweel, Drag coefficient for axisymmetric flow around individual spheroidal particles, *Powder Technology*, Vol. 57, 1989, pp. 193–195.

38. A. Bejan, How nature takes shape: extensions of constructal theory to ducts, rivers, turbulence, cracks, dendritic crystals, and spatial economics, *Int. J. Therm. Sci. (Rev. Gen. Therm.)*, Vol. 38, 1999, pp. 653–663.

39. P. W. Webb and D. Weihs, eds., *Fish Biomechanics*, Praeger, New York, 1983, Chap. 11.

40. L. Maddock, Q. Bone, and J. M. V. Rayner, eds., *Mechanics and Physiology of Animal Swimming*, Cambridge Univ. Press, Cambridge, UK, 1994.

41. D. C. Agrawal, Work and heat expenditure during swimming, *Phys. Educ.*, Vol. 34, 1999, pp. 220–226.

42. J. S. Lim and A. Bejan, Two fundamental problems of refrigerator thermal insulation design, *Heat Transfer Eng.*, Vol. 15, No. 3, 1994, pp. 35–40.

CHAPTER TEN

STRUCTURE IN TIME: RHYTHM

The generation of structure in space (geometric form) is the principle that united all the flow systems discussed until now. Nature impresses us not only with structure in space but also with structure in time – rhythm, or characteristic periodicity. The same holds for many engineering systems.

Best known among the natural flows that exhibit rhythm are the breathing and heart beating of all animals. These intermittent flows are the basis for some of the most puzzling allometric laws, for example, the observed proportionality between frequency and animal body mass raised to the power of approximately −0.25. An elephant breaths less frequently than the mouse, and the human heart beats faster than the heart of the cow.

The volume of theoretical material published on pulsating physiological processes is staggering. The preferred approach has been to take pulsating flows as given, i.e., not to question why nature has opted for periodic flow and not for steady flow. This is similar to the equally staggering amount of work done on tree networks in physiology and geophysics: The existence of trees is assumed, not questioned. Frequencies and trees are assumed empirically, as they are first observed and then analyzed. Optimized features of the fluid-flow path, such as Murray's law (Section 5.7 and Table 5.3), are then added based on flow-resistance minimization. This gives the impression that the work is theoretical, i.e., deduced entirely from principle. It is not.

What has not been rationalized until now is the *existence* of finely tuned pulsating flows in nature. To achieve this by invoking the constructal principle is the main objective of this chapter. The presentation follows closely the material disclosed recently in Refs. 1–3.

Breathing and heart beating have a lot in common with engineered processes that are governed by time-dependent transport by diffusion. When we manufacture ice by freezing water on a cooled surface, the rate of ice formation decreases as the ice layer thickens. To maximize the time-averaged rate of ice production, we must terminate the freezing process, scrape the surface clean, and start over. In the first section of this chapter we learn that there is an optimal, characteristic periodicity associated with any ice-making device or diffusion-controlled process.

Optimized intermittent operation is a characteristic of many other engineered systems, some very complicated. Frost and ice grow on the evaporator surface of a refrigerator. An automatically defrosting refrigerator interrupts its cooling cycle and cleans the evaporator surface by melting the ice (Section 10.2). Power plants are shut down periodically so that the layers of dirt (scale, fouling) are scraped off the surfaces of their heat exchangers (Section 10.3). All these techniques are made necessary by the pursuit of objectives subject to constraints: They all generate structure in time.

The lung and the vascularized tissue are systems with similar objectives and constraints. The fluid (air, blood) that is just admitted into the smaller passages (alveoli, capillaries) triggers mass diffusion into the surrounding tissue. The effectiveness of this transport mechanism decreases in time. The maximization of the mass transfer rate demands the purging of the old charge of fluid, so that high fluxes are triggered again by a fresh charge. Next to the spatial structures of the preceding chapters (trees, round ducts, river cross sections, eddies), the temporal structure of natural pulsating flows adds to the already overwhelming evidence that supports the constructal principle.

10.1 Intermittent Heat Transfer

Consider the production of a solid material, which is periodically solidified as a layer on a cooled wall and removed by contact melting by heating the wall. For example, ice is manufactured in this way on plane surfaces, on the outside of tubes, or inside tubes. The fundamental question is this: How should we select the on–off (freezing–removal) cycle, such that the time-averaged rate of ice production is maximum?

To show that an optimal freezing time exists, it is a good idea to use the simplest freezing model possible [4]. We assume therefore that the ice layer is sufficiently thin relative to the radius of curvature of the wall so that the wall may be regarded as plane (Fig. 10.1). Water at the freezing point T_m is held in a channel whose wall is suddenly cooled to a lower temperature T_w, starting with the time $t = 0$. An ice layer grows on the cooled wall until the time $t = t_1$, when the cooling is interrupted, and the ice is removed. The time interval required for removing the ice, t_2, is assumed constant and known. This freezing and ice removal cycle of total duration $(t_1 + t_2)$ is repeated many times.

The thickness of the ice layer at the end of the freezing interval is

$$\delta_1 = b\,t_1^{1/2}, \tag{10.1}$$

where the constant b is shorthand for $[2k(T_m - T_w)/\rho h_{sf}]^{1/2}$, and h_{sf} is the latent heat of solidification. This solution is exact when the Stefan number $c(T_m - T_w)/h_{sf}$ is smaller than 1 [5]. The properties ρ, c, and k are the density, specific heat, and thermal conductivity of ice, respectively. The evolution of the ice thickness $\delta(t)$ and the freezing–removal production cycle are shown qualitatively in Fig. 10.1.

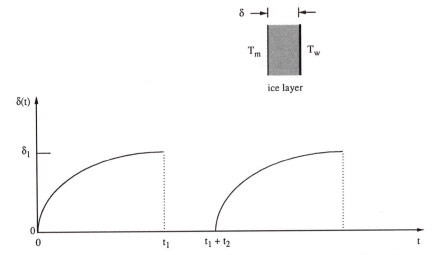

FIGURE 10.1. The intermittent production of ice: the freezing time t_1, followed by the ice removal time t_2.

The objective is to maximize the amount of ice produced over the entire duration of one cycle, namely,

$$\bar{\delta} = \frac{\delta_1}{t_1 + t_2} = \frac{b}{t_2^{1/2}} \frac{\tau^{1/2}}{\tau + 1}. \tag{10.2}$$

The lone degree of freedom is the freezing time t_1 or its dimensionless counterpart $\tau = t_1/t_2$. By solving $d\delta/d\tau = 0$, we find that $\tau_{\text{opt}} = 1$, or

$$t_{1,\text{opt}} = t_2. \tag{10.3}$$

The optimal freezing time interval of the cycle is as long as the ice-removal interval. This conclusion will change somewhat as we include in the model additional features such as the wall curvature (e.g. freezing inside a tube), the finite heat transfer coefficient between the coolant and the outer surface of the water container, and the dependence between the ice-removal time and the final thickness of the ice layer [6, 7].

The essential point made by this simple analysis is that an optimal time of heat transfer exists and that as a first approximation that time is equal to the time required for removing the ice layer. Furthermore, this conclusion is fundamental: An optimal freezing time exists in any installation (or more realistic model) for ice making based on freezing and contact melting. This optimization principle is relevant to the production by contact solidification of other solid materials, not just ice.

The same principle governs the optimization of other diffusion time intervals in the absence of phase change. The thickness of a material (solid, fluid) penetrated by thermal diffusion is proportional to $t^{1/2}$, as in Eq. (10.1). The time-averaged rate of heat transfer assumes a form similar to that of Eq. (10.2) and serves as the basis for the optimization of the periodic heating process. This optimization

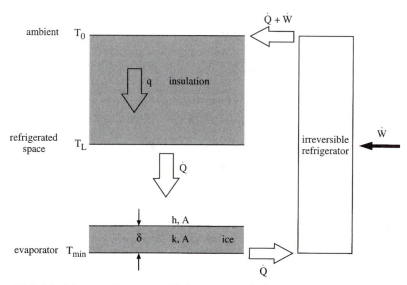

FIGURE 10.2. Model of a refrigerator with frost accumulation on the evaporator surface [13].

opportunity was demonstrated for natural convection [8, 9], forced convection [10], boiling [11], and condensation [12]. It is one of the most basic, simple, and general optimization principles in engineering. The next two sections illustrate two classes of large-scale applications.

10.2 Defrosting Refrigerators

Most of us are familiar with the frosting and icing on the coils of household refrigerators, particularly when these operate in climates of high humidity. A layer of frost and, later, ice builds up on the visible surface (e.g., bare or finned tubes) of the evaporator, which is surrounded by the cold air trapped in the cold space of the refrigerator. The growth of the ice layer leads to an increase in the thermal resistance between the cold space and the even colder evaporator surface. As this resistance increases, the temperature of the evaporator surface must decrease to continue to pull the appropriate refrigeration load out of the cold space. At the same time, the refrigerator works harder as the compressor uses more electric power to maintain the cold space at the prescribed temperature level.

Figure 10.2 shows the simplest features of a refrigerator that experiences icing on its evaporator surface (T_{min}) [13]. The function of the refrigerator is to maintain the temperature of the refrigerated space (cold box) at a low level T_L in spite of the average heat transfer rate q that leaks steadily through the insulation surrounding the cold space. The heat leak q reaches the cold box T_L and is removed as $\dot{Q}$ during the interval t_1 when the refrigerator is turned on:

$$\dot{Q} = \frac{q(t_1 + t_2)}{t_1}. \tag{10.4}$$

Each t_1 interval is followed by a time interval t_2 during which the refrigerator is turned off and the evaporator surface is being deiced. In this simplest treatment of the problem we assume that the defrosting time interval t_2 is known and fixed and that the electric power used for melting the ice layer is negligible. The time interval in which the refrigerator must be left running, t_1, is the chief unknown in this problem. The refrigerator load $\dot{Q}$ is greater than the heat leak q, because the latter is being accumulated in the thermal inertia of the cold box during the time interval when the refrigeration cycle is turned off. We are assuming for simplicity that the cold-box thermal inertia is large enough so that the temperature T_L is practically constant during each on–off cycle of total duration $(t_1 + t_2)$.

The refrigeration cycle that operates between the ambient (T_0) and the inner evaporator surface (T_{min}) is irreversible, with a second law efficiency η_{II} assumed constant and known. The second law efficiency accounts for all the irreversibility features other than the finite temperature difference $T_L - T_{min}$; these features are lumped into a constant second law efficiency [14] η_{II} between 0 and 1 because they are peripheral to the optimization question posed in this problem.

The feature that is affected by the formation of ice is the temperature gap $(T_L - T_{min})$, which increases as the ice layer becomes thicker. The refrigeration load is driven into the evaporator by the temperature difference $(T_L - T_{min})$:

$$\dot{Q} = \left(\frac{1}{hA} + \frac{\delta}{kA}\right)^{-1}(T_L - T_{min}). \tag{10.5}$$

In this expression, A, h, δ, and k are the evaporator surface, the convective heat transfer coefficient between the cooled material and the ice surface, the ice thickness, and the ice thermal conductivity, respectively. We further assume that A and h are constant – that is, independent of δ.

Experimental measurements with household refrigerators reveal that the ice-layer thickness increases almost proportionally with the time of operation [13]:

$$\delta = at. \tag{10.6}$$

The rate of ice formation a is assumed known from direct measurements. The power required by the refrigerator during the interval t_1 is

$$\dot{W} = \frac{1}{\eta_{II}}\dot{Q}\left(\frac{T_0}{T_{min}} - 1\right), \tag{10.7}$$

for which $\dot{Q}$ is furnished by Eq. (10.5). The total work required during the interval t_1 is

$$W = \int_0^{t_1} \dot{W}dt. \tag{10.8}$$

This work is used for the purpose of removing the corresponding heat leak $\dot{Q}t_1$ or $q(t_1 + t_2)$. In conclusion, the figure of merit of the complete on–off sequence of operation is the ratio $W/[q(t_1 + t_2)]$. We show next that this ratio can be minimized by choosing an optimal time of refrigerator operation, t_1.

When Eqs. (10.4)–(10.8) are combined, it is possible to express the average power requirement $W/(t_1 + t_2)$ in the following nondimensional form:

$$\frac{\eta_{II} W}{q(t_1 + t_2)} = \frac{(T_0/T_L)H}{\text{Bi}(1 + \tau)} \ln\left[\frac{H - 1 - 1/\tau}{H - 1 - (1/\tau) - \text{Bi}(1 + \tau)}\right] - 1. \qquad (10.9)$$

Here Bi is the Biot number [5] based on the ice thickness of size at_2,

$$\text{Bi} = \frac{hat_2}{k}, \qquad (10.10)$$

and H is the nondimensional counterpart of the convective heat transfer coefficient between the evaporator surface and the refrigerated space,

$$H = \frac{hAT_L}{q}. \qquad (10.11)$$

The Biot number may be viewed as a nondimensional rate of ice accumulation. The nondimensional time τ is proportional to the time interval of refrigerator operation:

$$\tau = \frac{t_1}{t_2}. \qquad (10.12)$$

Equation (10.9) shows that the ratio work/(heat leak) is a function of four dimensionless numbers, τ, H, Bi, and T_0/T_L. Only τ represents a degree of freedom in the operation of the system, while the other three numbers are fixed as soon as the apparatus is constructed. In Fig. 10.3 we see that there exists an optimal time τ (or t_1) for minimum work/(heat leak) ratio – that is, a best moment when the refrigeration should be interrupted to deice the evaporator. The small-time and large-time asymptotes of all the curves shown in Fig. 10.3 are quite steep and stress the importance of knowing the optimal τ value.

The optimal operating time derived from Fig. 10.3 is a function of only two parameters, Bi and H. The solution reported in Ref. 13 shows that the optimal time of refrigerator operation is almost proportional to the inverse of Bi. The effect

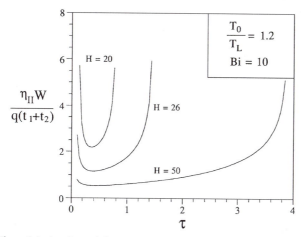

FIGURE 10.3. The minimization of the power requirement of a defrosting refrigerator [13].

of the electric power used during the t_2 interval to melt the ice layer is analyzed further in Ref. 13. An even more realistic model of a defrosting refrigerator based on the vapor compression cycle was optimized by use of the same approach in Ref. 15.

10.3 Cleaning Power Plants

We now turn our attention to a related fundamental problem that occurs in the operation of a power plant in which the working fluid fouls gradually (i.e., deposits scale on) the heat exchanger surfaces. The scale adds to the overall thermal resistance of the heat exchanger and causes a steady decrease in the energy-conversion efficiency of the power plant. The question is when to shut down the power plant for the purpose of removing the scale from the heat exchanger surfaces [13].

In Fig. 10.4 we see the simplest model in which we can examine this fundamental engineering question. The heat source (e.g., flame) temperature level T_H is fixed, while the working fluid (e.g., water) executes a cycle between the high temperature T_{max} and the ambient T_0. The heat exchanger that experiences fouling is assumed to be located at the hot end of the cycle, in the temperature gap $T_H - T_{max}$. The thermal resistance of this heat exchanger is due to its finite heat transfer A, the finite heat transfer coefficient h between the heat source and the surface of the scale, and the scale itself. The power plant between T_0 and T_{max} is operating irreversibly.

We assume that the scale thickness δ varies (increases) with the time of operation, $\delta = f(t)$. The rate of heat transfer that drives the power cycle is

$$\dot{Q} = \left(\frac{1}{hA} + \frac{\delta}{kA} \right)^{-1} (T_H - T_{max}). \tag{10.13}$$

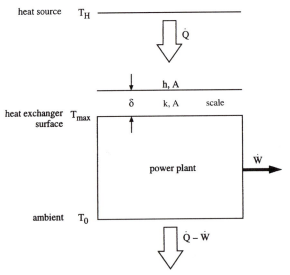

FIGURE 10.4. Model of power plant with scale deposition (fouling) on the surface of the hot-end heat exchanger [13].

For the sake of brevity and clarity we use the linear model

$$\delta = bt, \tag{10.14}$$

in which the fouling rate b is assumed to be a known empirical constant. Epstein's [16] review and the experiments described by Konings [17] show that the thickness measured experimentally can be fitted at small times with Eq. (10.14).

Let t_1 be the unknown time interval in which the power plant is on and t_2 be the known and fixed cleaning time interval that follows t_1. The power plant is shut down during t_2. If we are interested in producing maximum power averaged over time, we must maximize the ratio $W/(t_1 + t_2)$, where W is the work produced while the power plant is on:

$$W = \int_0^{t_1} \dot{W} dt. \tag{10.15}$$

In this problem two parameters, t_1 and T_{max}, can be chosen optimally such that the average power output $W/(t_1 + t_2)$ is maximized. The selection of the optimal T_{max} is now a classical result that was reviewed specifically in Refs. 4 and 14. For example, if the power cycle executed between T_{max} and T_0 is reversible, the instantaneous power output $\dot{W}$ is maximized if T_{max} has the constant value

$$T_{max,opt} = (T_H T_0)^{1/2}. \tag{10.16}$$

The fact that $T_{max,opt}$ is independent of t_1 is an important feature that must be kept in mind as we turn our attention to the optimal selection of the second free parameter, t_1.

The instantaneous power output maximized with respect to T_{max} is

$$\dot{W} = \eta_{II}\left(1 - \frac{T_0}{T_{max,opt}}\right) \dot{Q}, \tag{10.17}$$

where the second law efficiency constant η_{II} accounts for the assumed irreversibility of the power cycle and $T_{max,opt}$ is a constant depending on T_0, T_H, and η_{II}. When Eqs. (10.14)–(10.17) are combined, the time-averaged power output becomes [13]

$$\frac{W}{t_1 + t_2} = \eta_{II}\left(1 - \frac{T_0}{T_{max,opt}}\right)(T_H - T_{max,opt})hAF, \tag{10.18}$$

where

$$F(\theta, Bi_*) = \frac{\ln(1 + \theta Bi_*)}{(1 + \theta)Bi_*}, \tag{10.19}$$

$$\theta = \frac{t_1}{t_2}, \quad Bi_* = \frac{hbt_2}{k}. \tag{10.20}$$

Noteworthy in these definitions are the nondimensional time interval of power plant operation θ and the nondimensional rate of fouling Bi_*. The effect of θ and

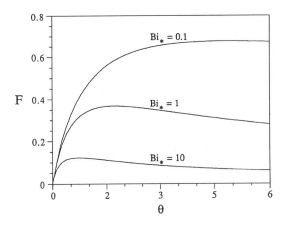

FIGURE 10.5. The effect of the operating time interval and the rate of fouling on the average power output [13].

Bi_* on the average power output is conveyed by the function $F(\theta, Bi_*)$, which is plotted in Fig. 10.5. The average power output reaches a maximum at a distinct operating time θ_{opt}. The Bi_* number may be seen as a way of nondimensionalizing the known cleaning time t_2. The figure shows that, as the cleaning time decreases, the maximum average power and the time ratio θ_{opt} increase. The optimal time interval of power plant operation, $t_{1,opt}$, is proportional to the product $\theta_{opt}Bi_*$. Results reported in Ref. 13 show that this time interval decreases as the cleaning time decreases.

Similar behavior and results are found when the deposition of scale is occurring on the surface of the cold-end heat exchanger [13].

To summarize this section, we relied once again on the most basic principles of heat transfer and thermodynamics to prove that an optimal on–off sequence exists in power plant design. Intervals of power plant operation must be interspaced optimally with intervals for the removal of scale from heat exchanger surfaces so that the time-averaged power output of the plant is maximum. The power plant model used in this demonstration was the simplest. An optimal on–off sequence of the type described in this section will be present in the design of any power plant that experiences fouling, regardless of the complexity of the design and regardless of whether fouling occurs at the hot end of the cycle or at the cold end.

Additional manifestations of this principle are found in the optimal operating lifetime of power plants driven by energy extracted from hot-dry-rock deposits [18]. The rock layers cool down by time-dependent diffusion, as the working fluid of the power plant is circulated through manmade fissures.

10.4 Breathing

In this section we shift the focus to living systems and show how the same power minimization or maximization leads to optimal frequencies for on–off flows. Unlike in the latest work that is being done in physiology and fluid mechanics, here I rely on the simplest possible models and a minimum of algebra. The models of the

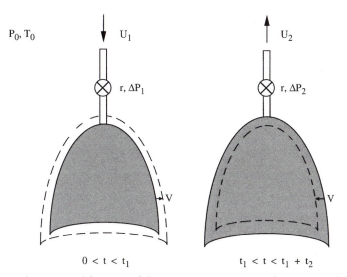

FIGURE 10.6. The principal features of the respiratory system and its two-stroke cycle [1].

organs (lungs, circulatory system) account in nakedly simple terms for irreversibility features such as flow with friction and mass transfer [1].

We begin with respiration. Consider the inhaling and exhaling processes sketched in Fig. 10.6. During the inhaling time t_1 the chest cavity experiences the volume increase V, while drawing in atmospheric air (T_0, P_0) with the time-averaged mean velocity U_1. This flow is made possible by the lower-than-atmospheric pressure $(P_0 - \Delta P_1)$ maintained inside the lungs during the inhaling process. The pressure difference varies monotonically with the mean inhaling velocity,

$$\Delta P_1 = r U_1^n, \tag{10.21}$$

where r is the fluid resistance of the air passages and n varies from $n \cong 1$ in laminar flow to approximately $n \cong 2$ in turbulent flow. Mass conservation during the t_1-long process requires that

$$\rho_0 U_1 A_f t_1 = \rho_1 V, \tag{10.22}$$

where A_f is the effective cross-sectional area of the flow path, ρ_0 is the air density at atmospheric conditions, and ρ_1 is the density of the inhaled air (evaluated at $P_0 - \Delta P_1$ and body temperature T_b).

The exhaling process is characterized by the excess pressure ΔP_2 inside the chest cavity:

$$\Delta P_2 = r U_2^n, \tag{10.23}$$

where r is the same flow resistance and U_2 is the mean exhaling velocity. The duration of the exhaling process is t_2; therefore mass conservation requires that

$$\rho_1 U_2 A_f t_2 = \rho_1 V. \tag{10.24}$$

The total work done by the thorax muscles during the inhaling and exhaling cycle $(t_1 + t_2)$ is

$$W = \oint (P_0 - P_{\text{cavity}}) \mathrm{d}V_{\text{cavity}} = (\Delta P_1 + \Delta P_2)V. \qquad (10.25)$$

The cycle-averaged power consumed by the thorax is $\dot{W} = W/(t_1 + t_2)$. By combining Eqs. (10.21)–(10.25) and the assumption that $\rho_0 \approx \rho_1$, we find that the average power reduces to

$$\dot{W} = r \frac{V^{n+1}}{A_f^n} \frac{t_1^{-n} + t_2^{-n}}{t_1 + t_2}. \qquad (10.26)$$

Equation (10.26) shows that the power requirement decreases monotonically as either t_1 or t_2 increases. The conclusion we reach at this stage is that effortless breathing corresponds to the longest inhaling and exhaling strokes possible. It turns out, however, that both t_1 and t_2 must be finite, as required by the chief function of the breathing process. That function is to facilitate the transfer of oxygen between the freshly inhaled air and the vascularized tissue beneath the surface of the pulmonary passage. The transfer of O_2 is by mass diffusion on both sides of the surface, because of the small size of both the terminal ramifications of the air passages (alveoli) and the blood capillaries.

The mass diffusion analysis begins with writing that during the inhaling interval t_1 the mass flux of O_2 is of the order of [19]

$$j \sim \frac{D\Delta C}{\delta}, \quad \delta \sim (Dt_1)^{1/2}, \qquad (10.27)$$

where δ is the mass diffusion distance associated with t_1. Although a pair of equations such as (10.27) can be written for each side of the pulmonary passage surface (air side and blood side), in the present analysis Eqs. (10.27) refer to the blood side because it is more restrictive to mass transfer. As a numerical example, consider the breathing of mammals in the human size range characterized by the time scale $t_1 \sim 1$ s. The orders of magnitude of the mass diffusivities of O_2 on the air and the blood sides are, respectively, 10^{-5} and 10^{-9} m²/s (Ref. 5, pp. 588–589). The second of Eqs. (10.27) indicates that after 1 s the mass penetration distance would be of the order of 3 mm (i.e., 2 orders of magnitude greater than the alveolus scale of 50 μm (see Ref. 20), while on the blood side δ is of the order of 30 μm, which is of the same order as the capillary diameter scale, 10 μm (Ref. 21, p. 118).

Combining Eqs. (10.27), we find that the oxygen mass transferred during the t_1 interval is $m = j A t_1$, or

$$m \sim At_1 \frac{D\Delta C}{(Dt_1)^{1/2}}, \qquad (10.28)$$

where A is the total contact (mass transfer) area of the air passages. An important concept in this chapter is the mass transfer rate averaged over one complete cycle,

$\dot{m} = m/(t_1 + t_2)$, or

$$\dot{m} \sim AD^{1/2}\Delta C \frac{t_1^{1/2}}{t_1 + t_2}. \tag{10.29}$$

To make this mass transfer possible is the function of the lungs and their muscular system. In conclusion, approximation (10.29) places a constraint on the inhaling and the exhaling time intervals, forcing both to be finite:

$$\frac{t_1^{1/2}}{t_1 + t_2} \sim \frac{\dot{m}}{AD^{1/2}\Delta C} = K \quad \text{(constant)}. \tag{10.30}$$

Finally, consider the behavior of the average power requirement $\dot{W}$, Eq. (10.26), subject to constraint (10.30). Eliminating t_2 and solving $\partial \dot{W}/\partial t_1 = 0$, we obtain the optimal inhaling time $(t_{1,\text{opt}})$ for minimum breathing power consumption. This result is given implicitly by [1]

$$\left(1 - Kt_{1,\text{opt}}^{1/2}\right)\left[\left(\frac{1}{Kt_{1,\text{opt}}^{1/2}} - 1\right)^n + 1\right] \sim \frac{n}{2n + 1}. \tag{10.31}$$

This shows that periodic flow of a certain frequency is a necessity. It is demanded by the minimization of the mechanical power required for breathing. The constructal principle that generated spatial structure (Chap. 9) and temporal structure (Sections 10.1–10.3) in power systems also generates rhythmic breathing in animals.

The flow regime (n) has no effect on the scale of $t_{1,\text{opt}}$: The scale of $t_{1,\text{opt}}$ is K^{-2}. For example, when the animal is small such that the air passages are very tight and U_1 is small enough that the flow is laminar $(n = 1)$, the optimal inhaling time given by approximation (10.31) is $t_{1,\text{opt}} \sim (9/16)K^{-2}$. Substituting this estimate into constraint (10.30), we obtain the optimal exhaling time $t_{2,\text{opt}} \sim (3/16)K^{-2}$. A theoretical milestone is the conclusion that the breathing time intervals are of the same order of magnitude:

$$(t_1, t_2)_{\text{opt}} \sim K^{-2} = \left(\frac{AD^{1/2}\Delta C}{\dot{m}}\right)^2. \tag{10.32}$$

This is in agreement with the very large volume of observations catalogued and correlated in the biology literature [21–23].

In Section 10.6 we will show that the theoretical breathing time (10.32) increases with the animal body size raised to a power of approximately 0.25, which is again in excellent agreement with data collected in the biology literature.

10.5 Heart Beating

The existence of a characteristic (finely tuned) heartbeat frequency that decreases as the body size increases can be anticipated based on the same constructal principle: the minimization of pumping power subject to a global mass transfer constraint.

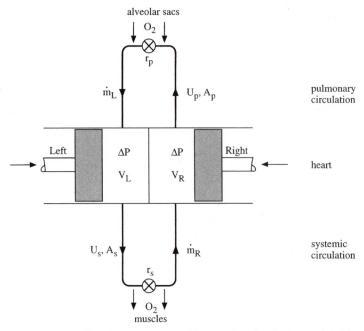

FIGURE 10.7. Circulatory system with two-chamber heart model [1].

Consider the circulatory system model shown in Fig. 10.7. The system consists of three main parts: the heart, the pulmonary circulation, and the systemic circulation. The pulmonary circulation begins with the arteries that guide the oxygen-depleted blood to the lungs, and it ends with the pulmonary veins that return the oxygenated blood to the left side of the heart. The systemic circulation loop accounts for the arteries that take the oxygenated blood to the muscles, where O_2 is released, and the veins that return the oxygen-depleted blood to the heart. The left versus right arrangement shown in Fig. 10.7 corresponds to the physiology of the human circulatory system.

The simplest heart model has only two chambers: the working chambers (left, L; right, R), which force the blood away from the heart and into the capillaries of the systemic and pulmonary circuits. These two chambers contract at the same time during the time interval t_1. The heart wall squeezes both chambers, raising the pressure to $P_0 + \Delta P$ relative to the background pressure (P_0) of the blood that returns to the heart. The pressure rise is related to the flow resistances of the systemic and pulmonary paths (r_s, r_p):

$$\Delta P = r_s U_s, \qquad \Delta P = r_p U_p, \tag{10.33}$$

where U_s and U_p are the average velocities of the blood that leaves the heart. Linear relations (10.33) between pressure drop and velocity are justified because each flow resistance is dominated by the contributions made by the smaller vessels and capillaries. In other words, unlike in the general equations (10.21) and (10.23), here we model the flow as laminar, or as Darcy flow if we view the vascularized tissues as porous media saturated with fluid, e.g., Refs. 19 and 24.

Mass conservation requires that the average flow rates issuing from the two chambers be equal:

$$\dot{m}_L = \dot{m}_R = \dot{m},$$ (10.34)

where

$$\dot{m}_L = \frac{\rho V_L}{t_1} = \rho A_s U_s, \quad \dot{m}_R = \frac{\rho V_R}{t_1} = \rho A_p U_p.$$ (10.35)

In these equations ρ is the blood density, and A_s and A_p are the effective flow-path cross-sectional areas corresponding to U_s and U_p. Equations (10.34) and (10.35) show that the contractions of the two chambers are the same: $V_L = V_R = V$.

Let t_2 be the resting time between two consecutive heart beats. The cycle-averaged power required by the heart muscle is then

$$\dot{W} = \frac{2 V \Delta P}{t_1 + t_2}.$$ (10.36)

Eliminating ΔP and U_s from Eqs. (10.33)–(10.35), we find that $\dot{W}$ is proportional to the inverse of $t_1(t_1 + t_2)$:

$$\dot{W} = \frac{2 r_s V^2}{A_s t_1(t_1 + t_2)}.$$ (10.37)

The time interval t_2 is when oxygen diffuses from alveoli into pulmonary capillaries and from systemic capillaries into muscles. Each of these mass diffusion processes is governed by the scale analysis (10.27) and (10.28) with t_2 in place of t_1. In conclusion, the mass of oxygen transferred to and from the circulatory system during the t_2 interval is proportional to $t_2^{1/2}$. The time-averaged mass transfer rate is proportional to $t_2^{1/2}/(t_1 + t_2)$; this group is also proportional to the metabolic rate of the animal and acts as a constraint:

$$\frac{t_2^{1/2}}{t_1 + t_2} \sim K \quad \text{(constant)}.$$ (10.38)

The constant K has the same form as in Eq. (10.32): Now A represents the total mass transfer surface of all the capillaries, pulmonary and systemic.

Mass transfer constraint (10.38) can be used to eliminate t_1 from Eq. (10.37) and to conclude that the heart power consumption is proportional to the inverse of the group $t_2(1 - K t_2^{1/2})$. This shows that $\dot{W}$ is infinite at $t_2 = 0$ and $t_2 = K^{-2}$ and has a relatively sharp minimum at the intermediate value $t_{2,\text{opt}} \sim (4/9) K^{-2}$. Mass transfer constraint (10.38) delivers the corresponding contraction time $t_{1,\text{opt}} \sim (2/9) K^{-2}$ or the time ratio $(t_1/t_2)_{\text{opt}} \sim 1/2$. These results are meant to be valid in only an order-of-magnitude sense: Both $t_{1,\text{opt}}$ and $t_{2,\text{opt}}$ are of the same order as K^{-2}; see Eq. (10.32).

In conclusion, the existence of a unique (characteristic) heartbeat frequency in an animal can be rationalized based on the minimization of heart power consumption subject to contact area and oxygen mass transfer (or metabolic) rate constraints.

The predicted equality of the orders of magnitude of $t_{1,\text{opt}}$ and $t_{2,\text{opt}}$ agrees with the measured time intervals of animals over a wide range of body sizes [21–23].

It pays to look back at the engineering examples that opened this chapter, to see similarities not only in the invoked principle but also in the ensuing mathematics. Note, for example, the function of type $t_1^{1/2}/(t_1 + t_2)$, which stems from the averaging of a diffusive process over a complete on–off cycle. This analytical form can be seen in Eqs. (10.2) and (10.19) and approximations (10.29) and (10.38). The maximization of this function with respect to the time interval of active diffusion is the source of all the optimal pulsations determined for engineered and natural systems in this chapter.

10.6 The Effect of Animal Body Size

Another contribution of the preceding theory is that it explains why the breathing interval increases with the size of the animal. Specifically, Eq. (10.32) predicts that the breathing time intervals must vary as $(A/\dot{m})^2$. To predict the relation between breathing time and body size, we need

1. the relation between metabolic rate (or $\dot{m}$) and body size (mass M)
2. the relation between the mass transfer contact area (A) and body size (M).

We question first the allometric law (1). Allometric laws are widely recognized power-law relations between geometric and functional (flow) parameters of living bodies. They are accurate over wide ranges of body size [21–23]. Predicting these relations from a purely theoretical standpoint has been a real challenge. One of the most challenging of all allometric laws is (1) the proportionality between metabolic rate and body mass (or volume V) raised to the power 3/4.

The pre-1984 history of the theoretical attempts to predict this relation was recounted by Schmidt-Nielsen [21]. From this history, what is most relevant to the present work is the oldest theory: The metabolic rate must be proportional to the heat loss from the body to the ambient. Since the convective heat loss is proportional to the body surface, the metabolic rate must be proportional to the length scale ($V^{1/3}$) squared, i.e., the body mass or volume raised to power 2/3 [14]. I refer to this earliest explanation as the heat transfer theory.

The heat transfer theory was effectively discredited in this century by mounting observations of birds and mammals, indicating an exponent much closer to 3/4 than 2/3. Deviations from this 3/4-power trend are known to occur in the limit of small body sizes. I return to this trend in the closing paragraphs of this section.

The heat transfer theory appears to have been pushed aside completely by the tree model of West et al. [25], which drew attention to a class of interesting relations between geometric and flow parameters in an optimized tree network for fluid flow. To discuss the finely tuned model of West et al. [25] is not the objective of this book, because the fine-tuning activity of modeling what is being observed does not represent theory. It is sufficient to note that, as in earlier optimizations of fluid tree networks in physiology and river morphology [26–28], the optimization of West et al. was based on invoking the minimization of pumping power, this among numerous ad hoc assumptions.

My objective is to propose a much more direct and familiar explanation for the 3/4-power exponent [29]. I submit this as an engineering alternative to the modern explanations that have been proposed in biology and physics. My explanation is based on the discarded heat transfer theory and, because of this, it closes the loop and brings under the same theory the metabolic rates of all the vertebrates, the warm blooded (3/4-power law) and the cold blooded (2/3-power law).

Why should anyone question the apparent success of Ref. 25 by resurrecting an abandoned theory? I have two reasons. First, the invocation by West et al. of the minimization of pumping power makes sense, because reduced power consumption makes sense to the animal (the animal would need less food to survive). Less food means a lower metabolic rate, i.e., a lower rate of heat loss to the ambient. This is why in my view the minimization of pumping power goes hand in hand with the heat loss doctrine, not against it. Minimum pumping power consumption and minimum loss of body heat are parts of the same optimization principle: the constructal principle, or *how to be* the fittest.

The second reason is that by minimizing pumping power alone, the analysis [25] of the fluid tree leads to the 3/4-power law that correlates mammals and birds. The 3/4 exponent does not correlate cold-blooded vertebrates, even though the bodies of reptiles, amphibians, and fish are equally dominated by flow structures shaped as trees.

Heat transfer is the new element introduced by the following analysis. Unlike in the heat loss theory of the 19th century, here heat transfer is not considered in isolation. It is combined with the more recent progress made based on pumping power minimization in constructal fluid trees (Chap. 5). It is not necessary to repeat here the optimized features of the three-dimensional fluid tree developed in Table 5.3. In Problem 5.1 we found that the main features of that structure can be illustrated much more briefly by optimizing a plane construct consisting of a T-shaped junction (Fig. 10.8A) [30]. Contrary to the direction used in Chap. 5, in the following analysis the counting of constructs (levels of assembly) starts with the largest tube ($i = 0$) and ends with the elemental ($i = n$).

The stream $\dot{m}_i$ encounters the flow resistance of two L_{i+1} tubes in parallel, which are later connected in series with one L_i tube. When the resistance is minimized by fixing the total tube volume, we find the optimal diameter ratio $D_{i+1}/D_1 = 2^{-1/3}$. This old result (Murray's law) is extremely robust because it is independent of the lengths (L_i, L_{i+1}) and the relative position of the three tubes.

Newer is the constructal optimization of the lengths when the space allocated to the construct is fixed. In Fig. 10.8A the space constraint is $2L_{i+1}L_i =$ constant. This second minimization of the flow resistance yields the optimal length ratio $L_{i+1}/L_i = f = 2^{-1/3}$ (Problem 5.1), which happens to match the optimal diameter ratio. The optimized diameter and length ratios are drawn to scale in Fig. 10.8A.

In the tree that was optimized step by step into three-dimensional parallelepipedic constructs (Fig. 5.18 and Table 5.3), the tube lengths increase by factors in

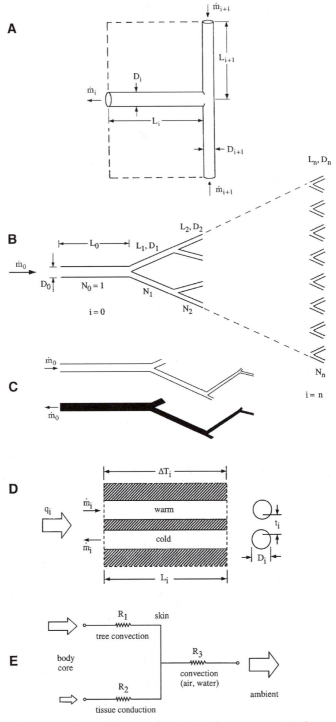

FIGURE 10.8. The construction of the tree of convective heat currents: A, the constrained optimization of the geometry of a T-shaped construct; B, the stretched tree of optimized constructs; C, the superposition of two identical fluid trees oriented in counterflow; D, the convective heat flow along a pair of tubes in counterflow; and E, the three resistances for heat flow from the animal body core to the ambient [29].

the cyclical sequence $2, 1, 1, 2, 1, 1, \ldots$. The average of this factor for one step is $2^{1/3}$; therefore the optimization of the plane construct of Fig. 10.8A is a condensed summary of the optimized three-dimensional construct averaged over each three-step cycle. The averaged tree is laid out (stretched) in Fig. 10.8B, so that we may see all the tubes and levels (i) of construction. The number of tubes at level i is $N_i = 2^i$ and the total number of levels is n.

This concludes the introductory analysis of the theoretical fluid tree. It is not even necessary to accept the constructal method as the origin of this structure. The following heat transfer analysis may also be started right here by accepting the tree network heuristically, i.e., as an assumption, as in the first assumption acknowledged by West et al. [25] in their model.

We now turn our attention to the flow of heat between the root and the canopy of the tree. This proposal is new [29]. Of interest is the heat lost by a warm-blooded animal through the volume situated under its skin. The trees of blood vessels are one geometric feature in this region, but not the only one. The other is the *superposition* of the arterial and venous trees, so closely and regularly that tube i of one tree is in counterflow with tube i of the other (Fig. 10.8C). The existence of counterflows of blood and other fluids is widely recognized in physiology [21–23].

The counterflow formed by two tubes of level i is shown in the detail drawing of Fig. 10.8D. The arterial stream is warmer than the venous stream: Heat is transferred transversally, from stream to stream (Section 9.4). Consider now the adiabatic control surface drawn with a dashed line around the counterflow. Since the enthalpy of the warmer stream is greater than that of the colder stream, the counterflow convects longitudinally the energy current $q_i = \dot{m}_i c_p \Delta T_{t,i}$, where c_p is the blood specific heat and $\Delta T_{t,i}$ is the stream-to-stream temperature difference at level i. It was shown first in heat transfer engineering [31] and, more recently, in bioengineering [32] that such a counterflow sustains a longitudinal temperature gradient $\Delta T_i / L_i$ and that the convective heat current is proportional to this gradient [see Eq. (9.33)]:

$$q_i = \frac{(\dot{m}_i c_p)^2}{h_i p_i} \frac{\Delta T_i}{L_i}. \tag{10.39}$$

In heat transfer terminology [5], h_i and p_i are the overall stream-to-stream heat transfer coefficient and the perimeter of contact between the two streams. In the case of blood counterflow, the stream-to-stream thermal resistance h_i^{-1} is the sum of two resistances: the resistance through the fluid in the duct ($\sim D_i / k_f$, where k_f is the fluid thermal conductivity) plus the resistance through the solid tissue that separates two tubes ($\sim t_i / k$, where k is the tissue thermal conductivity). Even when the tubes touch, t_i is of the same order as D_i. In addition, since $k_f \sim k$, we conclude that $h_i \sim k / D_i$, and Eq. (10.39) becomes

$$\Delta T_i \sim \frac{q_i L_i}{\dot{m}_i^2} \frac{k}{c_p^2}. \tag{10.40}$$

The double-tree structure of fluid streams is a single tree of convective heat streams with zero net mass flow. The convective tree stretches from the core

temperature of the animal (at $i = 0$) to the skin temperature. The latter is registered in many of the elemental volumes ($i = n$) that happen to be near the skin. The
many counterflows of the double tree sustain the overall temperature difference
ΔT (constant) associated with warm bloodedness:

$$\Delta T = \sum_{i=0}^{n} \Delta T_i \sim \frac{q_0 k}{\dot{m}_0^2 c_p^2} \sum_{i=0}^{n} N_i L_i. \tag{10.41}$$

In going from approximation (10.40) to Eq. (10.41) we used the continuity relations
for fluid flow ($N_i \dot{m}_i = \dot{m}_0$, constant) and heat flow ($N_i q_i = q_0$, constant). Recalling
that $L_{i+1}/L_i = f$, we substitute $L_i = L_0 f^i$, $L_n = L_0 f^n$ and $N_i = 2^i$ into Eq. (10.41),
and, after rearranging, we obtain

$$q_0 \sim \left(\frac{q_0}{\dot{m}_0}\right)^2 \frac{k L_n f^{-n}[(2f)^{n+1} - 1]}{c_p^2 \Delta T (2f - 1)}. \tag{10.42}$$

Separated on the right-hand side are the quantities that are constant and the quantities that depend on n (the number of construction steps). Note that the ratio $q_0/\dot{m}_0$
is independent of body size (n) because both q_0 and $\dot{m}_0$ are proportional to the
metabolic rate. Recall also that the elemental volume length scale L_n is treated as
a constant, in accordance with the constructal method.

We estimate the volume scale by regarding the stretched tree as a cone in
Fig. 10.8B. The base of the cone (at $i = n$) has an area of size $N_n L_n^2 \sim 2^n L_n^2$.
The height of the cone is of the same order as the sum of all the tube lengths,
$L_0 + L_1 + \cdots + L_n = L_0(1 - f^{n+1})/(1 - f)$. In conclusion, the volume scale is

$$V \sim L_n^3 \left(\frac{2}{f}\right)^n \frac{1 - f^{n+1}}{1 - f}. \tag{10.43}$$

We obtain the relation between metabolic rate and total volume by eliminating n
between approximations (10.42) and (10.43). The result is visible in closed form if
we assume that n is sufficiently large so that $(2f)^{n+1} \gg 1$ in approximation (10.42)
and $f^{n+1} \ll 1$ in approximation (10.43). In this limit q_0 is proportional to 2^n and
V is proportional to $(2/f)^n$. From this follows the conclusion

$$\frac{\log q_0}{\log V} = \frac{3}{4}, \tag{10.44}$$

which means that q_0 increases as $V^{3/4}$. The proportionality between metabolic
rate and body size raised to the power 3/4 has just been predicted based on pure
theory.

It can be verified numerically that Eq. (10.44) is accurate even for small n. The
3/4 exponent that has been so puzzling over the years is a reflection of the optimized
ratio of successive tube lengths, in a *fixed space*, $L_{i+1}/L_i = f = 2^{-1/3}$. Specifically, if
we use $f = 2^{-a}$ in the derivation of Eq. (10.44), instead of 3/4 we obtain $(1 + a)^{-1}$,
where $a = 1/3$. The 3/4 exponent is intimately tied to the optimization that generated the tube lengths ratio subject to the total volume constraint, after the ratio of

tube diameters had been optimized subject to the tube volume constraint. This double geometric optimization, with the two constraints and the pairing of tubes into constructs larger than the fixed elemental volume, is the essence of the constructal method.

We cannot dismiss geometry. Counting tubes and optimizing diameters regardless of spatial arrangement (e.g. Ref. 25) is one thing, and optimizing the lengths and relative positions of these tubes is *constrained* spaces is an entirely different thing. Constructal theory combines these two ideas, and brings *geometry* where it belongs and works, in physiology, river morphology and every domain in which macroscopic shape and structure defines the nonequilibrium (flow) system.

In conclusion, what had been missing was the combination of (1) the tree architecture optimized for minimum pumping power subject to spatial constraints, and (2) the convective heat transfer (or, better, thermal insulation) characteristics of two identical fluid trees superimposed in counterflow. Putting (1) and (2) together in a heat transfer theory is the contribution of this section.

The convective thermal resistance posed by the trees in counterflow (R_1 in Fig. 10.8E) resides inside the animal. This resistance runs in parallel with a second internal resistance (R_2) associated with the conductive heat leak through the solid tissue (R_2 was neglected in the preceding analysis). On the outside of the animal the heat current flows through the convective resistance associated with the body surface exposed to the ambient (air, water).

In cold-blooded vertebrates the temperature drop across the internal resistances (R_1, R_2) is minimal, and, when environmental temperature changes occur, the dominant resistance is R_3. Consequently, the heat loss and metabolic rates follow closely $V^{2/3}$, as shown by the convection analysis reported in Ref. 14.

In warm-blooded animals a significant thermal resistance ($R_1^{-1} + R_2^{-1}$) is located on the body side of the skin. In larger animals R_1 is less than R_2, the heat current is carried convectively by the double tree (R_1), and the observed metabolic rate follows the predicted $V^{3/4}$ trend.

The conductive resistance R_2 is proportional to the body thickness scale $V^{1/3}$ divided by the body surface $V^{2/3}$; hence $R_2 \sim V^{-1/3}$. The tree resistance R_1 is proportional to $V^{-3/4}$. The ratio $R_2/R_1 \sim V^{5/12}$ shows that R_2 becomes progressively weaker (i.e., the preferred path) as the body size decreases. In that limit the exponent in the power law between heat loss and body size becomes 1/3. In other words, from pure heat transfer theory we should expect a gradual decrease in the power-law exponent as the body size decreases. At the other end, 3/4 is the asymptotic value of the exponent for large body sizes.

Even the large-body asymptote may change, because, at least in principle, when tubes and fluid-flow rates become large enough the Hagen–Poiseuille regime is replaced with the turbulent regime. In Problem 6.1, we repeated the double geometric optimization of the T-shaped construct of Fig. 10.8A by assuming fully developed turbulent flow in the fully rough regime [30] through each tube, i.e., a proportionality between pressure drop and mean velocity squared. In this regime the optimized ratios are $D_{i+1}/D_i = 2^{-3/7}$ and $L_{i+1}/L_i = 2^{-1/7}$. The exponent calculated on the right-hand side of Eq. (10.44) becomes $(1 + 1/7)^{-1} = 7/8$, which is greater than 3/4.

The lung is also a tree of convective currents, which results from the superposition of two air-flow trees: the inhaling flow and the exhaling flow. The convective tree is made up of currents of heat, and constitutes a heat flow path of the same type as the tree analyzed above. Cold air warms up gradually along the air passages during inhaling. Warm air cools down gradually along the same passages during exhaling. The air passage (its wall tissue) acts as a regenerative heat exchanger in the proper engineering sense. In addition to the convective tree for thermal insulation, and, based on the same in–out mechanism, the lung works as a tree-shaped path for minimizing the loss of water.

Another way to summarize the theoretical progress that we just made is to recognize that the body surface of warm-blooded animals does not serve the same heat transfer function as the surface of cold-blooded animals. In mammals and birds the surface (or, better, the vascularized tissue under the surface) serves a thermal insulation function. This is why it has also been possible to predict from the geometric minimization of heat transfer the well-correlated proportionality between hair-strand diameter and body-size length scale raised to the power 1/2 [24].

In lizards and salamanders the surface has the opposite mission: It must facilitate the transfer of heat between body and ambient, and vice versa. For the past three decades in thermal engineering, Bergles [33] argued that the dorsal protuberances of large lizards (the stegosaurus was Bergles' favorite) are fins in the proper engineering sense: extended surfaces optimized for augmenting thermal contact [34, 35]. The present analysis supports this view, and goes one step further to suggest that the entire cold-blooded body is elongated (finlike), because it must permanently maximize its contact with the ambient.

We made so much progress on heat transfer aspects of animal body design that we almost forgot the problem stated at the start of this section. So far, we predicted the allometric law (1), namely,

$$\dot{m} \sim M^{3/4}. \tag{10.45}$$

For the allometric law (2) between contact area and body size, I repeat the argument proposed in Ref. 14, pp. 786–787. The thickness of the tissue penetrated by mass diffusion during the breathing or heart beating time t is proportional to $t^{1/2}$. The body volume (or mass) of the tissue penetrated by mass diffusion during this time obeys the proportionality $M \sim At^{1/2}$. Eliminating t between $M \sim At^{1/2}$ and $t \sim (A/\dot{m})^2$ [see Eq. (10.32)] and using approximation (10.45), we conclude that the contact area should be almost proportional to the body mass:

$$A \sim M^{7/8}. \tag{10.46}$$

This trend is comparable with what we found theoretically in approximations (5.58) and (5.60). Finally, by substituting approximations (10.45) and (10.46) into $t \sim (A/\dot{m})^2$, we conclude that the time intervals should vary as

$$t \sim M^{1/4}. \tag{10.47}$$

This allometric law is supported convincingly by the large volume of observations accumulated in the physiology literature [21–23].

PROBLEMS

10.1 Ice forms on the inner surface of a cylindrical wall cooled on the outside by convection. In the beginning the cylinder is filled with water at the freezing point, T_m. The outer wall of the cylinder is exposed to a coolant $(T_\infty < T_m)$ when $t > 0$. Furthermore, the heat transfer coefficient between the wall and the coolant, h, is assumed constant. The water movement can be neglected because the buoyancy effect is zero (the liquid is isothermal at T_m). The heat transfer through the ice shell and the tube wall are by quasi-steady conduction. The formation of an ice column with annular cross section lasts t_1 seconds. Determine the relation between the ice inner radius r and t_1 (as a guide, use Ref. 5, p. 187). Assume that the time associated with the removal of the ice column is

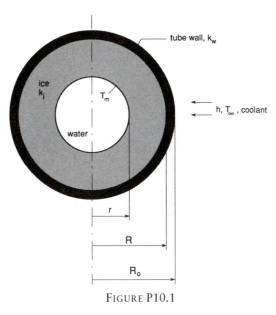

FIGURE P10.1

fixed, t_2. Determine the optimal freezing time $t_{1,opt}$, or optimal inner radius r_{opt}, for maximum time-averaged rate of ice production.

10.2 Ice is being made on the outside of a tube cooled internally by convection. The coolant temperature is T_∞, and the wall-coolant heat transfer coefficient is h. The wall has radii R and R_0 and thermal conductivity k_w. The tube is immersed in water at the freezing point T_m. Ice grows to the radius r during the time t_1. The time needed to remove the ice column is fixed, t_2. Determine the relation between r and the freezing time t_1 (e.g., Ref. 5, p. 187). Maximize the time-averaged rate of ice production during a complete freezing–removal cycle. Report formulas for calculating the optimal freezing time and the optimal outer radius of the ice column.

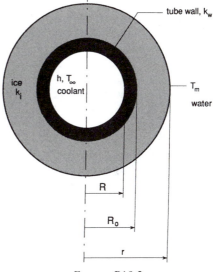

FIGURE P10.2

10.3 A power plant experiences fouling on the surface of its cold-end heat exchanger. The cycle that operates between T_H and T_{min} is irreversible, and its second law efficiency η_{II} is known. In time, the fouling thickness increases, $\delta = bt$, and so does the thermal resistance between the working fluid ($T_{\min}$) and the ambient (T_0). The power plant runs during the time interval t_1, after which it is shut down for cleaning the heat exchanger surface. The cleaning time t_2 is a known constant. Determine the optimal time of operation such that the time-averaged rate of power generation is maximum.

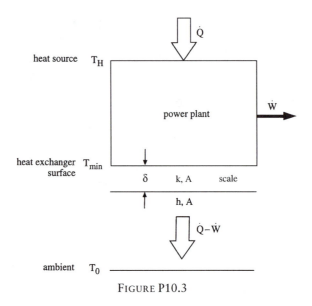

FIGURE P10.3

REFERENCES

1. A. Bejan, Theory of organization in nature: pulsating physiological processes, *Int. J. Heat Mass Transfer*, Vol. 40, 1997, pp. 2097–2104.

2. A. Bejan, Fundamental optima in thermal science, in *The Symposium on Thermal Science and Engineering in Honor of Chancellor Chang-Lin Tien*, R. O. Buckius, ed., University of California, Berkeley, 1995, pp. 523–530.

3. A. Bejan, Fundamental optima in thermal science: time-dependent (on & off) processes, in *Process, Enhanced, and Multiphase Heat Transfer: A Festschift for A. E. Bergles*, R. M. Manglik and A. D. Kraus, eds., Begell House, New York, 1996, pp. 51–57.

4. A. Bejan, *Entropy Generation Minimization*, CRC, Boca Raton, FL, 1996, pp. 341–343.

5. A. Bejan, *Heat Transfer*, Wiley, New York, 1993, p. 188.

6. J. V. C. Vargas and A. Bejan, Fundamentals of ice making by convection cooling followed by contact melting, *Int. J. Heat Mass Transfer*, Vol. 38, 1995, pp. 2833–2841.

7. J. V. C. Vargas, A. Bejan, and A. Dobrovicescu, The melting of an ice shell on a heated horizontal cylinder, *J. Heat Transfer*, Vol. 116, 1995, pp. 702–708.

8. J. V. C. Vargas and A. Bejan, Optimization principle for natural convection pulsating heating, *J. Heat Transfer*, Vol. 117, 1995, pp. 942–947.

9. J. L. Lage and A. Bejan, The resonance of natural convection in an enclosure heated periodically from the side, *Int. J. Heat Mass Transfer*, Vol. 36, 1993, pp. 2027–2038.

10. A. M. Morega, J. V. C. Vargas, and A. Bejan, Optimization of pulsating heaters in forced convection, *Int. J. Heat Mass Transfer*, Vol. 38, 1995, pp. 2925–2934.

11. J. V. C. Vargas and A. Bejan, Optimization of pulsating heating in pool boiling, *J. Heat Transfer*, Vol. 119, 1997, pp. 298–304.

12. J. V. C. Vargas and A. Bejan, Optimization of film condensation with periodic wall cleaning, *Int. J. Therm. Sci. (Rev. Gen. Therm.)*, Vol. 38, 1999, pp. 113–120.

13. A. Bejan, J. V. C. Vargas, and J. S. Lim, When to defrost a refrigerator, and when to remove the scale from the heat exchanger of a power plant, *Int. J. Heat Mass Transfer*, Vol. 37, 1994, pp. 523–532.

14. A. Bejan, *Advanced Engineering Thermodynamics*, 2nd ed., Wiley, New York, 1997.

15. V. Radcenco, J. V. C. Vargas, A. Bejan, and J. S. Lim, Two design aspects of defrosting refrigerators, *Int. J. Refrig.*, Vol. 18, 1995, pp. 76–86.

16. N. Epstein, Fouling in heat exchangers, in *Heat Transfer 1978—Proceedings of the 6th International Heat Transfer Conference*, Hemisphere, New York, 1979, Vol. 6, pp. 235–253.

17. A. M. Konings, Guide values for the fouling resistances of cooling water with different types of treatment for design of shell-and-tube heat exchangers, *Heat Transfer Eng.*, Vol. 10, 1989, pp. 54–61.

18. J. S. Lim, A. Bejan, and J. H. Kim, Thermodynamics of energy extraction from fractured hot dry rock, *Int. J. Heat Fluid Flow*, Vol. 13, 1992, pp. 71–77.

19. A. Bejan, *Convection Heat Transfer*, 2nd ed., Wiley, New York, 1995.

20. J. B. Grotberg, Pulmonary flow and transport phenomena, *Annu. Rev. Fluid Mech.*, Vol. 26, 1994, pp. 529–571.

21. K. Schmidt-Nielsen, *Scaling (Why is Animal Size So Important?)*, Cambridge Univ. Press, Cambridge, UK, 1984.

22. R. H. Peters, *The Ecological Implications of Body Size*, Cambridge Univ. Press, Cambridge, UK, 1983.

23. S. Vogel, *Life's Devices*, Princeton Univ. Press, Princeton, NJ, 1988.

24. D. A. Nield and A. Bejan, *Convection in Porous Media*, 2nd ed., Springer-Verlag, New York, 1999.

25. G. B. West, J. H. Brown, and B. J. Enquist, A general model for the origin of allometric scaling laws in biology, *Science*, Vol. 276, 1997, pp. 122–126.

26. D'A. W. Thompson, *On Growth and Form*, Cambridge Univ. Press, Cambridge, UK, 1942.

27. E. R. Weibel, *Morphometry of the Human Lung*, Academic, New York, 1963.

28. I. Rodriguez-Iturbe and A. Rinaldo, *Fractal River Basins*, Cambridge Univ. Press, Cambridge, UK, 1997.

29. A. Bejan, The tree of convective streams: Its thermal insulation function and the predicted 3/4-relation between body heat loss and body size, *Int. J. Heat Mass Transfer*, Vol. 43, 2000, to appear.

30. A. Bejan, L. A. O. Rocha, and S. Lorente, Thermodynamic optimization of geometry: T- and Y-shaped constructs of fluid streams, *International Journal of Thermal Sciences*, Vol. 40, 2001, to appear.

31. A. Bejan, A general variational principle for thermal insulation system design, *Int. J. Heat Mass Transfer*, Vol. 22, 1979, pp. 219–228.

32. S. Weinbaum and L. M. Jiji, A new simplified bioheat equation for the effect of blood flow on local average tissue temperature, *J. Biomech. Eng.*, Vol. 107, 1985, pp. 131–139.

33. R. M. Manglik and A. D. Kraus, eds., *Process, Enhanced, and Multiphase Heat Transfer: A Festschrift for A. E. Bergles*, Begell House, New York, 1996, p. xv.

34. J. O. Farlow, C. V. Thompson, and D. E. Rosner, Plates of the Dinosaur Stegosaurus: Forced convection heat loss fins?, *Science*, Vol. 192, 1976, pp. 1123–1124.

35. K. J. Bell, Early development in enhanced surfaces, *Heat Transfer Eng.*, Vol. 4, 1983, pp. 15, 106.

CHAPTER ELEVEN

TRANSPORTATION AND ECONOMICS STRUCTURE

In the preceding chapters we started from engineering and showed how the constructal principle allows us to anticipate the generation of geometric form in nature. The architectures that emerged cover a very wide domain populated by inanimate and animate flow systems. Along the way, we generated entirely new classes of results for engineering design – optimal trees for heat conduction, fluid flow, and combined heat and fluid flow.

Natural self-organization is even more widespread. It strikes closest to home in the form of connections, links, and networks that constitute the "societies" that we know. These patterns are everywhere we look: transportation routes, urban growth, and economics and business structure distributed in clear patterns all over the globe. Simpler, but equally fascinating, structures are exhibited by spatially distributed populations of other living systems, from domesticated and wild animals all the way to bacterial colonies.

What generates structure in living groups? In this chapter we see that the constructal principle is an effective and legitimate way to formulate answers. The engine that drives the generation of form is the objective, or the purpose: to gather more food, to cover a larger area faster, to earn more revenue, etc. The concept of objective, which was invoked consistently throughout this book, is used most clearly and most openly in this chapter. Objective and purpose are an integral part of our common understanding of life and a part of our language.

Constraints are also clear when we think of living groups. Territory is limited, food is scarce, time is finite, and so on. Whenever the constraint is absent or temporarily removed, the living group spreads over an increasingly larger territory, with a flow structure that becomes more and more complex in time (see Section 4.4). In a living group, flow means the movement of individuals, food and other goods, money, and information. Constraints and objective are so clear and so close to the thinking that goes on in engineering that quite often urban and economics structures are thought to be "engineered." There may be some truth in this today, in today's technological world, but not in earlier times, and certainly not in living groups other than our own. It is fair to say that social structure is a natural phenomenon of geometric extensions of the individual (self-organization), which is driven by the purpose felt by the individual (see Section 1.3).

11.1 Minimum Travel Time

The simplest and oldest problem of social-structure generation is the minimization of travel time between one point (the individual home, den, or burrow) and a finite-size territory. This is not only how streets, roads, and cities (civilization) began, but also how constructal theory was first exposed to the nonengineering public [1, 2].

The problem is how to connect a finite area A to a single point (M), Fig. 11.1. It is important to note that area A contains an *infinite number of points* and that every single one of these points must be taken into account when we are optimizing the access from A to M and back. Time has shown that this problem was a lot tougher than the empirical game of connecting many points, that is, a *finite number* of points distributed over an area. The many-points problem can be solved on the computer by brute-force methods (random walk or Monte Carlo – more points on better computers), which are not theory. This approach is discussed further in Section 11.6.

The fundamental problem of Fig. 11.1 was stated above in the most general and abstract terms because its solution and its diverse manifestations benefit from such a formulation. It helps, however, to see a real-life problem in this statement before attempting a solution. The area A could be a flat piece of farm land populated uniformly with M as its central market or harbor. It also helps to think *in time*, by beginning with the most ancient type of community that faced this access-optimization problem. The oldest solution to this problem was also the simplest: Unite with a straight line each point P and the common destination M, and you will minimize the total time spent by the population en route to M.

The straight-line solution was, most likely, the preferred pattern as long as man, his load, and his ox, had only one mode of locomotion: walking, with the average speed V_0. The farmer and the hunter would walk straight to the point (farm, village, river) where the market was located. This radial pattern of access paths can still be seen today, especially in perfectly flat and uniformly rural areas such as the plain of the lower Danube, in Romania (Fig. 1.1). The once-ancient (Roman times) settlement is now a larger town, and the surrounding farmers have become

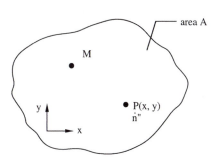

FIGURE 11.1. Finite-size area A covered by a uniformly distributed population $\dot{n}''$ traveling to a common destination M [1].

a constellation of almost equidistant tiny villages. As a matter of fact, six seems to be the usual number of surrounding communities, such that the plain is now covered by hexagonal cells with larger villages in the center. When we look at the map, the honeycomb of Bénard convection cells comes to mind (Sections 7.5–7.8). The radial length of any such "wheel" was set in antiquity by the distance that a pedestrian (or an ox) could cover in a few hours, such that a round trip to the market place could be made during daylight. The order of magnitude of that distance is 10 km, and is what we see printed on the map today.

The radial pattern disappeared naturally in areas where settlements were becoming too dense to permit straight-line access to everyone. Why the radial pattern disappeared "naturally" is the heart of the present problem. Another important development was the horse-driven carriage: With it people had *two modes* of locomotion, walking (V_0) and riding in a carriage with an average velocity V_1 that was significantly greater than V_0. It is as if the area A became a composite material with two conductivities, V_0 and V_1 (recall the thermal conductivities k_0 and k_p of Chap. 4). Clearly, it would be faster for every inhabitant (P, in Fig. 11.1) to travel in straight lines to M with the speed V_1. This would be impossible, however, because the area A would end up being covered by beaten tracks, leaving no space for the inhabitants and their land properties.

The real, more modern problem then is one of bringing the street near a small but finite-size group of inhabitants; this group would have to walk in order to reach the street. The problem is one of allocating a finite length of street to each finite patch of area (A_1), Fig. 11.2, where $A_1 \ll A$. The problem is also one of connecting these street lengths in an optimal way such that the time of travel of the entire population is minimum.

Let us consider this optimization and construction sequence in detail in order to see the constructal steps and the emergence of optimal shapes and tree networks. In Fig. 11.2 travelers have access to more than one mode of locomotion, starting with the slowest speed (V_0, walking) and proceeding toward faster modes ($V_1 < V_2 < \cdots$). The given area (A) is covered in steps of increasingly larger constructs ($A_1, A_2, A_3, \ldots$). The shape of the area and the angle between the new (collecting) path and its tributaries are optimized at each level.

The size of the smallest area element is fixed ($A_1 = H_1 L_1$) but the shape H_1/L_1 varies. The smallest size A_1 is not arbitrary: It is known because it is dictated by the land property of the individual. To minimize the time of travel between A_1 (all its points) and the boundary point M_1, it is sufficient to consider the travel between the farthest point P and M_1. We obtain

$$t_1 = \frac{L_1}{V_1} + \frac{f_1 H_1}{2 V_0}, \tag{11.1}$$

where

$$f_1 = \frac{1}{\cos \alpha_1} - \frac{V_0}{V_1} \tan \alpha_1. \tag{11.2}$$

The angle α_1 is formed between the V_0 path and the perpendicular to the V_1 stream

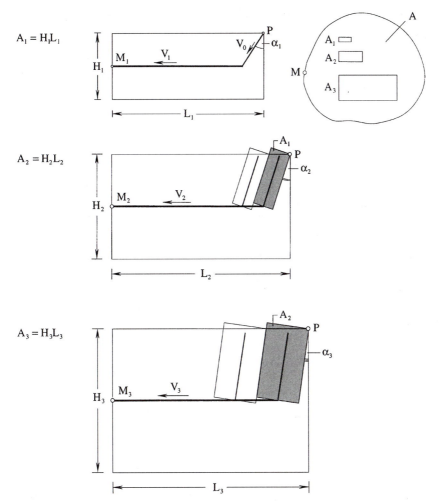

FIGURE 11.2. The constructal approach to the minimal-time route between one point M and the infinity of points of a finite-size area A [2].

(the first street). Equation (11.1) is valid when $\tan \alpha_1 < 2L_1/H_1$. Next, we eliminate L_1 from Eq. (11.1),

$$t_1 = \frac{A_1}{V_1 H_1} + \frac{f_1 H_1}{2 V_0}, \tag{11.3}$$

and note that t_1 has a minimum with respect to the overall shape (H_1):

$$H_{1,\text{opt}} = \left(\frac{2 A_1 V_0}{f_1 V_1} \right)^{1/2}, \tag{11.4}$$

$$\left(\frac{H_1}{L_1} \right)_{\text{opt}} = \frac{2 V_0}{f_1 V_1}, \tag{11.5}$$

$$t_{1,\text{min}} = \left(\frac{2 f_1 A_1}{V_0 V_1} \right)^{1/2}. \tag{11.6}$$

We can minimize the minimal time (11.6) or the f_1 expression (11.2) further by varying the angle α_1. This second step yields

$$\alpha_{1,\text{opt}} = \sin^{-1}\left(\frac{V_0}{V_1}\right), \qquad f_{1,\text{min}} = \cos\alpha_{1,\text{opt}}. \qquad (11.7)$$

When V_1 is at least three times larger than V_0, the optimal angle α_1 is close to zero and $f_1 \cong 1$. To visualize mentally the construction and repetition that follows, it helps to think that $V_0 \ll V_1 \ll V_2 \ll, \ldots,$ such that each new path is perpendicular to its tributaries. This choice was also made in the first paper on constructal streets [1, 3]. The optimized angles reported in this section are from a subsequent generalization of the construction [2].

That the V_0 path shown in the A_1 frame of Fig. 11.2 is a line of zero thickness does not mean that the rest of A_1 remains not covered, i.e., not connected to M_1. Every single point of the white area A_1 is connected to the V_1 path by a line (not shown) that is nearly parallel to V_0. The time of travel from any such point to M_1 is less than the time from P to M_1. This is why the optimal shape of Eq. (11.5) represents the geometric design for optimal access between the point M_1 and the infinity of points of A_1.

If instead of minimizing the longest travel time $[t_1,$ Eq. (11.1)] we minimize the average travel time between the points of A_1 and M_1 we arrive at the same optimal geometry as in Eq. (11.5); see Section 1.3 and Problem 11.1. In other words, what is good for the most disadvanted individual (point P, Fig. 11.2) is good for the community as a whole. This finding has profound implications in the spatial organization of all living groups, from bacterial colonies to our own societies:

The urge to organize is an expression of selfish behavior.

Consider next the fixed area $A_2 = H_2 L_2$, which is larger than the optimized element A_1. This system is shown in the second frame of Fig. 11.2. The objective, again, is to optimize the access from an arbitrary point of A_2 to a common destination point M_2. Access to every point that belongs to A_2 is ensured if we fill A_2 with a number n_2 of A_1 elements, the size and shape of which are preserved:

$$A_2 \cong n_2 A_1. \qquad (11.8)$$

Relation (11.8) is correct but not exact, because inclined A_1 elements do not fit perfectly inside A_2. This slight imprecision is an important facet of constructal theory. A certain degree of imperfection is to be expected in patterns that emerge naturally. These patterns are not identical or perfectly similar: This is consistent with the historic difficulty of attaching a theory to naturally organized systems. Natural patterns are *quasi* similar, but only in the same sense in which no two human faces are identical. Their performance, however, is practically the same as that of the mathematically optimized pattern. The contribution of constructal theory is that the performance and the main geometric features (structure, mechanism) of the organized system can be predicted in purely deterministic fashion.

The twice-optimized geometry of the A_2 construct is represented by

$$H_{2,\text{opt}} = \left(\frac{A_2 V_1}{f_2 V_2} \right)^{1/2},$$ (11.9)

$$\left(\frac{H_2}{L_2} \right)_{\text{opt}} = \frac{V_1}{f_2 V_2},$$ (11.10)

$$t_{2,\text{min}} = 2 \left(\frac{f_2 A_2}{V_1 V_2} \right)^{1/2}.$$ (11.11)

The angle factor

$$f_2 = \frac{1}{\cos \alpha_2} - \frac{V_1}{2 V_2} \tan \alpha_2$$ (11.12)

reaches its minimum at $\alpha_{2,\text{opt}} = \sin^{-1}(V_1/2 V_2)$. These results are valid provided that $\tan \alpha_2$ is less than approximately $2L_2/H_2$. The number of A_1 elements assembled in the optimized A_2 element is

$$n_2 \cong 2 f_1 f_2 \frac{V_2}{V_0} \left(1 - \frac{V_1^2}{4 V_2^2} \right).$$ (11.13)

The optimal features of assemblies of higher order ($A_i, i \geq 2$) fall into a pattern summarized as $(H_i/L_i)_{\text{opt}} = V_{i-1}/(f_i V_i)$, $\alpha_{i,\text{opt}} = \sin^{-1}(V_{i-1}/2 V_1)$, and $t_{i,\text{min}} = 2(f_i A_i / V_{i-1} V_i)^{1/2}$. If these theoretical recurrence formulas were to be repeated ad infinitum in both directions, then the resulting image would be a fractal. The image of Fig. 11.2 is not a fractal [3, 4]. In other words, geometric optimization of volume-to-point accounts for (1) why certain natural structures look like images generated by fractal algorithms truncated at a small length scale (inner cutoff), (2) why an inner cutoff exists, and (3) why in a natural structure the algorithm breaks down in the steps situated close to the cutoff (e.g., step $i = 1$ in Fig. 11.2).

In summary, a single design principle (the minimization of travel time) generates every geometric feature of the path (V_0, V_1, V_2, . . . ,) that connects one point (M_1, M_2, . . .) to the infinity of points that make up an area (A_1, A_2, . . .). The path that emerges is a tree network, i.e., a loopless pattern (not a net!) in which links do not cross. The upper row of Fig. 11.3 shows two examples drawn to scale with the same elemental area A_1 used as the area unit [2]. It is assumed that the speed increases by the same factor whenever a new assembly is formed. Note that the streets become closer to perpendicular as the velocity increase factor V_i/V_{i-1} increases, for example, as the vehicle on the first street becomes faster than the pedestrian. At the same time, the number of constituents (tributaries) in each new assembly increases and the construction sequence (A_2, $A_3 \cdots$) spreads faster over the given area A. The lower part of Fig. 11.3 illustrates one case in which the velocity increase factor V_i/V_{i-1} decreases as the assemblies become larger. The higher-order assemblies become less slender, and their number of constituents decrease as V_i/V_{i-1} decreases.

Three-dimensional volume-to-point access can be optimized in the same manner as in Fig. 11.2 [2]. It is convenient to use Fig. 11.2 once more, by replacing mentally

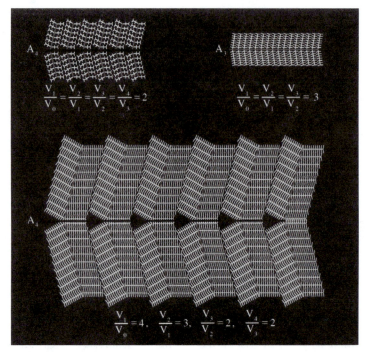

FIGURE 11.3. Three examples of the growth of street patterns as the minimization of travel time between a finite-size area and one point. Velocities increase as the constructs become larger. Each construct has been optimized twice, for shape and angle [2] (see Plate XII).

the area elements ($A_1 - A_3$) with cylindrically shaped volume elements ($B_1 - B_3$). Each volume element has two geometric features that can be optimized: the shape represented by the diameter/length ratio H_i/L_i in Fig. 11.2 and the angle α_i. Of these, the most important is the shape of the volume element, which is solely responsible for how the optimized building blocks are assembled into larger entities to fill the given space.

The optimal features of the smallest volume element (B_1) are $(H_1 L_1)_{opt} = 4V_0/(f_1 V_1)$, where $f_1 = (\cos \alpha_1)^{-1} - (V_0/V_1) \tan \alpha_1$ and $\alpha_{1,opt} = \sin^{-1}(V_0/V_1)$. For the B_2 volume element we find $(H_2/L_2)_{opt} = 4V_1/(3 f_2 V_2)$, where $f_2 = f_1^{2/3}(\cos \alpha_2)^{-1} - (V_1/3V_2) \tan \alpha_2$, and $\alpha_{2,opt} = \sin^{-1}(V_1/3V_2 f_1^{2/3})$. The results for B_2 and higher-order assemblies are correct only approximately because constant-diameter cylinders do not fit perfectly inside a larger cylinder. Finally, when $i \geq 3$ the geometric results follow the pattern $(H_i/L_i)_{opt} = 4V_{i-1}/(3 f_i V_i)$, where $f_i = (\cos \alpha_i)^{-1} - (V_{i-1}/3V_i) \tan \alpha_i$ and $\alpha_{i,opt} = \sin^{-1}(V_{i-1}/3V_i)$. In conclusion, tree networks in three dimensions have the same deterministic origin as tree networks in two dimensions.

One application of the optimal-access problem in three dimensions is the sizing and the shaping of the floor plan in a multistory building, and the selection and placement of the optimal number of elevator shafts and staircases. The floor-plan optimization proceeds according to the method illustrated based on A_1, Fig. 11.2, where V_0 corresponds to the travel through the rooms and V_1 is the travel along

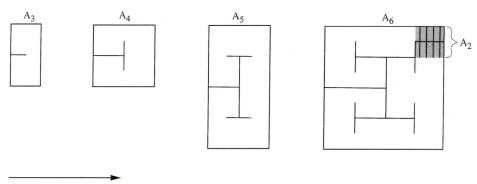

time, growth, evolution, development, purpose, life

FIGURE 11.4. Higher-order area elements in the constructal sequence [1].

the corridor. The unknown in this first building block is the shape of the floor area A_1. Next, we stack a number n_2 of A_1 elements on the vertical such that the next path V_2 is vertical and accounts for the elevator or the staircase. The optimization sequence may be taken to a third assembly (or even higher-order assemblies) if the towers optimized in the three-dimensional equivalent of Fig. 11.2 must be integrated into a larger building with several wings.

The network optimization and organization theory can be extended generally to areas that are populated unevenly, and to highways, railroads, telecommunications, and air routes (e.g., the organization of such connections into hubs, or centrals). Another extension is in operations research and manufacturing, in which the invention of the first auto assembly line is completely analogous to the appearance of the first street.

Another opportunity to optimize the street-patterning geometry arises in cases in which the vehicular speed (V_i, $i \geq 1$) increases with the street width and the paved surface of all the streets is constrained. This idea is explored in Refs. 1 and 3.

We end with another look at the two-dimensional, point-to-area travel, which occupied most of this section. Figure 11.4 was drawn on the assumption that the V_i / V_{i-1} ratios are such that, starting with A_3, the number of constituents is two.

The sequence shown in Fig. 11.4 is only for illustration, because it is unlikely to be repeated beyond the street of third generation. The reason is that as the community and the area inhabited by it grow, other common destinations (e.g., church, hospital, bank, school, train station) emerge in A, in addition to the original M point (Fig. 11.1). Some of the streets that were meant to provide access to only one end of the area element (e.g., A_3 in Fig. 11.4) must be extended all the way across the area to provide access to both ends of the street. As the destinations multiply and shift around the city, the dead ends of the streets of the first few generations disappear, and what replaces the "growth pattern" of Figs. 11.2 and 11.3 is a *grid* with access to both ends of each street. The multiple scales of this grid and the self-similar structure of certain areas (neighborhoods) of the grid, however, are the fingerprints of the deterministic, element-by-element optimization and organization principle employed throughout this book.

11.2 Minimum Cost

In this and the following sections constructal theory is extended into the economics realm [3, 5–7]. The economic activity that covers a given area is driven by the flow-access-optimization (objective and constraints) principle, and the economic and business structure is the result.

To see how constructal theory accounts for the origin of structure in economics and business, we consider a stream of goods that proceeds from one point (producer, distributor) to every point of a finite-size territory (consumers). The flow may also proceed in the opposite direction, from a finite area (e.g., grain, or carpets woven by individuals) to one point of collection.

Two types of economic activity are studied. In this section the objective is cost minimization, as in the design of a postal-service network for the delivery of mail over a given area. Section 11.3 focuses on the more common goal of revenue maximization, an example of which is the sale of fertilizer to farmers and, later, the harvesting of crops over a given area. This entire treatment is based on Ref. 7.

The fundamental contribution of this extension is that the *law of parsimony* [8] emerges as the economics analog of the resistance-minimization principle recognized in physics and engineering, and that constructal theory emerges as an even more general theory of parsimony in nature.

Consider a stream of goods that proceeds from one point (producer, distributor) to every point of a finite-size territory (consumers). The objective is to minimize the total cost associated with the given point-to-area or area-to-point stream. The economics principle of economies of scale tells us that the cost is lower when the goods move in the aggregate; that is, when they are organized into thicker streams. The cost is also proportional to the distance traveled. Clearly, the cost plays the same role as the local thermal resistance in heat trees (Chap. 4), the inverse of the travel speed in street trees (Section 11.1), or the local fluid-flow resistance in fluid trees (Chap. 5). The given business territory is covered naturally by trees, that is, links of decreasing cost, starting from the highest unit cost that is allocated to the smallest area scale (the individual), and continuing in a sequence of intermediaries (distributors, collectors) who handle increasingly larger fractions of the total stream of goods.

We illustrate this deterministic process by considering the area A in which goods must arrive at (or depart from) every point at the uniform rate γ [units/(m^2 s)], Fig. 11.5. The total stream of goods $m = \gamma A$ flows between A and a single point M (producer or collection point). For a given total stream m, it is advantageous to have higher rates near the collection point M and lower rates in distant areas dA. Such a nonuniform source of goods was studied within the framework of solar cells [9], in which the "good" to be collected is the photocurrent. However, in many cases, a uniform production rate density $\gamma = m/A$ is given, and this distribution cannot be changed. Therefore we restrict ourselves to this case.

Several means of transport are available, and they are represented by the sequence of costs K_i [$/(m unit)], $i = 0, 1, 2, \ldots$, such that $K_0 > K_1 > K_2 \ldots$. Each

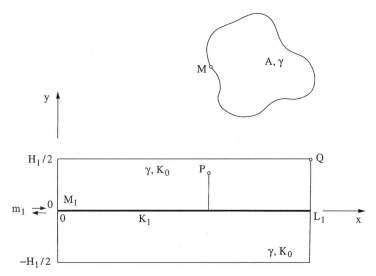

FIGURE 11.5. The area-to-point transport configuration (top) and the elemental area A_1 (bottom); the double arrow means that the flow of goods may proceed in either direction [7].

K_i represents the cost per unit of goods transported and per unit of distance traveled. The problem consists of covering A with trajectories of various unit costs such that the total cost required by the m stream is minimum. One way to approach this problem is by allocating an area element to each K_i link and connecting the links such that the entire area A is connected to M.

Let us assume that the smallest area element – the elemental area – over which the stream is delivered to (or collected from) every point is A_1 (see Fig. 11.5). In this section the size of A_1 is assumed known and fixed. The area A_1 is elemental because the stream associated with it ($m_1 = \gamma A_1$) is channeled along a single link (K_1). Every point P of the elemental area has access to the K_1 link along a direct (perpendicular) path of the highest unit cost, K_0. In other words, goods travel between A_1 and M_1 by two means: First, the most expensive mode of transport touches every point of the infinity of points found in A_1 and, second, the less expensive mode of transport K_1 makes the final connection with the exit point M_1.

The minimization of cost suggests that we shape A_1 as a rectangle ($H_1 L_1$) that surrounds the K_1 link, such that K_1 is placed on the longer of the two axes of the rectangle. The shape parameter H_1/L_1 is free to vary. The per-unit-time cost associated with the transport of one unit from P to M_1 is $K_0 y + K_1 x$. The cost of transporting the total stream between A_1 and M_1 is

$$C_1 = 2\gamma \int_0^{H_1/2} \int_0^{L_1} (K_0 y + K_1 x)\, dx\, dy = \gamma \left(\frac{1}{4} K_0 H_1^2 L_1 + \frac{1}{2} K_1 H_1 L_1^2 \right). \quad (11.14)$$

Since the elemental area and its total stream are fixed, we may express cost C_1 as

$$\frac{C_1}{m_1} = \frac{K_0 A_1}{4 L_1} + \frac{1}{2} K_1 L_1. \quad (11.15)$$

We reach the important conclusion that shape – optimal form – can be deduced from cost minimization. The optimal shape of the elemental area is independent of its size:

$$\left(\frac{H_1}{L_1}\right)_{\text{opt}} = 2\frac{K_1}{K_0} < 1. \tag{11.16}$$

Associated features of this geometric optimum are $H_{1,\text{opt}} = (2K_1 A_1/K_0)^{1/2}$, $L_{1,\text{opt}} = (K_0 A_1/2K_1)^{1/2}$, and the total minimum cost associated with the transport of the m_1 stream,

$$C_{1,\min} = m_1(A_1 K_0 K_1/2)^{1/2}. \tag{11.17}$$

Two observations define the course of the geometric optimization that will follow. First, instead of minimizing the cost integrated over A_1 [Eq. (11.14), or Problem 11.1 for minimum travel time], we could have minimized the unit cost faced by the most distant producer or consumer (point Q in Fig. 11.5). That cost is $K_0 H_1/2 + K_1 L_1$. Minimized subject to the constant elemental area $A_1 = H_1 L_1$, this cost leads to *exactly* the same geometric optimum as that in Eq. (11.16). When the elemental shape is optimal, the cost from Q to the bend in the path ($x = L_1$) equals the cost from the bend to M_1. Note that the unit cost faced by individual points P situated closer than Q relative to M_1 is always smaller than the unit cost from Q to M_1. This means that what is good for the individual, especially for the individual with the worst geographic location, is good for the entire group of producers or consumers that inhabits the elemental area. This sheds light on why the spatial organization of the flow of goods (the first route K_1) can occur *spontaneously*, that is without a discussion, plan, and agreement for the entire group. In the following analysis, we continue with calculations of total costs integrated over finite areas, as in Eq. (11.14).

The second observation is based on the minimum cost determined in Eq. (11.17). What happens when the goods must flow between the point M_1 and an increasingly larger territory A_1? Both the flow rate m_1 and the cost $C_{1,\min}$ increase, but the rate of increase in the total cost ($C_{1,\min}/m_1$) increases as well. It increases as $A_1^{1/2}$. If global cost minimization is the driving force behind the generation of spatial economics structure, then the population of producers or consumers (γ) will search for a geometric way of slowing down the cost increase that comes with territorial expansion.

The population is already organized in optimally shaped area elements of type A_1. The alternative to covering a larger territory (the alternative to increasing A_1) is to assemble a number of A_1 elements into a larger construct A_2, Fig. 11.6. The number of constituents $n_2 = A_2/A_1$, or the external shape H_2/L_2, is the unknown. The total stream of goods flowing from A_2 to the new single point M_2 (the larger producer, collection point, or user of goods) is $m_2 = \gamma A_2 = n_2 m_1$. The K_1 links of the A_1 elements are connected to M_2 through a new link of unit cost K_2. Since the new link (perhaps an advanced mode of transportation) facilitates the transport of a stream that is larger than that in each elemental area, its unit cost is lower than in the elements, $K_2 < K_1$.

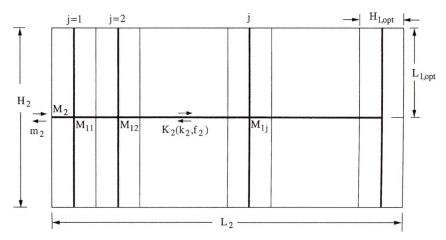

FIGURE 11.6. The second area A_2 as a construct of n_2 elemental areas [7].

We obtain the total cost C_2 required by the transport of m_2 between A_2 and M_2 by adding the $C_{1,\min}$ shares contributed by the n_2 constituents to the cost of transport along the central K_2 link. The latter is less than $m_2 K_2 L_2$, because the actual flow rate on this route is less than m_2: The flow rate varies from the full value m_2 at the root point M_2, all the way down to zero at the opposite end of the K_2 link. This variation is approximately linear when n_2 is sufficiently large. In this limit the total cost contributed by the K_2 link is $\frac{1}{2} m_2 K_2 L_2$, because the average flow rate along the K_2 route is $m_2/2$. In sum, the cost associated with the area-to-point configuration $A_2 - M_2$ is

$$C_2 = n_2 C_{1,\min} + \frac{1}{2} m_2 K_2 L_2. \tag{11.18}$$

We can rearrange this result by using Eq. (11.17), $A_1 = A_2/n_2$, and $L_2 = (n_2/2)H_1$:

$$\frac{C_2}{m_2} = \left(\frac{K_0 K_1 A_2}{2 n_2} \right)^{1/2} + \frac{K_2}{2} \left(\frac{K_1 A_2 n_2}{2 K_0} \right)^{1/2}. \tag{11.19}$$

An optimal elemental-area size ($A_{1,\mathrm{opt}} = A_2/n_{2,\mathrm{opt}}$) such that the total cost C_2 is minimum exists, i.e.,

$$n_{2,\mathrm{opt}} = 2 \frac{K_0}{K_2} > 1, \qquad C_{2,\min} = m_2 (K_1 K_2 A_2)^{1/2}. \tag{11.20}$$

The optimal external shape is described by $(H_2/L_2)_{\mathrm{opt}} = K_2/K_1$, $H_{2,\mathrm{opt}} = (A_2 K_2/K_1)^{1/2}$, and $L_{2,\mathrm{opt}} = (K_1 A_2/K_2)^{1/2}$. Note that what we have optimized is the shape of A_2, not the size of A_2. The optimal shape is fixed when K_2 and K_1 are specified. The A_2 area is not known because it is proportional to the elemental area A_1, which is not specified. Only when we maximize the revenue derived from the flow of goods (Section 11.3) do we optimize the shape and the size of each area construct.

It is useful to compare $C_{2,\min}$ with $C_{1,\min}$ to see that in the second-area structure (Fig. 11.6) the ratio $C_{2,\min}/m_2$ continues to be proportional to the total area A_2, but that the rate of increase has slowed down. The factor $(K_0 K_1/2)^{1/2}$ of Eq. (11.17)

TABLE 11.1. **The Optimized Structure of Point-to-Area or Area-to-Point Transport for Minimum Cost [7]**

i	$n_{i,\text{opt}}$	$(H_i/L_i)_{\text{opt}}$	$C_{i,\text{min}}/m_i$
1	—	$2K_1/K_0$	$(A_1 K_0 K_1/2)^{1/2}$
2	$2K_0/K_2$	K_2/K_1	$(A_2 K_1 K_2)^{1/2}$
≥ 3	$4K_{i-2}/K_i$	K_i/K_{i-1}	$(A_i K_{i-1} K_i)^{1/2}$

has been replaced with $(K_1 K_2)^{1/2}$ in Eqs. (11.20): The latter is smaller as soon as $K_2 < K_0/2$, which is true in the present case because of the first of Eqs. (11.20).

The more basic message of this geometric optimization is that cost minimization can be achieved (1) by optimizing the area and the structure and (2) by increasing the internal complexity of the structure. This method of cost minimization can be pursued further, toward more complex internal structures, as long as new and less costly modes of transport become available. For example, if the new mode has the unit cost K_3 that is sensibly smaller than K_2, then it makes sense to cover a larger territory, not by increasing A_2 indefinitely by use of the design of Fig. 11.6 and Eqs. (11.16) and (11.20), but by combining n_3 constructs of type A_2 into the new construct A_3 shown in Fig. 11.7. The n_3 streams of size m_2 are delivered or collected by a central route of length L_3 and unit price K_3. The total stream that reaches every point of the area A_3 is $m_3 = n_3 m_2$. When n_3 is sufficiently greater than 1, the total cost associated with the transport between A_3 and the root point M_3 is

$$C_3 = n_3 C_{2,\text{min}} + \frac{1}{2} m_3 K_3 L_3. \tag{11.21}$$

The optimal values of n_3 and H_3/L_3 that minimize the ratio C_2/m_3 are listed in Table 11.1. This optimization can be continued toward constructs of higher order: When $i \geq 3$, the results fall into the pattern summarized by the recurrence relations listed in the table.

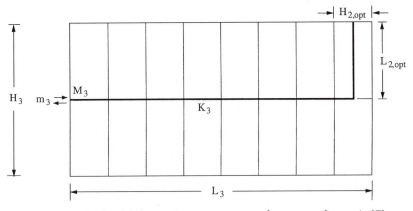

FIGURE 11.7. The third area A_3 as a construct of n_3 areas of type A_2 [7].

11.3 Maximum Revenue

In this section we solve a different problem of geometric optimization of point-to-area or area-to-point transport. We show that if, instead of minimizing cost, the objective is to maximize revenue, it is possible to predict not only the tree-shaped network of the transport routes but also the size of the elemental area A_1 [7].

Two features distinguish the new approach from the model used in Section 11.2. First, the unit cost associated with a certain link, c_i ($/unit), is assumed to be proportional to the distance traveled, d_i (m), and inversely proportional to the flow rate of goods along that link, f_i (units/s):

$$c_i = k_i \frac{d_i}{f_i},$$ (11.22)

where k_i [$/(s\ m)$] is a parameter that describes the interaction – the match – between producer and consumer. The ratio k_i/f_i is equivalent to the cost parameter K_i used in Section 11.2. The advantage of using k_i/f_i is one of increased flexibility: On the same route and at the same flow rate f_i, two different producers may face different costs (different k_i values). The cost analysis for the elemental area A_1 (Fig. 11.5) proceeds according to the steps outlined in Eqs. (11.14) and (11.15). In the first integrand of Eq. (11.14) we use $(k_0/f_0)y + (k_1/f_1)x$ in place of $K_0 y + K_1 x$ and arrive at the elemental cost formula

$$C_1 = \frac{\gamma}{2} A_1 \left(\frac{k_0 H_1}{2 f_0} + \frac{k_1 L_1}{f_1} \right).$$ (11.23)

The second new feature is the revenue obtained by the producer who distributes and sells a stream of goods over an area. If g ($/unit) is the price paid by the individual consumer who resides at a point (x, y) of the elemental area A_1, then the revenue cannot exceed $m_1 g$. From this ceiling value we subtract the transportation cost C_1 and obtain the net revenue

$$R_1 = m_1 \left(g - \frac{m_1 k_0}{4\gamma L_1 f_0} - \frac{L_1 k_1}{2 f_1} \right).$$ (11.24)

This expression shows that the net revenue vanishes in two extremes, when the flow of goods stops ($m_1 = 0$), and when the flow rate reaches the critical value,

$$m_{1,c} = 4\gamma L_1 \frac{f_0}{k_0} \left(g - \frac{L_1 k_1}{2 f_1} \right).$$ (11.25)

This means that R_1 has a maximum with respect to m_1. The net revenue has an additional maximum with respect to L_1, because R_1 tends to minus infinity in both limits, $L_1 \to 0$ and $L_1 \to \infty$. We obtain the maximum net revenue by solving simultaneously $\partial R_1/\partial m_1 = 0$ and $\partial R_1/\partial L_1 = 0$, and the results are

$$m_{1,\mathrm{opt}} = \frac{8}{9}\gamma g^2 \frac{f_0 f_1}{k_0 k_1}, \qquad L_{1,\mathrm{opt}} = \frac{2}{3} g \frac{f_1}{k_1}.$$ (11.26)

Since $m_1 = \gamma A_1$, the result obtained for $m_{1,\mathrm{opt}}$ prescribes the optimal size of the elemental territory that should be serviced by the producer:

$$A_{1,\mathrm{opt}} = \frac{8}{9} g^2 \frac{f_0 f_1}{k_0 k_1}. \qquad (11.27)$$

We obtain the optimal shape of the elemental area by combining $H_1/L_1 = A_1/L_1^2$ with Eqs. (11.26): The resulting expression is equivalent to Eq. (11.16), namely, $(H_1/L_1)_{\mathrm{opt}} = 2(k_1/f_1)/(k_0/f_0)$. The twice-maximized net revenue that corresponds to the two optimal parameters determined in Eqs. (11.26) is

$$R_{1,\mathrm{max,max}} = \frac{8}{27} \gamma g^3 \frac{f_0 f_1}{k_0 k_1}. \qquad (11.28)$$

To summarize, the geometric maximization of net revenue at the elemental level has 2 degrees of freedom, the size and the shape of A_1. What is new relative to the results of cost minimization (Section 11.2) is the optimal size of the elemental area. According to Eq. (11.27), the elemental area is large when the product is expensive (large g) and the transport is inexpensive (small k_0/f_0 and k_1/f_1). This optimal area size is independent of the surface density of flow of goods γ, because it is essentially a balance between the revenue generated by the stream of goods and the cost of transporting the same stream. The twice-maximized revenue per unit area is equal to $\gamma g/3$.

Consider next the second-order area element A_2 of Fig. 11.6, where $A_2 = n_2 A_1$. The root (storage) points of the elemental areas (M_{1j}, $j = 1, 2, \ldots, n_2/2$) are connected to the global root point M_2 by a new route of cost factor k_2 and flow rate f_2. The conservation of goods requires that $m_2 = n_2 m_1$. Each elemental area contributes to the net revenue maximized in accordance with Eq. (11.28). The question is how much of the total revenue $n_2 R_{1,\mathrm{max,max}}$ is used to offset the cost associated with the transport along the second-order route.

Along the short stretch between the points M_2 and M_{11} the flow rate is equal to the total flow rate for the second-order area, m_2. The unit cost is $(k_2/f_2)H_{1,\mathrm{opt}}/2$, and this means that the cost associated with the segment $M_2 M_{11}$ is

$$C_{2,1} = m_2 \frac{k_2}{f_2} \frac{H_{1,\mathrm{opt}}}{2}. \qquad (11.29)$$

Along the next segment ($M_{11} M_{12}$) the flow rate is $m_2 - 2m_{1,\mathrm{opt}}$, and the traveled distance is $H_{1,\mathrm{opt}}$. Together, these quantities pinpoint the total cost contributed by the segment,

$$C_{2,2} = (m_2 - 2m_{1,\mathrm{opt}}) \frac{k_2}{f_2} H_{1,\mathrm{opt}}. \qquad (11.30)$$

The remaining segments of the k_2/f_2 path are analyzed similarly, because the length of each is equal to $H_{1,\mathrm{opt}}$. For example, the cost associated with the segment $M_{1,j-1} M_{1j}$ is

$$C_{2,j} = [m_2 - 2(j-1)m_{1,\mathrm{opt}}] \frac{k_2}{f_2} H_{1,\mathrm{opt}}. \qquad (11.31)$$

TABLE 11.2. **The Optimized Spatial Structure of Point-to-Area or Area-to-Point Transactions for Maximum Revenue [7]**

i	$n_{i,\text{opt}}$	$(H_i/L_i)_{\text{opt}}$	$A_{i,\text{opt}}$	$R^*_{i,\text{max}}$	$R_{i,\text{max}}/A_{i,\text{opt}}$
1	—	$2\dfrac{k_1/f_1}{k_0/f_0}$	$\dfrac{8}{9}g^2\dfrac{f_0 f_1}{k_0 k_1}$	$\dfrac{8}{27}\gamma g^3\dfrac{f_0 f_1}{k_0 k_1}$	$\dfrac{1}{3}\gamma g$
2	$\dfrac{3k_0/f_0}{2k_2/f_2}$	$\dfrac{4k_2/f_2}{3k_1/f_1}$	$\dfrac{4}{3}g^2\dfrac{f_1 f_2}{k_1 k_2}$	$\dfrac{2}{3}\gamma g^3\dfrac{f_1 f_2}{k_1 k_2}$	$\dfrac{1}{2}\gamma g$
3	$\dfrac{3k_1/f_1}{2k_3/f_3}$	$2\dfrac{k_3/f_3}{k_2/f_2}$	$2g^2\dfrac{f_2 f_3}{k_2 k_3}$	$\gamma g^3\dfrac{f_2 f_3}{k_2 k_3}$	$\dfrac{1}{2}\gamma g$
≥ 4	$\dfrac{k_{i-2}/f_{i-2}}{2k_i/f_i}$	$2\dfrac{k_i/f_i}{k_{i-1}/f_{i-1}}$	$2g^2\dfrac{f_{i-1} f_i}{k_{i-1} k_i}$	$\gamma g^3\dfrac{f_{i-1} f_i}{k_{i-1} k_i}$	$\dfrac{1}{2}\gamma g$

* For $i = 1$, this parameter was maximized with respect to two variables, $R_{1,\text{max,max}}$.

In conclusion, the total cost that is due to transport along the central link that leads to M_2 is

$$C_2 = \sum_{j=1}^{n_2/2} C_{2,j}. \tag{11.32}$$

The net revenue collected over A_2 is equal to $R_2 = n_2 R_{1,\text{max,max}} - C_2$, which becomes, after some algebra,

$$R_2 = n_2 m_{1,\text{opt}}\left(g - \frac{n_2 k_2}{4 f_2}H_{1,\text{opt}}\right). \tag{11.33}$$

This quantity has a maximum with respect to n_2, because it vanishes at $n_2 = 0$ and at $n_2 = 4 f_2 g/(k_2 H_{1,\text{opt}})$. We obtain the optimal number of elements in the second-order area by solving $\partial R_2/\partial n_2 = 0$:

$$n_{2,\text{opt}} = \frac{3 k_0/f_0}{2 k_2/f_2}. \tag{11.34}$$

This solution is valid if $n_{2,\text{opt}} \geq 2$. In Eq. (11.34) we reached the important conclusion that optimal spatial growth (organization, assembly) can be reasoned on the basis of revenue maximization. A similar conclusion was reached based on cost minimization in Eq. (11.20). It is worth noting that Eqs. (11.34) and (11.20) do not prescribe the same number of elements in the A_2 assembly: If we recall the change in parameters made at the start of this section (namely, $K_0 = k_0/f_0$, $K_2 = k_2/f_2$), we see that revenue maximization recommends a smaller number of elements than cost minimization.

The remaining parameters of the optimized second-order area can be determined from $m_{2,\text{opt}} = n_{2,\text{opt}} m_{1,\text{opt}}$ and the geometric relations $H_{2,\text{opt}} = 2 L_{1,\text{opt}}$ and $L_{2,\text{opt}} = (n_{2,\text{opt}}/2) H_{1,\text{opt}}$. The resulting expressions are $H_{2,\text{opt}} = (4/3)g f_1/k_1$ and $L_{2,\text{opt}} = g f_2/k_2$, as shown in Table 11.2. At this optimum, net revenue expression

(11.33) becomes

$$R_{2,\max} = \frac{2}{3}\gamma g^3 \frac{f_1 f_2}{k_1 k_2},\qquad(11.35)$$

which reminds us that at the second level of assembly A_2 the net revenue was maximized with respect to a single variable. The corresponding area of the assembly is

$$A_{2,\text{opt}} = \frac{4}{3}g^2 \frac{f_1 f_2}{k_1 k_2},\qquad(11.36)$$

which means that the revenue per unit area at this level is $R_{2,\max}/A_{2,\text{opt}} = \gamma g/2$. This ratio is higher than the corresponding ratio at the elemental-area level ($\gamma g/3$). The stepwise increase in revenue per area is an additional incentive to continue the geometric-optimization sequence toward larger constructs. The other incentive is the increase in revenue, which proceeds in the direction of a larger area formed by the connection of the streams of optimized smaller areas. We examine this trend in the next section and Fig. 11.8.

The analytical results are listed in Table 11.2 and are valid, provided that $n_i \geq 2$. The results settle into a pattern when $i \geq 4$: This feature differs somewhat from the establishment of a repetitive pattern based on cost minimization, which occurred when $i \geq 3$ (Table 11.1). In the present solution, the routes (k_i, f_i) form a tree network that is completely deterministic. Every geometric detail is the result of invoking a single principle – the maximization of revenue.

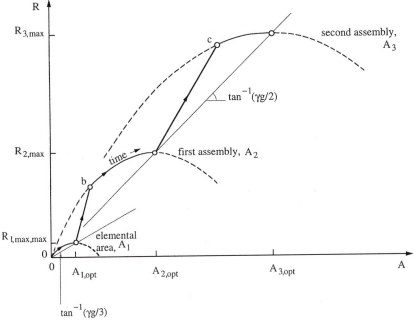

FIGURE 11.8. Increase in the size and complexity of the area-to-point flow as a result of the maximization of net revenue in time [7].

If the recurrence formulas listed for $i \geq 4$ in Table 11.2 were to be repeated ad infinitum in the opposite direction, until the area scale is of size zero, then the resulting tree structure would be a fractal. The present structure is not a fractal – it is a Euclidean figure – because the number of levels of assembly is finite [3, 4]. Access to the infinity of points of the given area is made by means of the costliest transport mode (k_0, f_0), which is placed over all the scales that are smaller than the elemental area $A_{1,\mathrm{opt}}$. Unlike in fractal tree networks, in which the smallest scale (the inner cutoff) is arbitrary and chosen for graphic impact, in the present construction, $A_{1,\mathrm{opt}}$ is finite and predictable based on the same principle of revenue maximization that generates the rest of the network structure.

11.4 Development of Economics Structure in Time

We should not read too much into the suggestion that the sequence outlined in Table 11.2 continues indefinitely. The passing of time, or better, the development of transport technology in time, dictates how far the structure spreads in two dimensions. In Table 11.2, the index i could be associated with the increase in time. For example, if the modes of transport (in order of decreasing cost per unit) are hand delivery (k_0, f_0), light auto transport (k_1, f_1), heavy trucks (k_2, f_2), and rail transport (k_3, f_3), then the final area-to-point connection is at the most an assembly of the third order. This does not mean that the area covered by the largest assembly ($A_{3,\mathrm{opt}}$) is small. The step changes in area size from one assembly to the next are described approximately by $(A_i/A_{i-1})_{\mathrm{opt}} \sim (k_{i-2}/f_{i-2})/(k_i/f_i) \gg 1$. The new assembly is much larger than its predecessor when its cost parameter (k_i/f_i) is much smaller than the cost parameter of the assembly formed two steps earlier.

The transition from one area assembly to the assembly of the next higher order is abrupt: Size and complexity change stepwise. Why should a producer or distributor opt for such abrupt changes? What drives these "transitions" in structure growth and development?

These questions are addressed in the rightmost column of Table 11.2. The transition from the elemental area ($A_{1,\mathrm{opt}}$) to the first assembly ($A_{2,\mathrm{opt}}$) is recommended by the promised increase in revenue per unit area. The ratio $\gamma g/3$ means that in a design of the elemental-area type the producer receives only one third of the money paid by all the consumers, while the transporter receives the remaining two thirds. In an area design of the first assembly type, the producer and the transporter receive equal shares of the revenue generated per unit area. This is another example of the principle of equipartition, or optimal allocation, which is a frequent feature in engineering, geophysical, and biophysical applications of constructal theory [3]. Equipartition of revenue between producer and transporter is preserved when $i \geq 2$, as shown in Table 11.2.

Another way to anticipate the stepwise increase in area size and complexity is to invoke the maximization of net revenue *in time*, as shown in Fig. 11.8. The producer begins servicing one or more areas of the elemental type A_1, which has one central route (k_1, f_1) and diffuse mode of transport (k_0, f_0) that reaches every point.

The elemental structure starts from $t = 0$ when $A = 0$ and expands in time until it reaches the optimal size and external shape represented by the point $(A_{1,\text{opt}}, R_{1,\text{max,max}})$. If the elemental structure would continue to expand beyond this optimum the net revenue would drop.

The route to higher revenue consists of assembling two or more structures of type $A_{1,\text{opt}}$ into a first assembly A_2, assuming that a new and less costly mode of transport is available (k_2, f_2). The resulting structure (point b in Fig. 11.8) is not optimal. However, it produces a revenue that is greater than the sum of the maximized revenues of the components of type $A_{1,\text{opt}}$. The A_2 structure grows from point b until it reaches its peak of revenue production $(R_{2,\text{max}}, A_{2,\text{opt}})$. The process repeats itself beyond this second peak: Sudden coalescence and jump in revenue are followed by gradual increase in revenue until a new maximum is reached. The horizontal distance between $A_{2,\text{opt}}$ and the abscissa of point c is at least as large as $A_{2,\text{opt}}$: In Fig. 11.8 this distance was drawn shorter because of page-width limitations.

11.5 Optimally Shaped Triangular Areas

In this section we show that costs can be minimized further by optimizing the shapes of the areas that make up the constructal transport structure. In Section 11.2 all the area elements were assumed rectangular: That assumption allowed us to assemble area elements into a larger area, and to cover the larger area *completely*. The downside of the rectangular shape is that it enforces a nonuniform distribution of unit cost, especially when we compare the points situated on its perimeter. In Fig. 11.5, for example, the largest unit cost is concentrated in the two corners of type Q, because these corners are the points P situated the farthest from the common destination, or origin, M_1.

Here we reconsider the area-to-point cost-minimization problem of Section 11.2 and begin the constructal sequence from the triangular area element shown in Fig. 11.9. The area $A_1 = B_1 L_1 / 2$ is fixed and is covered in the y direction by high-cost transport (K_0). The path of relatively low-cost transport coincides with the axis of symmetry x. The area shape parameter B_1 / L_1 is not fixed. The total cost required for transporting the stream of goods γA_1 between A_1 and M_1 is

$$C_1 = 2\gamma \int_0^{L_1} \left\{ \int_0^{H_1/2} [K_0 y + K_1(L_1 - x)] \mathrm{d}y \right\} \mathrm{d}x = 2\gamma \left(\frac{1}{24} K_0 B_1^2 L_1 + \frac{1}{12} K_1 B_1 L_1^2 \right),$$

(11.37)

where $H_1(x) = B_1 x / L_1$ is the local transversal dimension of the triangular area. The C_1 expression can be minimized with respect to the base/length ratio of the triangle, with the following results:

$$B_{1,\text{opt}} = 2 \left(\frac{K_1 A_1}{K_0} \right)^{1/2}, \qquad L_{1,\text{opt}} = \left(\frac{K_0 A_1}{K_1} \right)^{1/2}, \qquad (11.38)$$

$$\left(\frac{B_1}{L_1} \right)_{\text{opt}} = 2 \frac{K_1}{K_0}, \qquad C_{1,\text{min}} = \frac{2^{3/2}}{3} m_1 (A_1 K_0 K_1 / 2)^{1/2}. \qquad (11.39)$$

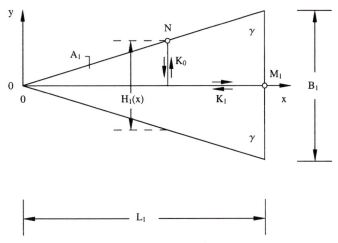

FIGURE 11.9. Triangular elemental area A_1 for area-to-point transport [7].

Equations (11.39) show, first, that the optimal slenderness ratio of the trian-gle is exactly the same as the optimal slenderness ratio for a rectangular element of the same area, Eq. (11.16). Second, the cost $C_{1,\min}$ is smaller than the corre-sponding cost in the rectangular design, Eq. (11.17). The cost-reduction factor associated with optimizing elemental triangles instead of elemental rectangles is $2^{3/2}/3 = 0.94$.

A very special quality of the optimized triangle becomes visible when we calculate the unit cost for transport between M_1 and an arbitrary point on its periphery, $N(x, H_1/2)$:

$$c_1 = K_0 H_1/2 + K_1(L_1 - x). \tag{11.40}$$

Noting that $H_1 = B_1 x/L_1$ and using the optimized aspect ratio $(B_1/L_1)_{\text{opt}}$ of Eq. (11.34), we conclude that the unit cost on the boundary is *constant*, i.e., inde-pendent of the position x of the peripheral point N:

$$c_1 = K_1 L_1 = \frac{1}{2} K_0 B_1 = (A_1 K_0 K_1)^{1/2}. \tag{11.41}$$

In this way we have deduced that the stretching of the points of largest unit cost into a continuous line [10] (as opposed to one or two points) is a mechanism that generates not only minimal total cost for A_1 but also geometric form: the complete architecture of A_1.

Because of the constant-c_1 property of the symmetric portion of the perimeter of area A_1, the best geometric figure for A_1 is the isosceles triangle, and the best of all such triangles is the one with the slenderness $(B_1/L_1)_{\text{opt}}$. When the triangle is more slender than the best, $B_1/L_1 < (B_1/L_1)_{\text{opt}}$, the peripheral point of highest unit cost is the sharp tip of the triangle. When the triangular shape is less slender than the optimal triangle, the points of highest unit cost are the two corners that define the base B_1. When the slenderness is optimal, the sharp tip and the two base corners (and the *lines* that connect them) have the highest unit cost.

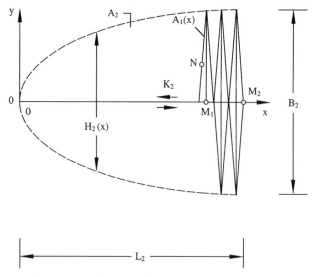

FIGURE 11.10. Second area A_2 with unspecified shape and covered incompletely by triangular elements [7].

The interior of the optimally shaped triangle is covered by a family of V-shaped lines of constant unit cost, which are terminated by the same baseline ($x = L_1$). The sides of each V-shaped figure are parallel to the $y = \pm H_1(x)/2$ sides of the A_1 triangle. The point M_1 is the internal "line" $c_1 = 0$, i.e., the center of this family of nested and geometrically similar triangles.

The cost reduction that was achieved by switching from elemental rectangles (Fig. 11.5) to elemental triangles (Fig. 11.9) is even more attractive at larger area scales. Consider the construct of area A_2 shown in Fig. 11.10, which is a substitute for Fig. 11.6. The shape of A_2, or the function $H_2(x)$, is not specified. The A_2 territory is covered incompletely by an infinity of geometrically similar triangular elements, $A_1(x)$, where $L_1(x) = H_2(x)/2$. Each elemental triangle has the slenderness ratio $B_1(x)/L_1(x) = 2K_1/K_0$ [see Eqs. (11.39)]; each has a constant-c_1 periphery; however, the c_1 value depends on the position x on the symmetry axis K_2, namely, $c_1(x) = K_1 L_1(x)$; see Eq. (11.41).

The optimal shape of A_2 follows from the requirement that the unit cost c_2 between a peripheral point N and new destination point M_2 be the same for every peripheral point, i.e., all along the contour of the toothy territory covered by transport:

$$c_2 = c_1(x) + K_2(L_2 - x), \qquad \text{constant.} \qquad (11.42)$$

Recalling that $c_1 = K_1 L_1(x) = K_1 H_2(x)/2$, we draw the conclusion that H_2 is linear in x, i.e., the optimal A_2 shape is an isosceles triangle of base B_2 and length L_2. We also find that $c_2 = K_1 B_2/2 = K_2 L_2$, and consequently

$$\left(\frac{B_2}{L_2}\right)_{\text{opt}} = 2\frac{K_2}{K_1}, \qquad L_2 = \left(A_2\frac{K_1}{K_2}\right)^{1/2}, \qquad B_2 = 2\left(A_2\frac{K_2}{K_1}\right)^{1/2}. \quad (11.43)$$

If $A_{1,max}$ is the largest elemental triangle used in the A_2 construct, namely, $A_{1,max} = A_1(x = L_2)$, then Eqs. (11.43) also mean that $A_2/A_{1,max} = K_0/K_2 \gg 1$.

To calculate the total cost associated with the transport between the A_2 construct and the point M_2, we analyze the stream of goods that arrives at the location x, from above and below the K_2 axis. There are two elemental triangles of size $A_1(x)$ that share the same base $B_1(x)$. They produce the stream $2m_1$ and, per unit length (dx), the stream $m_1' = 2m_1/B_1$, where $m_1 = \gamma A_1$. The total stream that flows between the A_2 construct and the point M_2 is

$$m_2 = \int_0^{L_2} m_1'\, dx = \frac{\gamma}{2} A_2. \tag{11.44}$$

Note that the size of m_2 is only half of γA_2, because the triangular elements cover exactly half of the area A_2. The other half of A_2 is the sum of all the triangular interstices situated between adjacent $A_1(x)$ elements.

The cost associated with the stream $m_1'\, dx$ has two components. Along the K_2 axis, the cost component is $K_2(L_2 - x)m_1'\, dx$. Off the K_2 axis, over the vertical stripe of width dx, the cost is $C_1'\, dx$, where $C_1' = 2C_{1,min}/B_1$, and $C_{1,min}$ is given by Eq. (11.39). In sum, the total cost is delivered by the integral

$$C_{2,min} = \int_0^{L_2} [C_1' + m_1' K_2(L_2 - x)]\, dx, \tag{11.45}$$

which, after some algebra and the use of Eq. (11.44), yields

$$C_{2,min} = \frac{7}{9} m_2 (A_2 K_1 K_2)^{1/2}. \tag{11.46}$$

Next to Eqs. (11.20), this result shows that the constant-cost shaping of the periphery of the transport territory inscribed in A_2 produces a significant reduction in the total cost. The cost-reduction factor $(7/9 = 0.78)$ is more significant than the reduction registered at the elemental level.

The analysis and the optimization of larger constructs follow the steps that we just outlined for the A_2 construct. We find that each new (larger) area must be an isosceles triangle of a certain slenderness ratio, so that each point on its toothy periphery is characterized by the same unit cost. This conclusion was drawn directly on Fig. 11.11, which shows the A_3 triangle covered by an infinite number of geometrically similar $A_2(x)$ triangles. The stream of goods that arrives on the K_3 axis at the location x is $m_2' = 2m_2(x)/B_2(x)$. The cost associated with this stream in regions off the K_3 axis is $C_2' = 2C_{2,min}(x)/B_1(x)$. We obtain the following results, in this order:

$$\left(\frac{B_3}{L_3}\right)_{opt} = 2\frac{K_3}{K_2}, \qquad L_3 = \left(A_3 \frac{K_2}{K_3}\right)^{1/2}, \qquad B_3 = 2\left(A_3 \frac{K_3}{K_2}\right)^{1/2}, \tag{11.47}$$

$$m_3 = \frac{\gamma}{2^2} A_3, \qquad C_{3,min} = \frac{23}{27} m_3 (A_3 K_2 K_3)^{1/2}. \tag{11.48}$$

The second of Eqs. (11.48) shows that relative to the corresponding rectangular

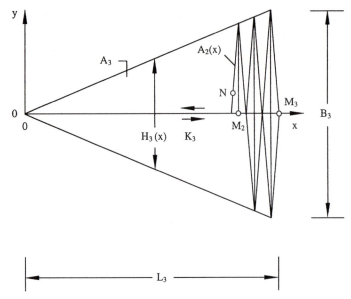

FIGURE 11.11. Triangular third area A_3, covered incompletely by triangular constructs A_2 [7].

design ($i = 3$, Table 11.1) the total cost has been reduced by the factor $23/27 = 0.85$. This reduction is comparable with the reduction obtained at the A_2 level, Eq. (11.46). The relative size of the A_3 construct is $A_3/A_{2,\text{max}} = K_1/K_3 > 1$, where $A_{2,\text{max}}$ is the largest $A_2(x)$ construct used in the A_3 internal structure, namely, $A_{2,\text{max}} = A_2(x = L_3)$.

Continuing this analysis, it can be shown that the fourth construct is another triangle (A_4), the shape of which is described by relations that fit the pattern visible already in Eqs. (11.43) and (11.47). The total flow rate of goods is $m_4 = \gamma A_4/2^3$, and the associated minimized cost is $C_{4,\text{min}} = (73/81)m_4(A_4 K_3 K_4)^{1/2}$.

The paragraph written at the start of Section 11.4 applies here as well. Although the sequence analyzed in this section can be continued to areas larger and more complicated than A_4, what we have presented is sufficient for drawing several important conclusions. The newest is that the triangle-in-triangle constructs (Figs. 11.10 and 11.11) look more fractallike in comparison with their rectangle-based counterparts (Fig. 11.6). The reason is that, in principle, in each larger triangle we could fit an infinite number of slender and geometrically similar triangles. On the other hand, in constructal theory and its earlier extensions covered in this book the smallest element size (volume, area) is finite. In the elemental unit the flow is ruled by laws that differ from those that govern the flows at larger scales. Constructal theory is an atomistic theory.

Constructs with triangular components happen to look more "natural," i.e., more like the dendritic patterns found in nature (e.g., leaves, young urban growth, fingers, dendritic crystals). The constructs based on rectangular components fill the allotted space better (completely). This observation suggests that "the urge to optimize" is why natural structures tend to look more and more fractallike. In other words, it is the refining of the performance of a rough design (e.g., the

Euclidean structure of Fig. 11.6) that pushes the design toward a fractallike structure. This tendency has more general implications, not just in spatial economics (see Section 11.8). It applies to all natural tree-shaped structures, which are always imperfect and incomplete (Euclidean), and in which the trend toward consistent refinements and fractal-looking structures is evident. We return to this observation in Chap. 12.

11.6 Older Methods in Spatial Economics

The preceding analyses were based on two important simplifying assumptions adopted from the earlier work on heat and fluid tree networks (Chaps. 4 and 5). These assumptions allowed us to show most clearly how the flow structure is deduced consistently from a single principle. One assumption is that the optimization results (the architecture) determined at one level of assembly are preserved for reuse as internal features of the next, larger area construct. The other assumption is that smaller areas assembled into a larger area communicate with each other only through the root points M_i, while the rest of their rectangular perimeters are impermeable. Additional, less pivotal simplifying assumptions such as the rectangular shape of area elements and the right angles between successive transport paths were also used.

These assumptions simplified the analysis because in each new construct they kept the number of degrees of freedom to a minimum. Their impact on the goodness of the optimized architecture was evaluated systematically in several studies dealing with tree-shaped heat flows [11–13]. The evaluation consisted of repeating the same work without making these assumptions, by coping with the large number of degrees of freedom, and solving the problem numerically in a continuum. A similar numerical approach is being used in the simulation and optimization of river basins and vascularized tissues [14, 15]. The numerical work on heat trees showed that the questioned assumptions have a relatively minor impact on the optimized shape and structure, and practically no effect on the optimized performance (the minimized global resistance). The work on river basins and vascularization [14, 15] also showed that in fully numerical simulations that begin with postulating an initial, nonoptimal tree network, the final architecture is also influenced by the initial features of the postulated network. If the initial network is chosen randomly, the final optimized architecture is not reproducible.

In sum, the constructal method provides a most transparent route – a shortcut – to the optimal performance, i.e., the optimal access (minimum resistance, time, cost) between an area and one point. The constructal tree is not the mathematically optimal design: That structure can be determined only numerically, based on the most refined formulation, involving the largest possible number of degrees of freedom. The constructal solution is the simplest structure that (1) is a tree, (2) is deterministic, i.e., is deduced by invoking a single principle, and (3) performs practically as well as the much more refined (and more "natural" looking) tree structures optimized by black-box numerical simulations.

This is also a good place to comment on how constructal theory fits in the research that is being conducted today in the field of spatial economics (this discussion continues in Sections 11.7 and 11.8). This field is highly developed because of the computer age and the limitless applications of the access optimization problem. On such a busy and important background, it may seem that the ideas of this chapter were developed in complete isolation from what goes on in the mainstream. This is true in the real sense: The ideas originate from a completely separate field (thermodynamics and engineering design). The challenge now is to communicate these ideas to the spatial economics community and to explain their purely theoretical origin. They were not derived from or built on developments found by reading of the spatial economics literature.

For example, today considerable work is being devoted in spatial economics to the so-called *p-median* problem. Some of the most representative and recent references in this area are Refs. 16–22. According to the recent review by Church and Sorensen [23], "the *p*-median problem involves the location of a fixed number of facilities in such a manner that the total weighted distance of all users assigned to their closest facility is minimized. The *p*-median problem was first posed by Hakimi [24] on a network of nodes and arcs. Each node was considered a place of demand as well as a potential facility location. The arcs represented transportation or accessibility linkages, and could be used for facility locations as well."

Constructal theory differs fundamentally from the *p*-median framework. First, in constructal theory we do not assume (postulate) a network: Our objective is to *deduce* the structure, i.e., to predict the very existence of structure on the basis of principle. Second, we do not simulate a territory by using a finite number of points (nodes): We optimize the access to a finite-size area, i.e., to an *infinity* of points. Third, we do not assume lines (links) between nodes, so that we may neglect the white area that separates two neighboring nodes. In constructal theory the transport over a white area (K_0) is as important as the transport along the route (K_1) that belongs to the area. Fourth, there is only one facility in the constructal problem: The point that serves as sink or source for the flow that reaches completely the given area.

In agreement with other area-to-point or volume-to-point flow examples from physics, biology, and engineering, in this chapter constructal theory led us theoretically to tree-shaped paths as optimal routes for transport. The tree image does not have loops and seems to contradict the "grids" – the networks with loops – that characterize urban areas, commerce, and communications. There is no contradiction, because the constructal tree is *the structure of the flow*: the path, or paths, followed by the area-to-point stream. The constructal tree is not a collection of immobile pieces of hardware (ducts, cables, or paved roads): It is the flow that happens to inhabit the necessary links when the area and point are specified. Take any urban grid, pick a single source or sink point, and try to reach the infinity of points of the given area, and then you will visualize mentally the tree-shaped stream that flows through some portions of the grid. Grids of streets develop because in real life the many inhabitants of an area must have access to several facilities, not one. Each facility point (source, sink) is the root of its own tree.

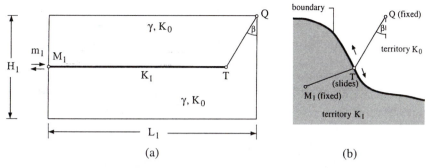

FIGURE 11.12. (a) Elemental area with variable angle of confluence between the K_0 and K_1 paths, (b) the refraction principle of minimizing the cost of transport between two fixed points [7].

11.7 The Law of Refraction

In the constructions of this chapter we made the simplifying assumption that the angle between two successive paths of transport is 90°. For example, in the cost-minimization analysis based on Fig. 11.5, we started with the assumption that the K_0 path is perpendicular to the K_1 path. In Fig. 11.6, K_2 was perpendicular to K_1, and so on. These angles of confluence too can be optimized to decrease further the cost per assembly or to maximize the revenue per assembly. The angle-optimization opportunity was illustrated for minimum-time travel in Section 11.1.

Consider again the elemental area A_1 of Fig. 11.5, but this time allow the most distant corner (Q) to be reached along a K_0 path that makes an unspecified angle β with the line perpendicular to K_1. This more general situation is shown in Fig. 11.12. The per-unit-time cost required for transporting one unit of goods from point Q to M_1 via the turning point T is

$$c = \frac{K_0 A_1}{2L_1}\left(\frac{1}{\cos \beta} - \frac{K_1}{K_0}\tan \beta\right) + K_1 L_1. \qquad (11.49)$$

We can minimize this cost not only by choosing the external shape H_1/L_1 [as in Eq. (11.16)], but also by selecting the angle β. The results of this two-variable minimization procedure are

$$\left(\frac{H_1}{L_1}\right)_{\text{opt}} = 2 \tan \beta_{\text{opt}}, \qquad \beta_{\text{opt}} = \sin^{-1}(K_1/K_0). \qquad (11.50)$$

When K_1 is sensibly smaller than K_0, β_{opt} is negligibly small, the K_0 and K_1 paths become perpendicular, and Eq. (11.50) approaches Eq. (11.16).

In general, there is an optimal angle of confluence, or an optimal angle of refraction if we liken the broken line QTM_1 as a ray of light that passes from a low-speed medium (K_0) into a high-speed medium (K_1). The analogy between this angle optimization and Fermat's principle of light refraction has been noted independently in two fields [3, 8]. In constructal theory, it was shown that the confluence angle can be optimized at every subsequent level of assembly, as in the minimum-time constructs for point-to-area travel (Section 11.1). In economics, beginning with Lösch's treatise [8], the refraction principle is a recognized deterministic method of transport route maximization [25, 26]. Lösch's angle optimization is different than the constructal version because, first, it does not recognize the opportunity to optimize the shape of the territory in which refraction occurs, and, second, the break point T is constrained to slide along a curve that serves as boundary between territory K_0 and territory K_1 (see the right-hand side of Fig. 11.12).

Lösch's angle optimization is simply an analog of Fermat's, because it is about point-to-point travel across a given boundary. In constructal theory, angle optimization is one feature in a more complex geometric construction for point-to-area or area-to-point transport. In this chapter angle optimization was not emphasized because its impact is minor. Routes are nearly perpendicular when the unit cost sequence is steep ($K_{i-1} \gg K_i$), and the other degree of freedom – the optimization of external area shape and numbers of constituents in new assemblies – is solely responsible for the formation and growth of structure in space and time.

11.8 The Law of Parsimony

In this chapter we saw that by minimizing cost in point-to-area or area-to-point transport it is possible to anticipate the formation and growth of dendritic routes over a growing territory. The generation of structure is a reflection of the optimization of area at each area scale. In addition, we saw that by maximizing revenue in point-to-area or area-to-point transactions it is possible to anticipate not only the expanding (compounding) dendrites of transport routes, but also the size of the smallest area element that is obtained by means of the highest unit cost available (K_0). Every geometric detail of the structure is deterministic. It is the result of invoking only one principle.

This principle is known and accepted in economics as the law of parsimony [8, 25, 26]. In physics, biology, and engineering, constructal theory unmasks this principle as a general law of access optimization for internal currents. This chapter showed that the law that generates structure in natural flow systems far from internal equilibrium also generates structure in economics.

The wide applicability of the law of parsimony in economics is acclaimed [8, 25, 26], but its manifestations outside the realm of economics are not emphasized. On the contrary, they are disclaimed. For example, Haggett and Chorley (Ref. 26, p. v) write

> "We are too conscious of the dangers of easy analogy and strained metaphor to claim that, for example, stream systems and transport systems are geographically 'the same'; to do so would force us to ignore aspects of network structure and evolution that are intrinsically important to physical and human geographers respectively."

The conclusion reached here and in Ref. 7 is to claim precisely what Haggett and Chorley have rejected. The similarities between the spatial structures of physical and economic flows are not mathematical coincidences. These structures are generated by the same deterministic principle. The constructal law or its earlier statements (law of parsimony, Fermat's and Heron of Alexandria's principle [3]) is the universal *law of nature* that accounts for the generation of shape and structure in heterogeneous flow systems subjected to constraints.

PROBLEMS

11.1 In the optimization of the elemental area A_1 of Fig. 11.2, we minimized the travel time between the most distant corner of the rectangle (P) and the entrance or exit point M_1. Perform the alternative analysis in which (1) the travel time between M_1 and an arbitrary point $P(x, y)$ situated inside A_1 is averaged over A_1 and (2) the area-averaged travel time is minimized. For simplicity assume that $V_1 \gg V_0$ so that the length of the V_1 path is equal to L_1. Compare the optimized shape of A_1 with the result developed in the text, Eq. (11.5).

11.2 Consider the symmetric elemental area A_1 of length L_1 and width $H_1(x)$, where $x = L_1$ marks the point M that serves as destination or origin for the flow between A_1 and M. This area is shown in Fig. P11.2 (top). The path of high speed V_1 coincides with the axis of symmetry x. Travel at the lower speed V_0 covers the territory above and below the x axis. Assume that V_0 is sufficiently smaller than V_1 so that the optimal angle between each V_0 path and the V_1 street is $90°$.

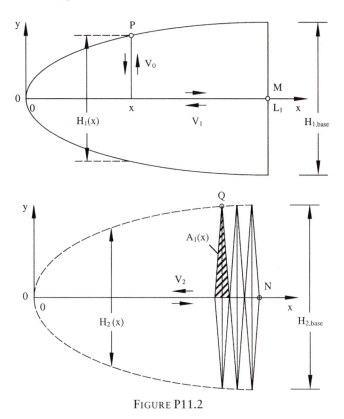

FIGURE P11.2

Determine the shape $H_1(x)$ such that the travel time between M and a point P (situated on the periphery of A_1) is a constant (t_{c1}), i.e., independent of x. Show that the A_1 area is an isosceles triangle with the base $H_{1,\text{base}}$ and height L_1:

$$H_{1,\text{base}} = 2\left(\frac{A_1 V_0}{V_1}\right)^{1/2} \qquad L_1 = \left(\frac{A_1 V_1}{V_0}\right)^{1/2}.$$

Show also that the constant time t_{c1} is shorter by the factor $2^{-1/2}$ than the corresponding time given for rectangular elements by Eq. (11.6).

We may cover a larger area A_2 by assembling on both sides of a new axis (new street of speed V_2) a large number of elemental triangles that are geometrically similar to the A_1 triangle analyzed until now. The size of each elemental triangle depends on its position on the new x axis, $A_1(x)$. The tips of the A_1 triangles fall on the perimeter of a new territory, which the triangles cover incompletely (Fig. P11.2, bottom). The length of this territory is L_2, and its total area is $A_2 = \int_0^{L_2} H_2 \, dx$.

Determine the shape $H_2(x)$ such that the travel time (t_{c2}) between the periphery of each triangle $A_1(x)$ and the destination (or origin) point $N(L_2, 0)$ is independent of x. Show that the A_2 territory must also be an isosceles triangle, with the following base and height:

$$H_{2,\text{base}} = 2\left(\frac{A_2 V_1}{V_2}\right)^{1/2} \qquad L_2 = \left(\frac{A_2 V_2}{V_1}\right)^{1/2}.$$

Derive the formula for t_{c2}, compare it with the t_{c1} formula, and comment on which design you would choose (A_1 or A_2) if your objective is to minimize the travel time per unit of $A_{1,2}^{1/2}$.

REFERENCES

1. A. Bejan, Street network theory of organization in nature, *J. Adv. Transp.*, Vol. 30, No. 7, 1996, pp. 85–107.
2. A. Bejan and G. A. Ledezma, Streets tree networks and urban growth: optimal geometry for quickest access between a finite-size volume and one point, *Physica A*, Vol. 255, 1998, pp. 211–217.
3. A. Bejan, *Advanced Engineering Thermodynamics*, 2nd ed., Wiley, New York, 1997.
4. D. Avnir, O. Biham, D. Lidar, and O. Malcai, Is the geometry of nature fractal?, *Science*, Vol. 279, 1998, pp. 39–40.
5. A. Bejan, The constructal law of structure formation in natural systems with internal flows, in *Proceedings of the ASME Advanced Energy Systems Division*, AES-Vol. 37, American Society of Mechanical Engineers, New York, 1997, pp. 257–264.
6. A. Bejan, How nature takes shape: extensions of constructal theory to ducts, rivers, turbulence, cracks, dendritic crystals, and spatial economics, *Int. J. Therm. Sci. (Rev. Gen. Therm.)*, Vol. 38, 1999, pp. 653–663.
7. A. Bejan, V. Badescu, and A. De Vos, Constructal theory of economics structure generation in space and time, *Energy Convers. Manage.*, Vol. 41, 2000, pp. 1429–1451.
8. A. Lösch, *The Economics of Location*, Yale Univ. Press, New Haven, CT, 1954.
9. P. Benítez, R. Mohedano, and J. Minano, Conversion efficiency increase of concentration solar cells by means of non-uniform illumination, in *Proceedings of the 14th European Photovoltaic Solar Energy Conference*, Barcelona, 1997, p. 2378.

10. M. Neagu and A. Bejan, Constructal-theory tree networks of "constant" thermal resistance, *J. Appl. Phys.*, Vol. 86, 1999, pp. 1136–1144.

11. G. A. Ledezma, A. Bejan, and M. R. Errera, Constructal tree networks for heat transfer, *J. Appl. Phys.*, Vol. 82, 1997, pp. 89–100.

12. N. Dan and A. Bejan, Constructal tree networks for the time-dependent discharge of a finite-size volume to one point, *J. Appl. Phys.*, Vol. 84, 1998, pp. 3042–3050.

13. A. Bejan and N. Dan, Two constructal routes to minimal heat flow resistance via greater internal complexity, *J. Heat Transfer*, Vol. 121, 1999, pp. 6–14.

14. I. Rodriguez-Iturbe and A. Rinaldo, *Fractal River Basins*, Cambridge Univ. Press, Cambridge, UK, 1997.

15. P. Meakin, *Fractals, Scaling and Growth Far From Equilibrium*, Cambridge Univ. Press, Cambridge, UK, 1998.

16. B. C. Tansel, R. L. Francis, and T. J. Lowe, Location on networks: a survey. Part I. The p-center and p-median problems, *Manage. Sci.*, Vol. 29, 1983, pp. 482–497.

17. S. L. Hakimi, p-median theorems for competitive locations, *Ann. Oper. Res.*, Vol. 6, 1986, pp. 77–98.

18. M. L. Brandeau, S. S. Chiu, and R. Batta, Locating the two-median of a tree network with continuous link demands, *Ann. Oper. Res.*, Vol. 6, 1986, pp. 223–253.

19. H. Tamura, M. Sengoku, S. Shinoda, and T. Abe, Location problems on undirected flow networks, *IEICE Trans.*, Vol. E73, 1990, pp. 1989–1993.

20. S. S. Chaudhry, I.-C. Choi, and D. K. Smith, Facility location with and without maximum distance constraints through the p-median problem, *Int. J. Oper. Prod. Manage.*, Vol. 15, 1995, pp. 75–81.

21. K. Watanabe, H. Tamura, and M. Sengoku, The problem of where to locate p-sinks in a flow network: complexity approach, *IEICE Trans. Fundam.*, Vol. E79-A, 1996, pp. 1495–1503.

22. P. Longley and M. Batty, eds., *Spatial Analysis: Modelling in a GIS Environment*, GeoInformation International, Cambridge, UK, 1996.

23. R. L. Church and P. Sorensen, Integrating normative location models into GIS: problems and prospects with the p-median problem, in Ref. 22. Chap. 9.

24. S. L. Hakimi, Optimum locations of switching centers and the absolute centers and medians of a graph, *Oper. Res.*, Vol. 11, 1964, pp. 450–459.

25. P. Haggett, *Locational Analysis in Human Geography*, Edward Arnold, London, 1965.

26. P. Haggett and R. J. Chorley, *Network Analysis in Geography*, St. Martin's, New York, 1969.

CHAPTER TWELVE

SHAPES WITH CONSTANT RESISTANCE

12.1 How, Not What

As concluding remarks, in this chapter I propose to review the territory covered theoretically in this book. I use once again the language of conductive heat transfer in which constructal trees were first described (Chap. 4). What flows is of secondary importance. The key is how the flow must be organized macroscopically in space and in time. The unknown is the geometry, and this is what the constructal principle anticipates. What holds for heat flow also holds for other flows: macroscopic structure (architecture) springs out of the objective and constraints principle.

In the constructs optimized in this book we showed repeatedly how the form of the structure depends on the number of degrees of freedom allowed in the design. Certainly, more degrees of freedom demand more complex optimizations and more time. Two benefits come with the harder work, and they are both important from the point of view of constructal theory. One is that the performance of the more polished construct is better, but only marginally better relative to the performance of optimized rougher constructs. The other thought is that the more a construct becomes "more efficient," the more its architecture looks "more natural." This gives even more meaning to the time arrow Principle → Nature previewed in the lower part of Fig. 1.3.

Finally, there is the analogy between the *flow* structure derived from principle and the generation of *solid* mechanical structures from the same objective and constraints principle. This analogy served as starting point for this book (Chap. 2). In the next section we find that as the degrees of freedom of a structure multiply, all the shapes become more rounded – more like leaves, branches, and pine cones. I called these shapes of constant resistance. We also find that the optimally curved shapes are characterized by hot spots that are spread continuously through the volume: lines and surfaces of hot spots, instead of discrete points. This feature is similar to Galilei's beam of constant strength [Fig. 2.1(c)], in which the maximum stress is spread as much as possible through the body of the beam. The connection between solid and flow structures is now complete. The two worlds of design are united by the principle that generates shape and structure.

12.2 More Degrees of Freedom

In this section we focus on an interesting trend that becomes visible when we attempt to make the tree constructs more efficient, i.e., less resistive to volume-to-point flow. We accomplish this by relaxing each geometric-optimization problem and optimizing more degrees of freedom at every volume scale. We discuss comparatively only the geometric features that result. The details of these optimizations are documented in a sequence of four papers [1–4].

This trend is illustrated in Fig. 12.1, which shows chronologically the four ways in which elemental volumes for two-dimensional volume-to-point heat conduction have been optimized. The boundary is insulated except over the cooled (heat-sink) point at $T_{\min}$. Heat is generated at every point in the white area. Each design is optimal and drawn to scale. The total volume, the volume fraction ($\phi_0 = 0.04$) occupied by the high-conductivity material (black), and the ratio of thermal conductivities ($k_p/k_0 = \tilde{k} = 575$) are the same in each design.

The earliest design [Fig. 12.1(a)] was the simplest, and the roughest. The elemental volume was assumed to be rectangular, and the lone k_p insert was assumed to have uniform thickness D_0. The thermal resistance is minimum when the large rectangle has a certain shape – an optimal slenderness ratio. The hot spots are concentrated in the two corners ($T_{\max}$) that are situated the farthest relative to the heat sink. See also Section 4.2.

The resistance of the optimized elemental rectangle was lowered by 6% after the constant-D_0 feature was abandoned and the profile of the k_p channel was optimized. In the design shown in Fig. 12.1(b), 2 degrees of freedom have been optimized: the slenderness ratio of the rectangle, and the shape function of the high-conductivity blade, $D_0 \sim x^{1/2}$, where x is measured away from the thin end [see also Fig. 4.7(b)]. The hot spots continue to reside in two points – the two most distant corners ($T_{\max}$).

A more substantial reduction in volume-to-point resistance is registered when the hot spots are distributed continuously over an optimally curved portion of the boundary [Fig. 12.1(d)]. The design is said to have constant resistance because the temperature difference between each boundary point ($T_{\max}$) and the common heat sink ($T_{\min}$) is constant. Three degrees of freedom are optimized: the outer shape ($H_0 \sim x^{3/5}$), the slenderness ratio of the outer shape, and the shape of the k_p insert ($D_0 \sim x^{7/5}$). The global resistance of design (d) is only 2/3 of the resistance of design (a) and 71% of the resistance of design (b).

Figure 12.1(c) shows a triangle-in-triangle approximation of the most polished design (d). This approximation is also a constant-resistance design with continuous hot-spot lines ($T_{\max}$), but the external triangular shape is assumed, not optimized. The 2 degrees of freedom optimized are the aspect ratio of the outer triangle and the shape of the high-conductivity inserts ($D_0 \sim x$). The resistance of design (c) exceeds by only 6% the resistance of design (d).

In the triangle-in-triangle structure, the formation of continuous hot-spot lines has its origin in the optimization of the aspect ratio of the external triangle. The

Resistance

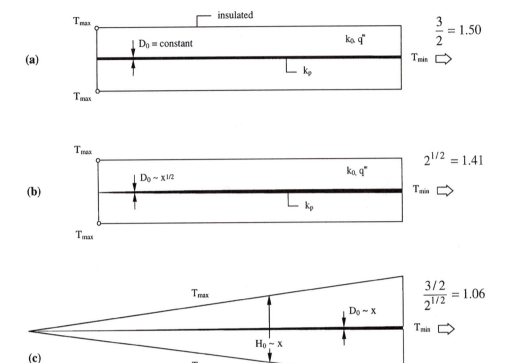

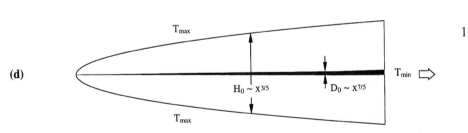

FIGURE 12.1. Evolution of the optimized elemental-volume design for minimum flow resistance between a volume and one point [3]. The numbers in the right column indicate the global volume-to-point resistance as a multiple of the resistance of design (d).

migration of the hot spots is indicated by the color red in Fig. 12.2. When the external triangle is too slender, the hot spot is concentrated in one point: the sharp tip of the triangle. In the opposite extreme, the hot spots are located in the two corners on the side with the heat sink. The optimal slenderness is in between, when the hot spot "jumps" from the tip of the triangle to the two base corners. At that moment the hot spot traces with T_{max} the two long sides of the triangle, and the design acquires its constant resistance.

    Viewed in ensemble, Fig. 12.1 shows that the performance improves at every step, from (a) to (d), and the improvement ranges from the significant to the marginal. When these elemental designs are assembled into larger constructs, they

FIGURE 12.2. The color red shows the migration of the hot spots as the aspect ratio of the triangle changes. There is an optimal, intermediate slenderness that marks the moment when the hot spot jumps from the tip to the two base corners. In this design the entire long sides of the triangle are at the hot-spot temperature (see Plate XIII).

cover the innermost scales of the tree structure. Designs based on elements (c) and (d) cover their allotted space incompletely.

Figure 12.3 shows the optimized first construct that results from using the best elements [Fig. 12.1(d)]. The optimal shape of the area allocated to the construct is such that the vertical dimension (H_1) of the structure increases as $x^{3/5}$, where x increases away from the left end. The optimal thickness (D_1) of the central blade (horizontal, black) increases as $x^{7/5}$. In this design the volume fraction of k_p material in the elemental volumes (ϕ_0) and the volume fraction averaged over the entire first construct (ϕ_1) satisfy the optimized proportion $\phi_0 = \frac{1}{2}\phi_1$.

The right-hand side of Fig. 12.3 shows that the isosceles triangle inscribed in the optimized first construct has a 90° tip angle. Unlike the slenderness of the elemental rectangles and triangles (Fig. 12.1), which decreases with $\tilde{k}\phi_0$, this tip angle is an invariant: it is independent of $\tilde{k}\phi_1$.

The increased internal complexity of the first construct has the effect of decreasing dramatically the resistance in volume-to-point flow. The global resistances of the optimized elemental volume [Fig. 12.1(d)] and first construct (Fig. 12.3) are

$$\frac{T_{\max} - T_{\min}}{q''' A_0 / k_0} = \frac{1}{3(\tilde{k}\phi_0)^{1/2}}, \tag{12.1}$$

$$\frac{T_{\max} - T_{\min}}{q''' A_1 / k_0} = \frac{8}{9\tilde{k}\phi_1}, \tag{12.2}$$

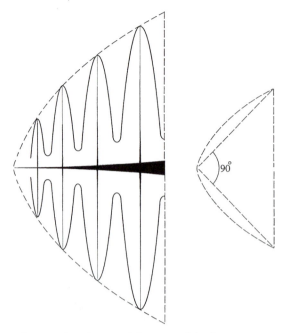

FIGURE 12.3. Optimal external and internal features of the first construct with constant thermal resistance [3].

where A_0 and A_1 are the actual areas covered by k_0 material and uniform heat generation (q''') in the two figures. Comparing the two resistances on an equal basis, namely, $A_0 = A_1$ and $\phi_0 = \phi_1 = \phi$, we see that the first-construct resistance is smaller as soon as the product $\tilde{k}\phi$ exceeds 64/9. This marks the transition from one configuration (elemental) to a more complex one (first construct) in the pursuit of better access for the volume-to-point current. As the product $\tilde{k}\phi$ increases above 64/9, the first-construct configuration becomes a considerably more effective path for the current, while its elemental volumes become more slender and numerous.

Simpler to draw is a first construct in which the constant-resistance triangular elements of Fig. 12.1(c) are mounted on a wedge-shaped high-conductivity blade D_1 and spread over a triangular space (Fig. 12.4). The optimization of this construct begins with the assumption that the internal and external shapes are triangular. The minimized global resistance is

$$\frac{T_{\max} - T_{\min}}{q''' A_1 / k_0} = \frac{1}{\tilde{k}\phi_1}, \tag{12.3}$$

which is only 12.5% larger than in Eq. (12.2) and Fig. 12.3. The slenderness of the outer shape and the relative thicknesses of the blades of high-conductivity material (or the ratio ϕ_1/ϕ_0) are optimized in this design. Each elemental triangle already comes into the assembly with constant temperature on its two long sides. The effect of optimizing the slenderness of the overall shape of the first construct is to make all the elemental triangles share the same temperature ($T_{\max}$) on their edges. In this way, the entire first construct becomes a structure with constant resistance.

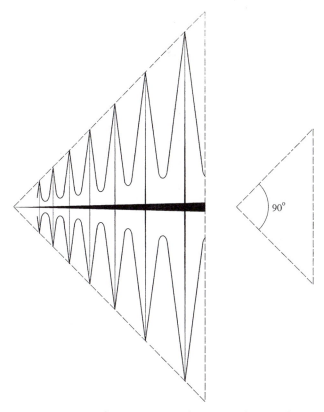

FIGURE 12.4. Constant-resistance first construct with optimized triangular external and internal shapes ($\phi_0 = 0.04$, $\phi_1 = 0.08$) [3].

The slenderness of the optimized first construct is independent of $\tilde{k}$ and ϕ_1, as its tip angle always equals 90°. Another feature of this structure is the presence of less k_p material at the elemental level, in the optimal proportion $\phi_0 = \frac{1}{2}\phi_1$. These features are robust, because they are exactly the same as in the best first construct with curved surfaces (Fig. 12.3).

Second constructs can be perfected in a similar way, for optimal external shape of the occupied territory, and for optimal shape of the new central k_p blade of thickness D_2. As in Section 4.4, it is found that in a second construct there is room for only two first constructs, because each first construct is far from being slender. The remaining role of the D_2 blade, which now has constant thickness, is to channel out of the second construct the two heat currents collected from the D_1 blades of the two first constructs. The optimized structure is drawn to scale on the left-hand side of Fig. 12.5. It is characterized by the optimal distribution of k_p material $\phi_1 = 0.513\phi_2$, where ϕ_2 represents the volume fraction allocated to the high-conductivity material (black).

The right-hand side of the same figure shows the corresponding structure optimized by starting with the assumption that the internal and external shapes are triangular. Here the optimal allocation of high-conductivity material is according to the rule $\phi_1 = \frac{1}{2}\phi_2$. Both sides of Fig. 12.5 have been drawn for $\phi_2 = 0.078$.

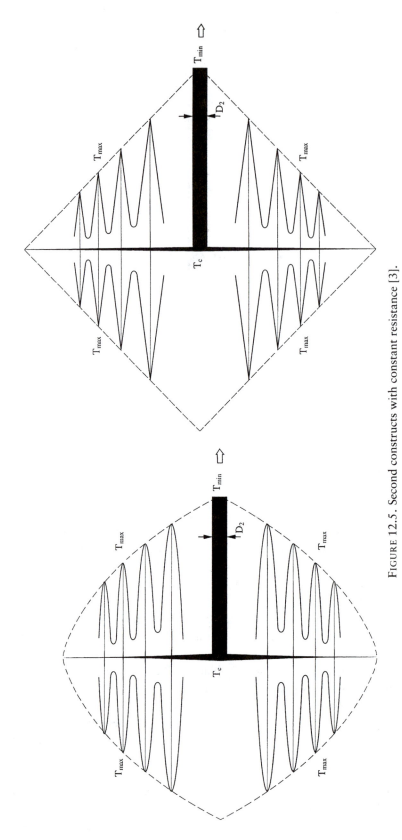

FIGURE 12.5. Second constructs with constant resistance [3].

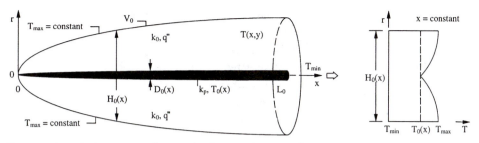

FIGURE 12.6. Axisymmetric elemental volume with unspecified external and internal shapes [4].

The global resistance of the triangle-in-triangle structure exceeds by 18% the resistance of the second construct with curved shapes.

Similar progress can be made in three dimensions. At the elemental volume level (Fig. 12.6), we begin with a body of revolution with unknown diameter variation, $H_0(x)$, and unknown diameter function $D_0(x)$ for the high-conductivity fiber. The total volume (V_0) and the volume of high-conductivity material ($\phi_0 V_0$) are constrained. The optimal shapes compatible with the presence of hot spots (T_{max}) all over the external surface of revolution are $H_0 \sim x$ and $D_0 \sim x$. In summary, the cone-in-cone design is the configuration for constant resistance at the elemental level. The elements become more slender as the product $\tilde{k}\phi_0$ increases. The minimized resistance is

$$R_0 = \frac{T_{max} - T_{min}}{q''' V_0^{2/3} / k_0} = \left(\frac{9}{2\pi}\right)^{2/3} \frac{(-1 + \phi_0 - \ln \phi_0)^{2/3}}{6\phi_0^{1/3} \tilde{k}^{1/3}}. \tag{12.4}$$

Figure 12.7 shows the route to the optimized first-construct architecture. The solid volume V_1 accommodates an infinite number of elemental constant-resistance volumes. The elemental volumes have the same $\tilde{k}$ and ϕ_0 and the same slenderness ratio. Their tips are situated on the unspecified surface $r = H_1(x)/2$. Each elemental cone generates a heat current, which is collected by a central fiber of unspecified

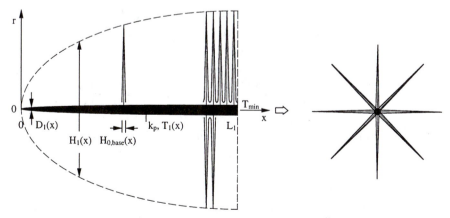

FIGURE 12.7. First construct with unspecified shapes, containing optimized cone-in-cone elemental volumes [4].

shape $D_1(x)$. Let $H_{0,\text{base}}(x)$ be the base diameter of the elemental cone, and take a thin slice (disk) of thickness $H_{0,\text{base}}$ perpendicular to the x axis. The number of elemental cones arranged radially in this slice is denoted by p (e.g., $p = 8$ in Fig. 12.7). Finally, there is the hot-spot temperature T_{max}, which must be spread uniformly over all the external surfaces of the elemental cones.

The optimized shapes of the first construct of Fig. 12.7 are represented by $H_1 \sim x^{1/2}$ and $D_1 \sim x$. This means that the central fiber is a cone. The outer shape is relatively round, almost hemispherical. In the limit $\phi_1 \to 0$, the external aspect ratio is $H_{1,\text{base}}/L_1 \to 2^{3/2}$, where $H_{1,\text{base}}$ is the largest (base) diameter of the external shape. The corresponding global resistance is

$$R_1 = \frac{T_{\text{max}} - T_{\text{min}}}{q''' V_1^{2/3}/k_0} = \frac{2^{1/3}}{(\pi \tilde{k} p)^{2/3}} \left[\frac{-1 + \phi_0 - \ln \phi_0}{\phi_0 (\phi_1 - \phi_0)} \right]^{1/3}. \tag{12.5}$$

The optimal distribution of k_p material ($\phi_1 V_1$) is such that the ratio ϕ_0/ϕ_1 increases little, from 0.54 to 0.6 as ϕ_1 increases from 0.001 to 0.1.

Once again, decreases in global resistance can be achieved by means of greater internal complexity. If we compare the elemental resistance (R_0) of Eq. (12.4) with the first-construct resistance (R_1) of Eq. (12.5) we find that the ratio R_1/R_0 is a function of ϕ_1 and $p\tilde{k}^{1/2}$. This comparison is done for the same total volume ($V_0 = V_1$) and the same amount of high-conductivity material ($\phi_0 V_0 = \phi_1 V_1$). The only thing that differentiates between R_0 and R_1 is the geometry of the internal space occupied by k_p material. We find that R_1 is less than R_0 when the product $p\tilde{k}^{1/2}$ is of the order of 10^2 and greater.

12.3 More Efficient Structures Look More "Natural"

The geometric designs developed in this article break new ground on several fronts. First, their performance is superior to that of previously optimized paths for minimum resistance to flow between a volume and one point. By allowing the external shape of the fixed-size domain to serve as a degree of freedom, in addition to the internal structure of the high-conductivity path, it is possible to optimize both the shape and structure to generate designs that maintain a constant resistance between every peripheral point (T_{max}) and the heat sink (T_{min}). The architectures that emerge are more efficient and look more like the tree structures found in nature.

The second new aspect is that the theoretical constant-resistance structures are fractals. They do not fill their allotted space completely because the addition of geometrically similar elemental volumes can, in principle, be continued indefinitely toward the tips of the first constructs (e.g., Figs. 12.3 and 12.7). These structures represent a theoretical limit. They are not real because they cannot be built or seen. Furthermore, their continuous hot-spot surfaces must be both isothermal (T_{max}) and adiabatic. The interstitial spaces that are left uncovered by the structure would have to be completely inactive. Then the uncovered spaces would be filled with useless material – zero heat generation rate and zero conductivity. It does not make sense to work so hard on maximizing the use of the volume fraction occupied by the dendritic heat-generating structure, when the uncovered volume is not used at all.

Indeed, if we reexamine the performance of the constant-resistance designs on the basis of the total volume allocated to the constructs, we find that the inactive interstices make the performance inferior in comparison with the performance of structures that fill their space completely. Take, for example, the elemental constant-resistance design of Fig. 12.1(d), which has the two-dimensional volume A_0 and the resistance (12.1). The rectangle circumscribed to the A_0 shape is larger, namely $A_{0C} = \frac{8}{5} A_0$. The heat current generated at the rate q''' in A_0 can also be generated uniformly at a lower rate ($\frac{5}{8} q'''$) in the A_{0C} rectangle. We obtain the resistance of the latter by multiplying Eq. (12.1) by $\frac{3}{2}$, because of the $\frac{3}{2}$ factor increase from (d) to (a) in Fig. 12.1, and by $\frac{5}{8}$ because of the weaker heat generation rate. The net effect is a factor of $\frac{15}{16}$ decrease in the resistance, meaning that the rectangular A_{0C} option is better.

The more sensible and realistic course is to use all the space available, as in the earliest applications of the constructal method (Chapter 4). There the construction began with finite-size rectangular elements that filled the space completely. There as well as in the tree networks of nature the structures are not fractals because they contain finite numbers of construction stages.

The constant-resistance structure – the fact that this geometric form represents the ultimate with respect to minimizing volume-to-point resistance – lends credence to the thesis that as natural structures evolve toward better performance in time, they also acquire shapes that look closer to fractal ones. The classical fractal approach postulates a priori self-similarity, that is, scale invariance of the geometry of the objects, or "invariance in the zoom direction." Constructal theory establishes instead relations between successive scales ("scale covariance," not invariance) as a "deterministic result" of constrained optimization [5]. The quotes indicate concepts introduced by Nottale [6], who wrote

> "One of the possible ways to understand fractals would be to look at the fractal behavior as the result of an optimization process.... Such a combination...may come from a process of optimization under constraint, or more generally of optimization of several quantities sometimes apparently contradictory (for example) maximizing surface while minimizing volume...."

Constructal theory is about the geometry-generating principle envisioned by Nottale.

Physicists, biologists, and engineers have a joint interest in the constructal theory and constant-resistance concepts. These methods provide (1) direct routes to resistance minimization by design, (2) rational explanations for the occurrence of flow structures in nature, and (3) a deterministic basis for the repeatability, regularity, and beauty of all such structures, natural and synthetic.

It is fitting that we end this book on point (3), which is also made in Fig. 12.8. This figure shows the design that the engineer would produce by covering a larger volume with four second-construct structures of the type optimized in Fig. 12.5. The heat current integrated over the four quadrants of the large square domain

FIGURE 12.8. The third construct obtained when four of the second constructs of Fig. 12.5 are joined (right) [3]. Red indicates the boundaries with hot-spot temperature. Blue indicates the high-conductivity inserts. This construct should be compared with Fig. 6.17, which shows the river drainage basin derived from a resistance-minimization erosion process (see Plate XIV).

is led by the diagonal D_2 channels to the center, which is cooled by a sufficiently strong stream that flows perpendicularly in Fig. 12.8. The reader should reexamine now Fig. 6.17, which shows the streaklines and the pattern of high-permeability channels developed numerically, in time, based on a fully deterministic erosion model of a river drainage basin [7]. Rain falls on every point of the rectangular (low-permeability) domain and is driven by pressure differences toward the central sink. The domain is made of 51×51 square grains (Section 6.6). At every time step, certain grains are selected and removed (washed downstream) based on the same constructal principle: the minimization of the global volume-to-point resistance subject to constraints. The river basin grows and marbleizes itself into nearly square territories that are served by central main streams.

The similarities (origin and image) between this unexpected structure (Fig. 6.17) and the designed version (Fig. 12.8) are worth contemplating. They reinforce a familiar view, which is new when seen from the reverse. Familiar is the empirical view that the more we look at natural systems the more we find that they operate most efficiently. The reverse is the theoretical direction of this book: The more we invoke the constructal law the more we discover that the deduced designs, our own drawings, look more and more "natural." The constructal principle works, in the inanimate, the animate, and the engineered.

The 'evolution' of engineering toward better designs is now anticipated by theory. To engineer is natural. I was reminded of this at an international thermodynamics conference in March 2000, where the invited speaker noted a key "difference" between a naturally optimized system (e.g., animal) and a system optimized in engineering (e.g., heat exchanger). According to him, the difference is that optimization makes a natural system better and bigger in time, while the engineered device is just that, a lifeless construct frozen in time (never mind the coincidence that the natural and the engineered are both *optimized*!). I think that the heat exchanger exists only as an extension of me – an extension of my mind, hands and all the other artifacts and humans that have been touched by me. Man today is a construct much larger and more complex than a human body examined in isolation. One individual today is a construct that covers the globe. Each of us is becoming a better and bigger spherical construct of flow networks, in time. The heat exchangers, which help us along this route, become better and bigger as well (think of air-conditioning, which spread all over the world in the twentieth century). They are accessories, parts of larger constructs. They should be likened to the toenails of the animal, not the animal.

12.4 More Material Where the Need is Greater

The flow of this book has been from engineering to nature. Along the way we developed a theory of how geometric form is generated in nature. Now the theory returns the favor to engineering: The same principle can be used to perfect the structure of engineering systems and to develop concepts for entirely new systems.

The evolution of tree designs toward the constant-resistance limit (Section 12.2). exemplifies the evolutionary work that can be done in engineering. Better global performance is achieved when more of the internal points are forced to work as hard as the few hardest working points. The system is destined to remain imperfect. The resistances to the internal flows cannot be eliminated because of the reality of design: Amounts and types of materials are given, volumes and matching (neighboring) systems are specified, and the time to contemplate changes is limited. In spite of these constraints, we can spread the imperfection around in optimal or near-optimal ways. The highs and the lows must be balanced. The optimal spreading of slopes and differences takes us to the architecture that serves the global purpose.

This process was illustrated several times in this book, most extensively in the construction of tree-shaped paths and beams of constant strength. Other structures were presented in rough forms, for example, the internal flow structures with one-size spacing (Chap. 3) and the convective fins with constant plate thickness (Chap. 8). In Fig. 3.3, the geometry can be improved by making the heat-generating plates somewhat denser near the inlet (blue). The reason is that the heat transfer coefficients are relatively larger near the leading edge of any structure in forced or natural convection. This is another way to say that in the one-spacing design of Fig. 3.3 the core regions of the channels near the bottom are still not used (note the blue color). Making these channels narrower would permit the packing of more heat generation rate into the given volume, while respecting the peak temperature constraint.

The improved shaping of a plate fin is illustrated in Fig. 12.9. Each line of heat flow encounters two resistances in series: the conduction along the fin, and the convection from the fin surface to the external fluid. The convective resistance is considerably lower near the leading edge of the plate. This means that we can tolerate a relatively larger conductive resistance in the fin material near the leading edge. When the total fin material is fixed, the minimization of the global resistance

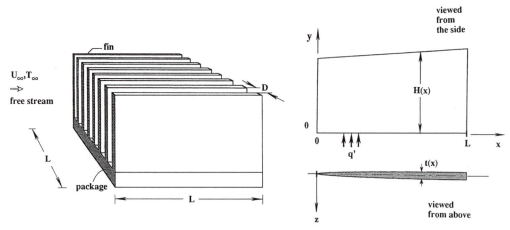

FIGURE 12.9. Electronic package with plate-fin heat sink and a single fin with variable thickness and height [8].

pushes the fin shape in two simultaneous directions: (1) smaller thicknesses near the leading edge and (2) larger fin lengths near the trailing edge. Optimized alone, effect (1) amounts to a plate thickness that varies as $x^{0.42}$, where x is measured from the leading edge, in the downstream direction. The reduction in global resistance that results from this optimal shaping of the fin profile is 15%. The impact of optimizing effect (2) alone is a reduction of 30% in the global resistance [8]. Note that the size of these improvements is consistent with the other improvements of the rough designs documented in this book.

Effect (2) – the tilting of the fin crest – is presented in color in Fig. 12.10. Yellow means cold, and orange means hot. Too much yellow in one region means that the plate is relatively cold, i.e., useless as a heat transfer device. It must be eliminated; hence the evolution to a crest with optimal inclination (the bottom figure). In the latter, the low temperature is distributed almost uniformly along the crest. We see that optimal form (constant resistance, etc.) can also be viewed as the uniform spreading of lazy material – the already cold regions of the plate, which do not transfer much heat to the fluid.

Pushed in the directions shown in Fig. 12.9 (right), the plate fin looks more natural. It looks more like the fin of a fish, which, incidentally, is why from the beginning this engineered system was named fin.

12.5 An Old and Prevalent Natural Phenomenon

In the engineering of flow systems the search for better global performance subject to present-day constraints is known as thermodynamic optimization, irreversibility minimization, or entropy generation minimization [9]. In this book I focused on the deterministic relationship between the improvement of performance (objective, constraints) and the generation of geometry (shape and structure) in the system.

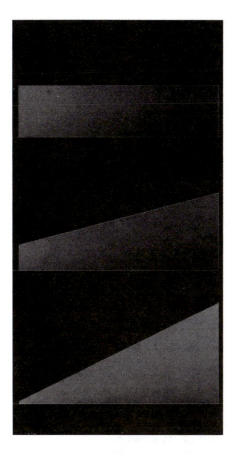

FIGURE 12.10. The temperature field in a plate fin with adjustable crest inclination [8] (see Plate XV).

Performance improvement, or optimization, is an old idea and an even older natural phenomenon. It has been with us throughout the history of engineering. Its reach, however, is much broader and more permanent: everything exhibits it. We can be sure that the performance of power plants – the performance of man, really – will continue to improve in time, in the same way that, in time, the rainfall will generate a more effective (dendritic) flow structure. Examples of geometric maximization of performance are everywhere, in the optimal folding of protein structures [10], the optimal configuration of reversed-field pinch plasma [11], the geometrical control of catalysis, adsorption and multiphase processes [12–15], the electromigration of metals [16], and geochemical patterns [17, 18]. Geophysicists and physiologists are converging on the conclusion that the optimization of performance has been the generating mechanism for geometric form in all plants, animals, and inanimate flow structures.

If this phenomenon is so old and prevalent, then what is new? New is the streamlining of its study into a single principle. This was my objective in this book, struggling, of course, under my own constraints! Most of this work could have been done one or two centuries ago, before thermodynamics. The geometric minimization of resistances to heat and fluid flow could have been accomplished based on Fourier's heat transmission and the hydrodynamics of Bernoulli, Poiseuille, and Darcy. It is a mystery that this was not done then, because that period was still influenced by Leibnitz's, Maupertuis', and Castigliano's intuition that, of all possible processes, the only ones that actually occur are those that involve minimum expenditure of "action." Instead, as Hildebrandt and Tromba [19] have noted, modern physics embarked on a course tailored to the principle of infinitesimal local effects. Constructal theory is a jolt the other way, a means to rationalize *macroscopic* features, objective, and behavior.

REFERENCES

1. A. Bejan, Constructal-theory network of conducting paths for cooling a heat generating volume, *Int. J. Heat Mass Transfer*, Vol. 40, 1997, pp. 799–816.

2. G. A. Ledezma, A. Bejan, and M. R. Errera, Constructal tree networks for heat transfer, *J. Appl. Phys.*, Vol. 82, 1997, pp. 89–100.

3. M. Neagu and A. Bejan, Constructal-theory tree networks of "constant" thermal resistance, *J. Appl. Phys.*, Vol. 86, 1999, pp. 1136–1144.

4. M. Neagu and A. Bejan, Three-dimensional tree constructs of "constant" thermal resistance, *J. Appl. Phys.*, Vol. 86, 1999, pp. 7107–7115.

5. A. Bejan and D. Tondeur, Equipartition, optimal allocation, and the constructal approach to predicting organization in nature, *Rev. Gen. Therm.*, Vol. 37, 1998, pp. 165–180.

6. L. Nottale, *Fractal Space-Time and Microphysics*, World Scientific, Singapore, 1993.

7. M. R. Errera and A. Bejan, Deterministic tree networks for river drainage basins, *Fractals*, Vol. 6, 1998, pp. 245–261.

8. M. Morega and A. Bejan, Plate fins with variable thickness and height for air-cooled electronic modules, *Int. J. Heat Mass Transfer*, Vol. 37, Suppl. 1, 1994, pp. 433–445.

9. A. Bejan, *Entropy Generation Minimization*, CRC, Boca Raton, FL, 1996.

10. C. Micheletti, J. R. Banavar, A. Maritan, and F. Seno, Protein structures and optimal folding emerging from a geometrical variational principle, *Phys. Rev. Lett.*, Vol. 82, 1999, pp. 3372–3375.

11. T. K. Chu, Turbulent steady-state configuration of an inductively driven reversed field pinch plasma, *Phys. Rev. Lett.*, Vol. 81, 1998, pp. 3148–3150.

12. M.-O. Coppens and G. F. Froment, The effectiveness of mass fractal catalysts, *Fractals*, Vol. 5, 1997, pp. 493–505.

13. M.-O. Coppens, Geometrical control of multiphase processes using a new fluid injection system, Paper 288c, AIChE Annual Meeting, Dallas, TX, Oct. 31–Nov. 5, 1999.

14. M.-O. Coppens, Y. Cheng, and C. M. van den Bleek, Controlling fluidized bed operation using a novel hierarchical gas injection system, Paper 304d, AIChE Annual Meeting, Dallas, TX, Oct. 31–Nov. 5, 1999.

15. M. Kearney, Control of fluid dynamics with engineered fractals – adsorption process applications, *Chem. Eng. Comm.*, Vol. 173, 1999, pp. 43–52.

16. S. J. Krumbein, Metallic electromigration phenomena, *IEEE Trans. Components, Hybrids, and Manufacturing Technology*, Vol. 11, 1988, pp. 5–15.

17. P. J. Ortoleva, *Geochemical Self-Organization*, Oxford University Press, Oxford, UK, 1994.

18. G. Israeli, Y. Zvirin, and Y. Zimmels, Evaporation and deposition processes in flows of aqueous concentrated solutions, *Heat Transfer 1998*, Proc. 11th Int. Heat Transfer Conf., Vol. 2, pp. 21–26, Aug. 23–28, 1998, Kyongju, Korea.

19. S. Hildebrandt and A. Tromba, *Mathematics and Optimal Form*, Scientific American Books, New York, 1985.

ABOUT THE AUTHOR

Adrian Bejan received his B.S. (1972, Honors Course), M.S. (1972, Honors Course), and Ph.D. (1975) degrees in mechanical engineering, all from the Massachusetts Institute of Technology. From 1976 until 1978 he was a Fellow of the Miller Institute for Basic Research in Science, at the University of California, Berkeley.

Professor Bejan's research covers a wide range of topics in thermal engineering: entropy generation minimization, exergy analysis, natural convection, combined heat and mass transfer, convection in porous media, transition to turbulence, melting, solidification, condensation, boiling, fouling, solar energy conversion, cryogenics, applied superconductivity, and tribology. More recently, he developed research projects on contaminant removal (ventilation) from enclosures, convection from surfaces covered with flexible fibers, the optimal geometry of electronics packages cooled by forced and natural convection, and the *constructal* theory of organization and complexity in nature and engineering.

Professor Bejan is the author of 10 books and over 300 journal articles. He is recipient of the Max Jakob Memorial Award of the American Institute of Chemical Engineers and the American Society of Mechanical Engineers. He was also awarded by the American Society of Mechanical Engineers the Worcester Reed Warner Medal, the James Harry Potter Gold Medal, the Gustus L. Larson Memorial Award, and the Heat Transfer Memorial Award in Science. He received the Ralph Coats Roe Award from the American Society of Engineering Education. He was awarded the J. A. Jones Chair at Duke (1989), and eight honorary doctorates at foreign universities.

AUTHOR INDEX

SUBJECT INDEX

Access optimization, 5, 58, 62, 158, 270–299; *see also* Constructal theory
Aggregation, *see* Construct
Aircraft, 234–242
Airways, *see* People
Allocation, 220–223, 237, 287, 311, 312; *see also* Equipartition
Allocation of imperfection, *see* Imperfection
Allometric laws, 3, 108–113, 238, 246, 257, 260–266
Alloys, 176
Altruism, 10
Angles, optimized, 11, 12, 62–65, 93, 99, 114, 115, 272–276, 295
Animate systems, xviii, 1, 44, 215, 230, 233–242, 265, 270, 277, 310, 311
Approach to equilibrium, *see* Equilibrium and nonequilibrium
Area moment of inertia, 15
Arteries, 235
Artificial systems, *see* Engineering
Assembly, *see* Construct
Asymptotes, intersection of, 32, 37, 40, 41
Atomistic theory, 292
Automobiles, *see* Vehicles
Avionics, 240

Basins, *see* River morphology
Be, pressure drop number, 37, 191, 200, 201
Beauty, 309
Bees, 44, 45
Behavior, 314
Bénard convection, 158–175
Bending, 14–22, 25, 26
Bernoulli, Daniel (1700–1782), 313
Better, the concept of, 24, 25, 226
Bi, Biot number, 182, 251, 253
Bifurcation, *see* Dichotomy
Bioengineering, xvi, 263
Biology, 3, 177, 178, 236, 240, 254, 257, 287,

294, 296, 309
Biomimetics, 7, 8, 19
Birds, 233–242, 260–266
Body size effect, xviii, 108–113, 238, 260–266
Boiling, 169, 174, 175, 249
Botany, xvi, 2, 19, 62, 66, 82–84, 215
Boundary layers, 31, 36, 42, 155, 165, 172
Boussinesq linearization, 170
Brancusi, Constantin (1876–1957), 25
Breathing, xviii, 102, 254–257, 266
Buckling
    elastic, 22–24
    inviscid fluid flow, 117–119, 152–158
Buoyancy, 160
Business, 270–299

Cantilever beam, 14–22
Castigliano, Carlo Alberto (1847–1884), 14, 314
Centrals, 277
Channel cross sections, *see* Duct cross sections
Channeling, 128–145, 160, 175, 203–212
Chaos, 159, 175
Circular fins, 198–201
Circulatory system, 258
Clausius, Rudolf Julius Emanuel (1822–1888), xviii
Cleaning, periodic, 252–254, 267, 268
Coalescence, 288; *see also* Growth
Coffee sediment experiment, 5, 121
Colburn analogy, 119, 156
Columbus, Christopher (1451–1506), 23, 24
Column in end compression, *see* Strut
Combustion, 240
Complexity, xvi, 62, 69–77, 126, 275, 277, 282, 296
Condensation, 155, 249
Conductance, thermal, 221
Conduction, 52–79, 161, 201–216; *see also* Diffusion
Conga lines, 144

320